PROGRESS IN ANIMAL BIOMETEOROLOGY
VOLUME 1 — PART II

PROGRESS IN BIOMETEOROLOGY

EDITOR-IN-CHIEF: S. W. TROMP EXECUTIVE EDITOR: JANNEKE J. BOUMA

BIOMETEOROLOGICAL RESEARCH CENTRE, LEIDEN
JAN STEENLAAN 3, OEGSTGEEST (LEIDEN)
THE NETHERLANDS

DIVISION B

PROGRESS IN ANIMAL BIOMETEOROLOGY

The Effect of Weather and Climate on Animals

VOLUME 1 PART II PERIOD 1963-1973

EFFECT OF LIGHT, HIGH ALTITUDE, NOISE, ELECTRIC, MAGNETIC AND ELECTRO-MAGNETIC FIELDS, IONIZATION, GRAVITY AND AIR POLLUTION ON ANIMALS

EDITED BY H. D. JOHNSON
DEPARTMENT OF DAIRY HUSBANDRY
UNIVERSITY OF MISSOURI
COLUMBIA (MO) U.S.A.
IN COOPERATION WITH 87 CONTRIBUTORS

SWETS & ZEITLINGER B.V. AMSTERDAM 1976

Printed by Drukkerij F. E. Mac Donald, Nijmegen

ISBN 90 265 02354 x

LIST OF CONTRIBUTORS OF DIVISION B, VOLUME 1

NAME	ADDRESS	CHAPTER and SECTION*
Y. ALARIE	Dept of Occupational Health, Graduate School of Public Health, Univ. of Pittsburgh, Pittsburgh, Pa 15261, U.S.A.	7.4
T. E. ALLEN	Div. of Animal Genetics, C.S.I.R.O., McMaster Field Station, Private Mail Bag, Liverpool, N.S.W., 2170, Australia.	2.2a
C. W. ALLISTON	Dept of Animal Sciences, The Roy W. Herrick Lab., Lilly Hall, Purdue Univ., West Lafayette, Indiana 47907, U.S.A.	1.15b(1)
P. D. ALTLAND	Physiological Section, Lab. of Chemical Physics, National Inst. of Arthritis, Metabolism and Digestive Diseases, National Inst. of Health, Bethesda, Maryland 20014, U.S.A.	3.2
A. ANTHONY	Dept of Biology, College of Science, The Pennsylvania State Univ., 208 Life Sciences I, Univ. Park, Pa 16802, U.S.A.	4.0
E. BEDRAK	Dept of Biology, Univ. of the Negev and Research and Development Authority, P.O.B. 1025, Beer-Sheva 84110, Israel	1.15d
S. C. BHATIA	Dept of Biochemistry, Vallabhohdi Patel Chest Inst., Univ. of Delhi, Delhi 110007, India Temporarily: Chef des Travaux at the Dept de Biochimie Médicale, Ecole de Médecine, Univ de Génève, 20, rue de l'Ecole de Médecine, 1211 Genève 4, Switzerland.	7.5
W. BIANCA	Institut f. Tierzucht, ETH, Universitätstr. 2, 8006 Zürich, Switserland	1.17
J. BLIGH	ARC Institute of Animal Physiology, Babraham, Cambridge, Great Britain.	1.4a
W. G. BRADLEY	Dept of Biological Sciences, Univ. of Nevada, Las Vegas, Nevada 89154, U.S.A.	1.18a
J. R. BRETT	Physiological Investigation Lab., Pacific Biological Station, Dept of the Environment, Nanaimo, B.C., Canada V9R 5K6	1.12c

* 1.5.b, 2.1.g means that the Contributor wrote Chapter 1, Section 5, Sub-Section b and Chapter 2, Section 1, sub-Section g.

G. D. BROWN	Div. of Wildlife Research, C.S.I.R.O., P.O.B. 84, Lyneham, A.C.T. 2602, Australia.	2.2a
W. E. CALDWELL	Dept of Psychology, The George Washington Univ., Washington, D.C. 20006, U.S.A.	5.2a
D. P. CARDINALI	Instituto Latino-Americano de Fisiología de la Reproducción (ILAFIR), Univ del Salvador, Casilla de Correo 10, San Miguel (P.B.A.), Argentina.	2.2b(2)
K. CENA	Academy of Physical Education, ul. Banacha 11, 51-617 Wrocław, Poland.	1.13a
R. R. CHAFFEE	Dept of Ergonomics, Univ. of California, Santa Barbara, Cal. 93106, U.S.A.	1.1a
J. W. DEATON	South Central Poultry Research Lab., Agricultural Res. Service, P.O.B. 5367, State College, Mississippi 39762, U.S.A.	1.12d
R. W. DOUGHERTY	National Animal Disease Lab., P.O.B. 70, Ames, Iowa 50010, U.S.A.	1.20.a
D. F. DOWLING	Dept of Animal Husbandry, Univ. of Queensland, St Lucia., Brisbane 4067, Australia.	1.12.a
T. W. DUKES	Animal Diseases Res. Inst., P.O.B. 11300, Station H, Ottawa K2H 8P9, Canada.	2.2c
G. E. FOLK Jr	Dept of Physiology and Biophysics, University of Iowa, College of Medicine, Iowa City, Iowa 52242, U.S.A.	1.11.b(1)
F. E. J. FRY	Ramsay Wright Zoological Labs, Univ. of Toronto, Harbord Str. 25, Toronto, Ontario, Canada, M5S 1A1.	1.4.e
M. GEBCZÝNSKI	Mammals Research Inst., 17-230 Bialowicza, Woj. Bialystok, Poland.	2.3
R. F. GROVER	Cardiovascular Pulmonary Research Lab., Div. of Cardiology, Dept of Medicine, Univ. of Colorado Medical Center, 4200 East-ninth Av., Denver, Col. 80220, U.S.A.	3.1
D. E. GRUBBS	Dept of Biology, California State Univ., Fresno, Calif., U.S.A.	1.3.c
G. L. HAHN	Dept of Agricultural Engineering, Building T-12, Univ. of Missouri-Columbia, Columbia, Mo, 65201, U.S.A.	1.21a
J. R. S. HALES	C.S.I.R.O., Div. of Animal Physiology, Ian Clunies Ross Animal Res. Lab., P.O.B. 239, Blacktown, N.S.W. 2148, Australia.	1.5.a 1.11.a(1)
C. L. HAMILTON	Dept of Physiology, Univ. of Pennsylvania, School of Medicine, Philadelphia, Pa 19104, U.S.A.	1.7
P. C. HARRISON	Dept of Animal Science, 124 Animal Sciences Lab., College of Agriculture, Univ. of Illinois, Urbana, Ill. 61801, U.S.A.	1.11.a(3) 1.11.b(2)

W. O. HAUFE	Veterinary and Medical Entomological Section, Canada Agriculture Research Station, Lethbridge, Alb. T1J-4B1, Canada.	1.20.d
F. L. HAYS	Eidgenössische Techn. Hochschule, Lehr- und Versuchsgut Chamau, Unterhünenberg (Zug), Bahnstation Cham, 6331 Unterhünenberg, Switzerland	7.1
J. C. HOFFMANN	Dept of Anatomy and Reproductive Biology, Univ. of Hawaii, School of Medicine, 1960 East-West Road, Honolulu, Hawaii 96822, U.S.A.	2.2.b(1)
M.E. HUGH-JONES	Pathological and Epidemiology Dept, Veterinary Lab., New Haw, Weybridge, Surrey, Great Britain.	1.20.b
J. C. D. HUTCHINSON	Division of Animal Physiology, Ian Clunies Ross Animal Research Lab., Prospect, CSIRO, P.O.B. 239, Blacktown, New South Wales 2148, Australia.	2.1.g 2.2.a
D. L. INGRAM	Agricultural Research Council, Institute of Animal Physiology, Babraham, Cambridge, Great Britain.	1.5.d 1.18.c(2)
G. L. JACKSON	Dept of Veterinary Physiology and Pharmacology, College of Veterinary Medicine, Univ. of Illinois at Urbana-Champaign, Ill. 61801, U.S.A.	2.1.c
L. JANSKÝ	Dept of Comparative Physiology, Charles University, Prague, Czechoslovakia.	1.8.b(4)
Elisabeth JOHNSON	Dept of Zoology, Univ. of Reading, Whiteknights, Reading RG6-2AJ, Great Britain.	2.1.f
H. D. JOHNSON	Dept of Dairy Husbandry, University of Missouri, Columbia, Missouri 65201, U.S.A.	1.2.a 1.14
K. G. JOHNSON	ARC Institute of Animal Physiology, Babraham, Cambridge, Great Britain.	1.5.c
T. H. KAMAL	Bioclimatology Unit, Radiobiology Division, Atomic Energy Establishment, Cairo, Egypt.	1.19
K. L. KNOX	Dept of Nutritional Sciences, College of Agriculture, The Univ. of Connecticut, Storrs, Conn. 06268, U.S.A.	1.8.a(2)
A. P. KRUEGER	School of Public Health, Earl Warren Hall, Univ. of Cal., Berkeley, Cal. 94720, U.S.A.	5.1
M. A. LESSLER	Dept of Physiology, Ohio State Univ., 333 W.10th Av., Columbus, Ohio 43210, U.S.A.	7.2
W. V. MACFARLANE	Dept of Physiology, Waite Institute, University of Adelaide, Private Bag 1, Glen Osmond, Adelaide (S.A.), Australia.	1.3.a 1.18.b

R. E. MACMILLEN	Dept of Population and Environmental Biology, Univ. of California, Irvine, Cal., U.S.A.	1.3.c
G. M. O. MALOIY	Dept of Animal Physiology, Univ. of Nairobi, P.O.B. 30197, Nairobi, Kenya, E. Africa.	1.3.b
E. J. MASORO	Dept of Physiology, The Univ. of Texas Health Center at San Antonio, 7703 Floyd Curl Drive, San Antonio, Texas 78284, U.S.A.	1.1.b
W. F. Y. MAWSON	Development Planning Branch, Dept of Primary Industries, Bribane, Qld 4000, Australia.	1.21.b
J. A. MCLEAN	Hannah Dairy Institute, Ayr, Scotland Great Britain	1.8.a(1)
J. L. MONTEITH	Dept of Physiology and Environmental Studies, Univ. of Nottingham, School of Agriculture, Sutton Bonington, Loughborough, Leics. LE12-5RD, Great Britain	1.13.a
P. R. MORRISON	Dept of Zoophysiology, Institute of Arctic Biology, Univ. of Alaska, Fairbanks, Alaska 99701, U.S.A.	3.3
L. E. MOUNT	Dept of Applied Biology, ARC Institute of Animal Physiology, Babraham, Cambridge, Great Britain.	1.8.b(3)
L. D. OLSON	Dept of Veterinary Pathology, University of Missouri, Columbia, Mo. 65201, U.S.A.	1.20.c
J. OYAMA	Aimes Research Center, NASA, Moffett Field, Cal. 94035, U.S.A.	6
Alice Y. S. P'AN	School of Public Health, Univ. of Montreal, 2375 Chemin Côte Ste-Catherine, Montreal 101, Quebec, Canada.	7.3
Pamela R. PENNYCUIK	Div. of Animal Genetics, CSIRO, P.O.B. 90, Epping, N.S.W. 2121, Australia.	1.18.c(3)
M. A. PERSINGER	Environmental Psychophysiological Lab., Dept of Psychology, Laurentian University, Sudbury, Ont., Canada.	5.2.b
V. M. PETROVIC	Inst. of Physiology and Biochemistry, Fac. Science, Univ. of Belgrado, Studentski/trg 16, Belgrado, Yugoslavia.	1.2.b(2)
W. S. PLATNER	Dept of Physiology, Univ. of Missouri, School of Medicine, Columbia, Missouri 65201, U.S.A.	1.9
R. S. POGRUND	Dept of Occupational and Environmental Medicine, School of Public Health, Univ. of Illinois at the Medical Centre, P.O.B. 6998, Chicago, Ill. 60680, U.S.A.	5.3
R. O. RAWSON	John B. Pierce Foundation, 290 Congress Av., New Haven, Connecticut 06519, U.S.A.	1.4.b

F. N. REECE	South Central Poultry Research Lab., Agricultural Research Service, P.O.B. 5367, Mississippi State College, 39762, U.S.A.	1.12.d
R. J. REITER	Dept of Anatomy, The Univ. of Texas, Health Science Center at San Antonio, 7703 Floyd Curl Drive, San Antonio, Texas 78284, U.S.A.	2.1.d
J. D. RISING	Ramsay Wright Zoological Lab., Univ. of Toronto, 25 Hardford Street, Toronto M5S-1A1, Ont., Canada.	1.4.f
D. ROBERTSHAW	Indiana Univ. School of Medicine, Medical Sciences Program, Meyers Hall, Bloomington, Ind. 47401, U.S.A.	1.5.b
F. RUSSO	Psychological Section, Behavioral Sciences Dept, Naval Medical Research Inst., Bethesda, Md 20014, U.S.A.	5.2.a
M. L. SCOTT	Dept of Poultry Science and Animal Nutrition, Rice Hall, New York State College of Agriculture, Cornell Univ. Ithaca, New York 14850, U.S.A.	1.10
H. S. SIEGEL	South East Poultry Research Lab., Univ. of Georgia, 934 College Station Road, Athens, Georgia 30601, U.S.A.	1.8.a(4) 1.8.b(5)
E. SIMENSEN	Dept of Veterinary Pathology, Univ. of Missouri, Columbia, Mo. 65201, U.S.A. Temporarily: Dept of Animal Husbandry and Genetics, Vet. College of Norway, P.O. Box 8146, Oslo-Dep, Oslo, Norway.	1.20.c
U. A. SOD-MORIAH	Dept of Biology, Univ. of the Negev and Research and Development Authority, P.O.B. 1025, Beer-Sheva 84110, Israel.	1.15.d
J. STEINBACH	Dept of Animal Science, Univ. of Ibadan, Ibadan, Nigeria.	1.12.b 1.15.c
M. B. TER HAAR	A.R.C. Inst. of Animal Physiology, Babraham, Cambridge, Great Britain.	2.1.b
G. E. THOMPSON	The Hannah Dairy Research Inst., Ayr, Scotland KA6 5HL, Great Britain.	1.8.b(1)
C. J. THWAITES	Dept of Livestock Production, Univ. of New England, Armidale, New South Wales 2351, Australia.	1.13.b
M. VAN KAMPEN	Lab. for Veterinary Physiology, Alex. Numankade 93, State Univ., Utrecht, The Netherlands.	1.5.e
H. A. W. VENTER	Dept of Animal Science, Univ. of Pretoria, Pretoria, South Africa.	1.15.a

J. E. VERCOE	C.S.I.R.O., Div. of Animal Genetics, Cattle Research Lab., Rockhampton, Qld, Australia.*	1.18.c(1)
A. J. F. WEBSTER	Rowett Research Inst., Bucksburn, Aberdeen, Scotland, AB2-9SB, Great Britain	1.2.b(1) 1.8.b(2)
G. C. WEST	Inst. of Arctic Biology, Lab. of Zoophysiology, Univ. of Alaska, Fairbanks, Alaska 99701, U.S.A.	1.18.d(2)
G. C. WHITTOW	Dept of Physiology, School of Medicine, Univ. of Hawaii, Honolulu, Hawaii 96822, U.S.A.	1.4.d 1.11.a(2)
W. O. WILSON	Dept of Avian Sciences, Univ. of California, Davis, Calif. 95616, U.S.A.	1.16
G. L. WRIGHT	Lab. of Cell Physiology, Dept of Physiology, 4196 Medical Science Building, 333 West Tenth Av., Columbus, Ohio 43210, U.S.A.	1.6 7.2
N. T. M. YEATES	Dept of Livestock Production, Univ. of New England, Armidale, New South Wales, Australia.	1.15.b(2)
M. K. YOUSEF	Dept of Biological Sciences, Univ. of Nevada, Las Vegas, Nevada 89154, U.S.A. and Lab. of Environmental Patho-Physiology, Desert Research Inst., Univ. of Nevada System, Boulder City, Nevada, U.S.A.	1.4.g 1.8.a(3) 1.18.a

* Temporary address: Animal Production and Health Section, Joint FAO/I.A.E.A. Division, Kärtnerring 11, P.O.B. 590, A–1101 Vienna, Austria.

PART II

PREFACE

Since 1963 Biometeorology has been developing rapidly due to the considerable increase in interest of the whole scientific world in man's physical environment. Also the recent developments in Space Research in connection with the landings on the moon and studies of the atmosphere of the planets have stimulated many research workers to enter this field of research.

The present book "PROGRESS IN ANIMAL BIOMETEOROLOGY" represents the first Volume of Division B in the series PROGRESS IN BIOMETEOROLOGY, following Volume 1 of Division A, PROGRESS IN HUMAN BIOMETEOROLOGY and Volume 1 of Division C, PROGRESS IN PLANT BIOMETEOROLOGY. The book PROGRESS IN HUMAN BIOMETEOROLOGY was prepared as a continuation of a previous textbook "MEDICAL BIOMETEOROLOGY" (Weather, Climate and the Living Organism) which was published by Elsevier Publishing Company, Amsterdam, 1963. It was written by the present Editor-in-Chief of the series PROGRESS IN BIOMETEOROLOGY.

Swets & Zeitlinger Publishing Company agreed to publish a continuation of "Medical Biometeorology" which covers the period 1963-1973. As the effects both on healthy man and on diseased subjects would be reviewed, the original title was changed to "PROGRESS IN HUMAN BIOMETEOROLOGY" with the subtitle "THE EFFECTS OF WEATHER AND CLIMATE ON MAN AND HIS LIVING ENVIRONMENT".

However, it was realized that not only "Human Biometeorology", needed a new up to date textbook, but also the recent developments in plant and animal biometeorology made it highly necessary to cover these developments.
For this reason it was decided to publish a completely new series which was called PROGRESS IN BIOMETEOROLOGY. It consists of three divisions: Division A: PROGRESS IN HUMAN BIOMETEOROLOGY; Division B: PROGRESS IN ANIMAL BIOMETEOROLOGY and Division C: PROGRESS IN PLANT BIOMETEOROLOGY. Each division has its own Editor. Division A is being prepared by Dr S. W. TROMP, Director of the Biometeorological Research Centre, Leiden, The Netherlands; Division B is being prepared by Prof. H. D. JOHNSON, Director of the Department of Dairy Husbandry, University of Missouri, Columbia, U.S.A.; Division C by Mr L. P. SMITH, former President of the Commission for Agricultural Meteorology of the World Meteorological Organization, Meteorological Office, Bracknell, Great Britain.
General Coordinator and Editor-in-Chief of the three divisions is Dr S. W. TROMP. Executive Editor of all volumes is Mrs Janneke J. BOUMA, Research Associate of the Biometeorological Research Centre, Leiden, The Netherlands. Depending on the amount of new information received by the Editors in their specific divisions, new volumes will be prepared every two to three years. For this reason the present book is called DIVISION B, VOLUME I (PERIOD 1963-1973).

Instead of giving a list of references after each Chapter or Section the same method was followed, previously used in MEDICAL BIOMETEOROLOGY and in Division A and C of PROGRESS IN HUMAN BIOMETEOROLOGY, which proved to be very satisfactory. All references are given at the end of each book arranged in the same sequence as the Chapters and Sections. At the end of each Section we briefly indicated the pages where the references of this particular Section can be found. An extensive Author and Subject Index will enable the reader to find each Section in the book and the Contributors of these Sections.

The Volumes 1 of Division A, B and C represent in a way the standard type of a Progress Series but differ from the usual publications in which a number of authors review only certain aspects of a whole science. In this Progress Series we have tried to prepare up to date textbooks on Human, Animal and Plant Biometeorology.

It is fully realized that the task ahead is a very difficult one, but we trust that the efforts of all the cooperating experts in the field of Biometeorology will enable us to keep abreast of the fast developments of this modern branch of science and will help future students in Meteorology, Biology, Medicine, Veterinary, Agricultural and many other sciences, in finding their way in this dense forest of rapidly growing scientific knowledge.

S. W. Tromp
Editor-in-Chief
Progress in Biometeorology

INTRODUCTION

PROGRESS IN ANIMAL BIOMETEOROLOGY was designed to provide a concise, up-to-date text of biometeorological effects on the mechanisms of physiological and behavioural responses of terrestrial vertebrates. Emphasis is on domesticated and wild mammalian and avian species; though, when possible other vertebrates and invertebrates are included. Most Sections will not cover all species; however, as information and authors become available, additional Sections will be added upon revision of the textbook.

This book will serve as a basic text for students of animal biometeorology and environmental physiology and related fields. The description and measurement of the physical and chemical environments and intensive ecological relationships must be provided from other sources of information.

Each Section has endeavored to include a concise concensus by the author of the present state of knowledge. When possible, figure(s) or diagram(s) have been used to illustrate the known or possible mode of action of the environmental effect and responses of the organisms.

Volume 1 consists of two Parts:
Part I comprises only Chapter 1, dealing with environmental temperature effects on animal functions and processes. It is the most extensive due to the greater availability of research on this subject.

Cell metabolism, endocrine functions, water and electrolyte balance, temperature regulation, evaporative and non-evaporative heat loss, energy metabolism, mineral and vitamin requirements and cardiovascular physiology have received major emphasis. Also the environmental effects on the process of growth, lactation, reproduction, fibre growth, egg production and aging are covered.
Adaptability and processes of engineering and management to alleviate or modify effects of heat and cold receive attention. Finally the effects of heat and cold on animal diseases are discussed.

In Part II a current evaluation is presented on the effects of light, high altitude, noise, electric, magnetic and electro-magnetic fields, ionization, gravity and chemical air pollution on animals.

Harold D. Johnson
Editor
Progress in Animal Biometeorology

DEFINITION AND CLASSIFICATION OF BIOMETEOROLOGY

Various definitions and classifications have been suggested since the Foundation of the International Society of Biometeorology (I.S.B.) in 1956. Biometeorology can be defined briefly as the study of the biological effects of weather and climate on living organisms, plants, animals and man and of their direct physico-chemical environment.

In 1956, during the first Biometeorological Symposium of the International Society of Biometeorology, at UNESCO Headquarters in Paris, the following definition was accepted by the members of ISB:

"Biometeorology comprises the study of the direct and indirect interrelations between the geophysical and geochemical environment of the atmosphere and living organisms, plants, animals and man. The term "environment" is broadly conceived and includes micro, macro and cosmic environments in so far as they affect the earth's atmosphere and the diverse physical and chemical factors which comprise these environments. The investigations in these disciplines should be conducted in nature or in the laboratory under as rigidly controlled conditions as possible to describe measurable and reproducible physical, chemical and biological factors which shows a sufficiently high statistical correlation with measurable physiological and pathological processes to suggest a valid cause and effect relationship between organism and environment."

Since 1969 the members of ISB were not satisfied anymore with this very long definition and the greatest majority of the members from 53 countries decided to use for official purposes the short definition only, as indicated above.

However, this short definition is insufficient if the entire field of biometeorological activities should be sharply defined. For this reason another slightly longer definition was suggested indicating the various types of phenomena studied by biometeorology. According to this definition biometeorology comprises the study of the direct and indirect effects (of an irregular, fluctuating or rhythmic nature) of the physical, chemical and physico-chemical micro- and macro-environments of both the earth atmosphere, and of similar extra-terrestrial environments, on physico-chemical systems in general and on living organisms (plants, animals and man) in particular.

The biometeorological sciences have been divided into six main groups: phytological, zoological, human, cosmic, space and paleo-biometeorology.

1. PHYTOLOGICAL BIOMETEOROLOGY studies the influence of weather and climate on the development and distribution of plants for general phytological, agricultural and forestry purposes, and also the effects on small plant organisms responsible for the development of piant, animal and human diseases.

2. ZOOLOGICAL BIOMETEOROLOGY studies the influence of weather and climate on animals in general, on milk production, reproduction and coating of cattle, etc. (*Veterinary biometeorology*), on egg-laying rhythm, etc., of birds (*Avian biometeorology*), but also the effect of weather and climate on insects and terrestrial arthropoda, their physiological activities and development and the outbreak of plant, animal and human diseases as a result of these developments, such as the abundant development of locusts, malaria mosquitoes, etc. (*Entomological biometeorology*).
3. HUMAN BIOMETEOROLOGY studies the influence of weather and climate on man. It is usually divided into five important sections:
   a. PHYSIOLOGICAL BIOMETEOROLOGY: the influence of weather and climate on physiological processes in normal healthy man;
   b. SOCIAL BIOMETEOROLOGY: use of favorable meteorological factors for the treatment of large population groups as preventive or curative measures;
   c. PATHOLOGICAL BIOMETEOROLOGY: the influence of weather and climate on the various physiological and pathological phenomena associated with the diseases of man, the period of outbreak, intensity and geographical distribution;
   d. ARCHITECTURAL AND URBAN BIOMETEOROLOGY: the influence of the microclimates in houses and cities on the health of man and the effect of architectural constructions and town planning on these microclimates;
   e. NAUTICAL BIOMETEOROLOGY: the influence of weather and climate on animals and man living on moving ships.

Of special importance for the general practitioner are the results obtained in *Acclimatization biometeorology* (a branch of physiological biometeorology which studies the various physiological processes involved during adaptation to extreme conditions of temperature, humidity, etc., e.g., in the tropics or polar regions); *Air pollution biometeorology* (influence of pollen, spores, dust and chemical air pollution on man) and the various methods applied in *Climatotherapy*.

4. COSMIC BIOMETEOROLOGY studies the influence of extraterrestrial factors such as variations in solar activity, cosmic radiation, etc., on living organisms on earth.
5. SPACE BIOMETEOROLOGY is closely related to space research. It studies the microclimatic conditions in space vehicles and the biological effects (both observed and expected) of extra terrestrial physical environments on man and animals on the moon and the planets.
6. PALEO-BIOMETEOROLOGY studies the influence of climatic conditions in the past (millions of years ago) on the evolutional development and geographic distribution of plants, animals and man.

Table 1. Classification of HUMAN BIOMETEOROLOGY

1. PHYSIOLOGICAL BIOMETEOROLOGY
   a. General physiological biometeorology
   b. Geographical biometeorology
   c. Ethnological biometeorology
   d. Acclimatization biometeorology
2. SOCIAL BIOMETEOROLOGY
   a. General social biometeorology
   b. Psychological biometeorology (including aestheto-biometeorology)
   c. Archaelogical biometeorology
3. PATHOLOGICAL BIOMETEOROLOGY
   a. General pathological biometeorology
   b. Meteorological pathology
   c. Climatological pathology
   d. Air pollution biometeorology
      (1) Pollution with organic particles (pollen, spores, etc.)
      (2) Pollution with inorganic particles (dust, etc.)
      (3) Chemical pollution
      (4) Aerosol biometeorology
   e. Geographical climatopathology
   f. Climatotherapy (or therapeutic biometeorology)
      (1) General climatotherapy
      (2) Thalassotherapy (climatological)
      (3) Heliotherapy
      (4) Thermotherapy
      (5) Aerosol (and ionization) therapy
      (6) Socio-climatotherapy
      (7) Climatic health resorts
      (8) Climatic chamber treatments
   g. Pharmacological biometeorology
4. ARCHITECTURAL AND URBAN BIOMETEOROLOGY
   a. General urban biometeorology
   b. Architectural biometeorology
   c. Sanatorium biometeorology
5. NAUTICAL BIOMETEOROLOGY
   a. General nautical biometeorology
   b. Cargo biometeorology

S. W. TROMP

CHAPTER 2

EFFECTS OF LIGHT ON ANIMALS

SECTION 1: PHYSIOLOGICAL EFFECTS OF PHOTOPERIODICITY

- (a) General Rhythmic Processes in Animals (J. L. CLOUDSLEY-THOMPSON) (see P.H.B.*, Part IB, Chapter 6, Sect. 21b and 22b (1))
- (b) Metabolic and Cellular Rhythmic Phenomena
  - (1) Circadian Rhythms of Protein Metabolism in the Rat Brain (M. B. TER HAAR)
  - (2) Circadian Rhythms of Food Intake and Control Processes Thereof in the Rat (M. B. TER HAAR)
- (c) Photoperiodic Effects on Neuro-Endocrine Functions (G. L. JACKSON)
- (d) Photoperiodic Effects on Mammalian Reproduction (R. J. REITER)
- (e) Seasonal Rhythms in Animals (J. L. CLOUDSLEY-THOMPSON) (see P.H.B.*, Part IB, Chapter 6, Sect. 22b (1))
- (f) Seasonal Molting Cycles in Wild Mammals (ELISABETH JOHNSON)
- (g) Photoperiodic Effects on Hair and Wool Growth of Domestic Animals (J. C. D. HUTCHINSON)

SECTION 2: DIRECT EFFECTS OF LIGHT

- (a) Effects of Solar Radiation on the Sensible Heat Exchange of Mammals (J. C. D. HUTCHINSON, G. D. BROWN and T. E. ALLEN)
- (b) Effects on Reproduction of Mammals and Birds
  - (1) Effects of Light Intensity on Reproduction of Mammals (J. C. HOFFMANN)
  - (2) Effects of Light Spectra on Reproduction of Mammals and Birds (D. P. CARDINALI)
- (c) Phototoxicity in Animals (T. W. DUKES)

SECTION 3: EFFECTS OF LIGHT AND TEMPERATURE ON CERTAIN PHYSIOLOGICAL FUNCTIONS OF SHREWS (M. GEBCZYŃSKI)

* P.H.B. = Progress in Human Biometeorology (Ed. S. W. Tromp).

Chapter 2, Section 1a

General Rhythmic Processes in Animals

by

J. L. Cloudsley-Thompson

See Progress in Human Biometeorology, Part IB, Chapter 6, Section 21b, pp. 496-499 and Section 22b (1), pp. 578-580.

Circadian Rhythms of Protein Metabolism in the Rat Brain

by

M. B. ter Haar

INTRODUCTION

Observations by Hemmingsen and Krarup (1937) that "the length of an oestrous cycle must always be a whole number of days", and that the reversal of the daylight cycle by 12 hrs led to a change of 12 hrs in the time of ovulation were early indications that ovulation in the rat was under the influence of circadian lighting conditions.

The presumption that light acted through the brain in exercising a circadian influence on ovulation was substantiated by three reports. Everett and Sawyer (1949) showed that progesterone administration to female rats showing a 5-day oestrous cycle could bring about the processes leading to ovulation a day earlier than would have been expected, and that this effect could only be produced at a certain time of day. Using drugs known to act on the nervous system, Everett et al. (1949) showed with Dibenamine and Atropine (adrenergic and cholinergic blockers respectively) that there was a period of neural activation leading to the release of an ovulatory quota of hormone from the anterior pituitary. This period of activation occurred between 14.00 and 16.00 hr on the afternoon of pro-oestrus. Furthermore, this "critical" period of neural activation could also be blocked by a barbiturate anaesthetic, Nembutal (Everett and Sawyer, 1950). Ovulation would occur 24 hrs later following this treatment, and further blocking of this delayed ovulatory stimulus could also be obtained if Nembutal was administered 24 hrs after the first injection over the same "critical" period of 14.00 to 16.00 hr.

Since the pioneer studies of Pfeiffer (1936) in the rat, it has become recognized that male and female brains become sexually differentiated by the action of testicular hormones produced very early in life. It has been shown that the sexual differentiation of the rat can be reversed from that of its genetic sex by altering the hormonal state of the animal within a few days of birth (see Barraclough, 1968; Harris, 1970). Hence, if androgens are injected into a neonatal female rat, the animal, when adult, will be anovulatory, areceptive to males, and will show persistent vaginal oestrus. Conversely, neonatal castration of a male rat permits the animal after puberty to maintain ovulatory cycles in implanted ovaries and vaginae. It is concluded from these experiments that the female and male brains differentiate in different directions so that the female brain can respond to elevated oestrogen levels in the blood in such a way that a surge of luteinizing hormone (LH) well above basal secretion levels occurs, and which can lead to ovulation. On the other hand, this "facilitation" of LH release is not generally obtained in the male, nor in the neonatally-androgenized female.

Numerous attempts have been made to detect variations in brain activity that might be involved in the correlation between circadian influences, and the induction of ovulation via a "facilitated" rise of LH in the female rat. That biochemical processes may be involved is suggested by the observation of Jackson (1973) who, through the use of inhibitors of RNA and protein synthesis, indicated that events involving these processes occurring over a closely defined period of a few hours are involved in the facilitation by oestrogen of an ovulatory rise of LH.

In order to establish a correlation between diurnal stimuli, biochemical events in the brain, sexual differentiation, and the control of ovulation, we have been investigating the incorporation of a radioactively labelled amino acid ([^{35}S]methionine) into protein in various neuroendocrine states of the rat (Ter Haar and MacKinnon, 1973; Ter Haar et al., 1974).

SEXUAL DIFFERENTIATION AND RHYTHMS OF PROTEIN METABOLISM

[^{35}S]methionine was chosen as a marker for investigations of brain protein metabolism since there is little incorporation of the [^{35}S] label into compounds other than proteins; and, as a result, it was relatively certain that the observations that were obtained were a function of protein metabolism alone.

From measurements of the incorporation of [^{35}S]methionine into protein in various areas* of rat brain at various times of day, it was observed (Ter Haar et al., 1974) that adult animals of either sex showed a circadian rhythm in this parameter, but that the phase was 8 hrs different between the sexes. Hence, whereas the male showed maximal levels of incorporation in the morning, the female showed peak levels late in the day (Fig. 1). The presence of circadian rhythm has been described for a number of biochemicals in the brain; and indeed, Richardson and Rose (1971) demonstrated a circadian rhythm of incorporation of another amino acid ([^{3}H]lysine) into protein of the lateral geniculate nucleus and cortex of male rats. Hiroshige, Abe et al. (1973) have observed a sex difference in a cerebral circadian rhythm in that they found the circadian rhythm of corticotrophin releasing hormone content of the hypothalamus of the male rat to be 180° out of phase with that of the female.

The treatment of genetic female rats with androgens at birth results in the animals being anovulatory when adult, as well as showing many male behavioural characteristics. Animals treated in this way showed a pattern of incorporation of [^{35}S]methionine into protein at adulthood similar to that normally observed in male rats i.e., with peak levels in the morning. Conversely, castration of genetic male rats gave rise to a female pattern in the adult (Fig. 2).

* The brain areas investigated were the amygdala, the preoptic area, and the area of the median eminence, since these have been implicated in the neural control of ovulation (Harris and Campbell, 1966), as well as thalamus and putamen which were taken as control areas. Since the rhythms of methionine incorporation described in this chapter were similar in all the areas investigated, only the patterns observed in the thalamus will be illustrated.

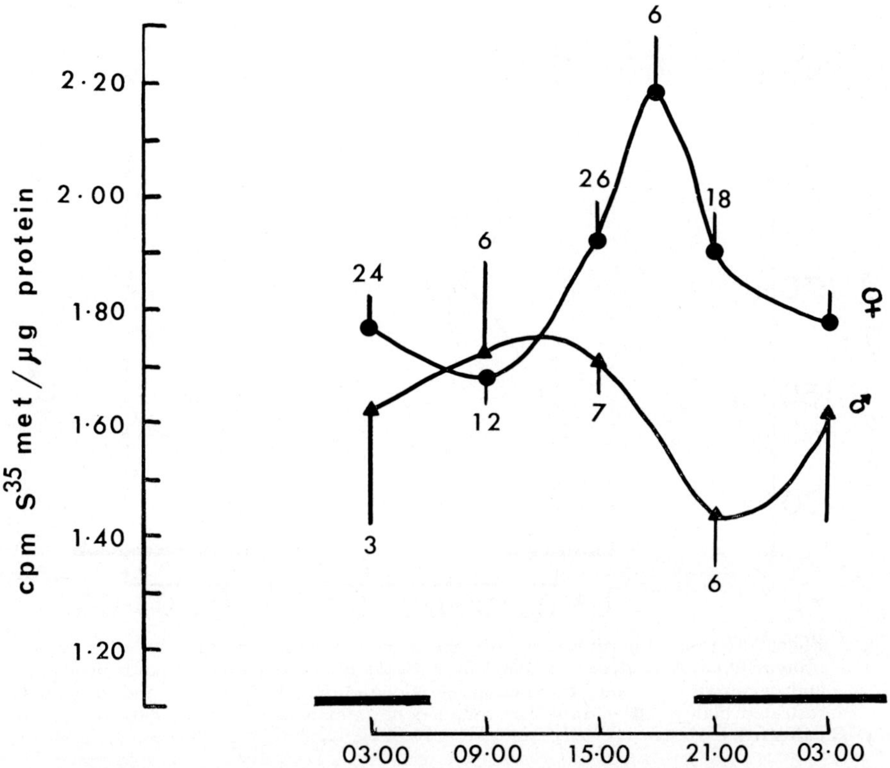

Fig. 1. The incorporation of [^{35}S] methionine into protein (cpm/μg protein $\pm$S.E.M.) over 24 hrs in the thalamus of normal adult male (—▲—) and female (—●—) rats. Values for the females are means for the same time of the day of each day of the 4-day oestrous cycle. The numbers of animals used and the dark periods are indicated.

The conclusions that are drawn from these observations are that the phase of the circadian rhythm demonstrated by the adult is determined by the hormonal environment at the critical period of development around the time of birth, and not by genetic determinants. Secondly, it appears that the phase demonstrated by the female is required for the potential to display oestrous cyclicity and ovulation. Thus the neonatally castrated male shows oestrous cyclicity in implanted vaginae and ovaries, as well as having a female phase of its circadian rhythm of protein metabolism.

ONTOGENY AND RHYTHMS OF PROTEIN METABOLISM

Following the experiments that led to the suggestion of a correlation between the presence of a female phase in the pattern of methionine incorporation in the brain and the potential of the brain to initiate the events leading to ovulation, a study was undertaken to find out how these rhythms developed during puberty and whether they still existed in the old anoestrous rat.

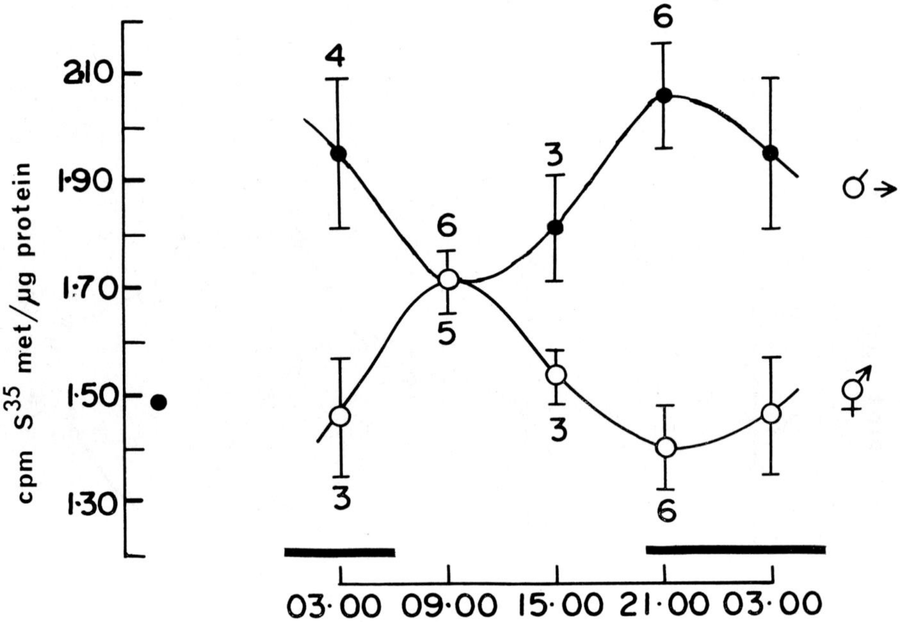

Fig. 2. Effects of neonatal hormonal environment. The incorporation of [35S] methionine into protein (cpm/μg protein ±S.E.M.) over 24 hrs in the thalamus of adult neonatally-androgenized (1.25 mg testosterone propionate) female (—○—) and neonatally-castrated male (—●—) rats. The numbers of animals used and the dark periods are indicated.

At 12 days of age, while the eyes of the rat are as yet unopened, no circadian rhythm of [35S]methionine incorporation can be detected in either male or female animals (Table 1), while at 30 days of age, though the animals are still prepubertal (vaginal opening occurs at around 40 days of age in our colony of rats), there appear to be circadian rhythms of incorporation which are out of phase between the sexes (Fig. 3).

Table 1. Incorporation of [35S]methionine into protein of the thalamus (cpm/μg protein ± S.E.M.) at 12 days of age in male and female rats

Time	03.00 hr	09.00 hr	15.00 hr	21.00 hr
Male	—	3.59 ± 0.06 (4)	—	3.83 ± 0.29 (4)
Female	3.30 ± 0.38 (4)	3.85 ± 0.05 (5)	3.23 ± 0.45 (4)	3.70 ± 0.30 (5)

The number of animals is indicated in parentheses.

However, the phase difference is slightly less and the time of each peak occurs somewhat earlier than in the adult.

It is of note that it is in fact possible to advance puberty to this age, or even a couple of days earlier (Donovan and Van der Werff ten Bosch, 1965; Critchlow and Bar

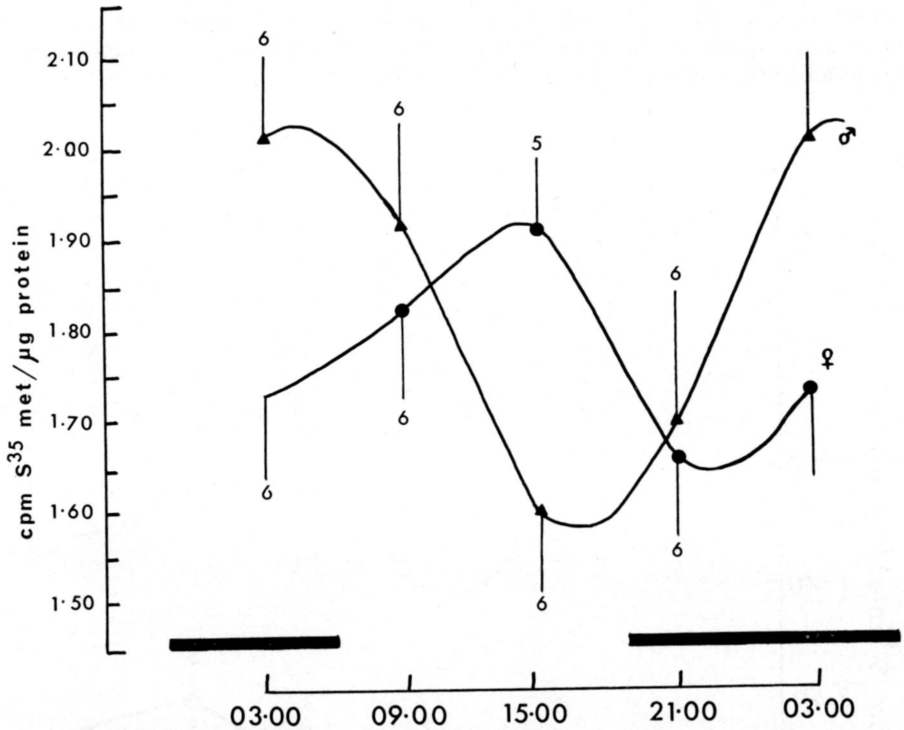

Fig. 3. Prepubertal rhythms. The incorporation of [^{35}S] methionine into protein (cpm/μg protein ±S.E.M.) over 24 hrs in the thalamus of 30-day old male (—▲—) and female (—●—) rats. The numbers of animals used and the dark periods are indicated.

Sela, 1967); and this may be related to the attainment by the female of some degree of its adult female pattern of protein metabolism by this age.

At the other end of the developmental scale, the female rat becomes acyclic with age, going into a state of persistent dioestrus, anoestrus, or oestrus. In such animals the female pattern of [^{35}S]methionine incorporation was found to be lacking as is shown in Figure 4. This loss of the full female circadian pattern of protein metabolism in the old female rat may be related to the loss of oestrous cyclicity.

EXTERNAL INFLUENCES AND RHYTHMS OF PROTEIN METABOLISM

Apart from the observation that the rhythms described above exhibit a period of 24 hrs, there is further evidence that these rhythms are under the direction of the light-dark cycle. This is suggested from observations that a 6 hr shift of the lighting regime gives rise to a 6 hr shift in the timing of the rhythms of methionine incorporation

(Burnet and MacKinnon, Personal Comm.). Secondly, animals placed under constant illumination appear to lose their "typical" rhythms of methionine incorporation, with the mean levels of incorporation occurring at somewhat higher levels than any of the

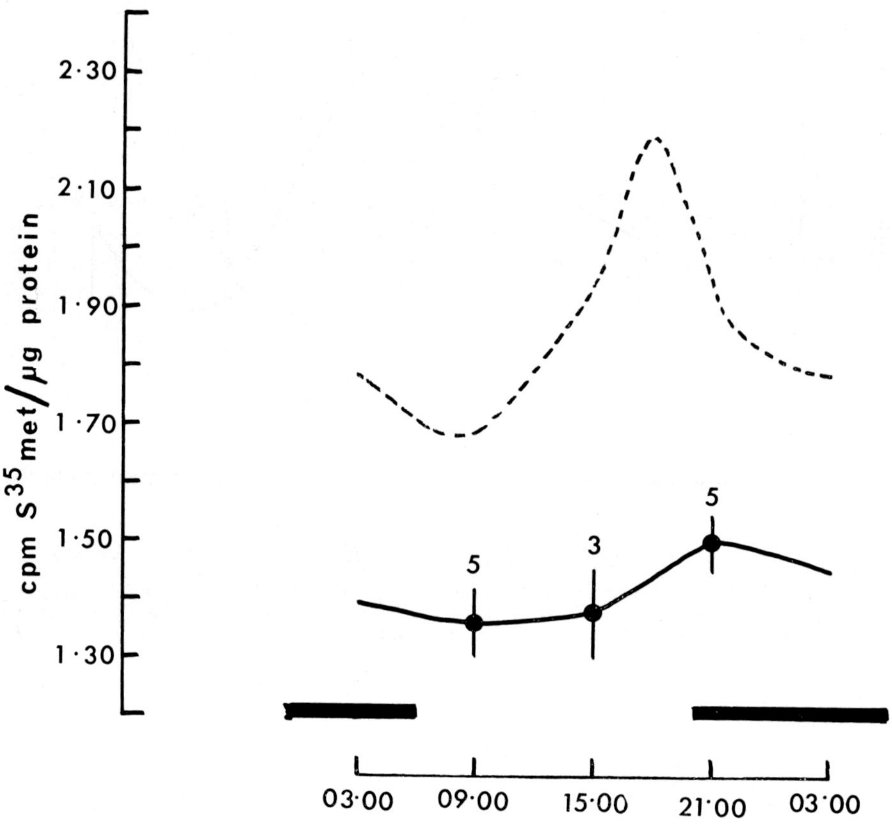

Fig. 4. Effect of old age. The incorporation of [^{35}S] methionine into protein (cpm/μg protein ±S.E.M.) in the thalamus of 25-30-month old acyclic female rats. The rhythm observed for normal cyclic females (......), the number of animals used, and the dark periods are indicated.

means observed in the normal light-dark regime (Fig. 5). Females kept under constant illumination become acyclic and enter a state of constant vaginal cornification, though they can exhibit reflex ovulation under these conditions (Brown-Grant et al., 1973). Thus it would appear that though the ability to ovulate does not disappear, the cyclic phenomena leading up to spontaneous ovulation do, together with the disappearance of the female phase of protein metabolism.

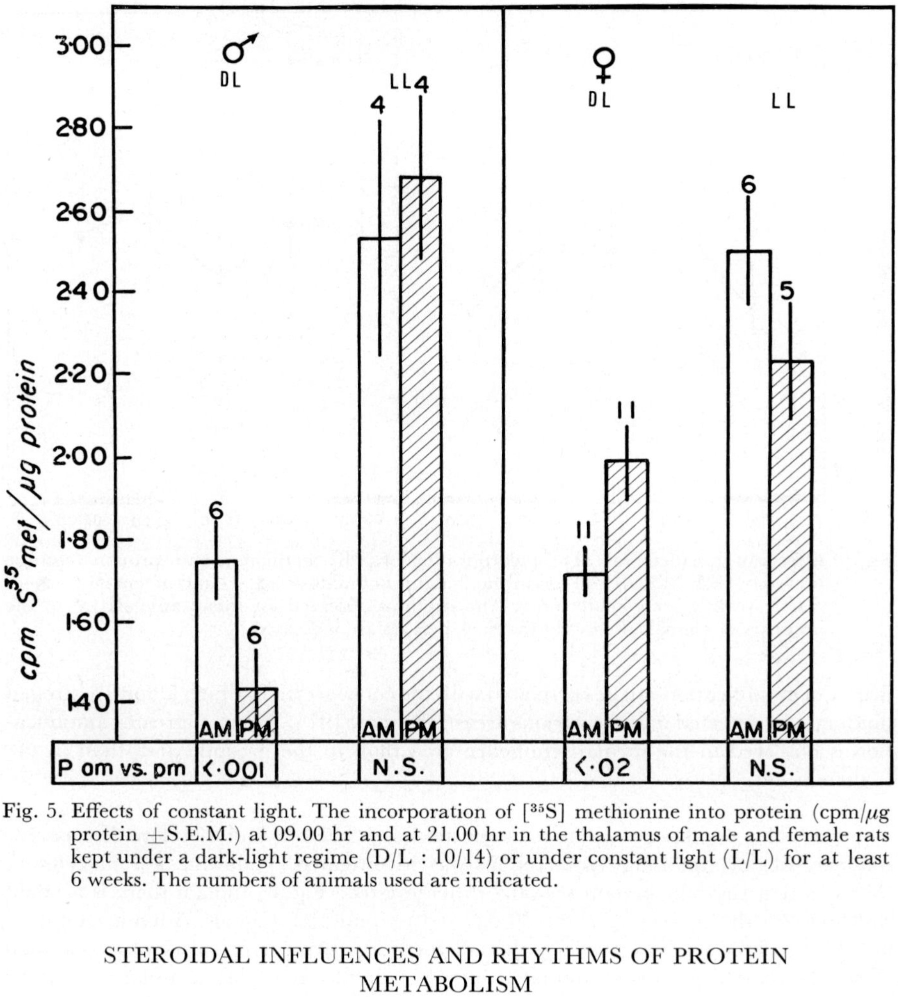

Fig. 5. Effects of constant light. The incorporation of [^{35}S] methionine into protein (cpm/μg protein $\pm$S.E.M.) at 09.00 hr and at 21.00 hr in the thalamus of male and female rats kept under a dark-light regime (D/L : 10/14) or under constant light (L/L) for at least 6 weeks. The numbers of animals used are indicated.

STEROIDAL INFLUENCES AND RHYTHMS OF PROTEIN METABOLISM

In the previous sections it has been demonstrated that adult circadian rhythms of methionine incorporation are dependent upon the neonatal hormonal environment and upon the adult external environment. These rhythms would also appear to be under the influence of the adult hormonal environment. Following removal of the gonads of the adult animal, the circadian rhythms of the adult animal disappear (Fig. 6). However, a single injection of oestradiol benzoate to ovariectomized females leads to the return of the female pattern (Burnet et al., 1974; Burnet and MacKinnon, 1975) together with the return of a daily facilitation of LH release (Caligaris et al., 1971).

The action of oestradiol administered to the intact animal, on the other hand, is to increase the amplitude of the female rhythm (Fig. 7). In the different brain areas the

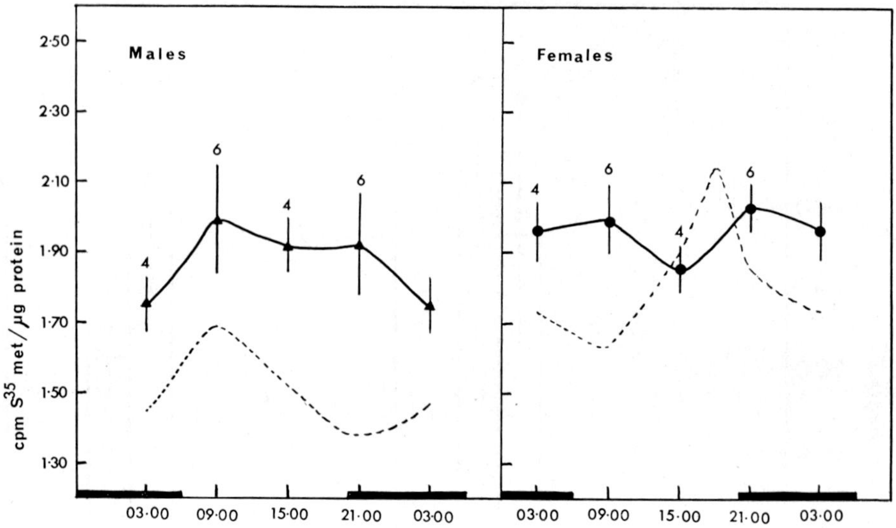

Fig. 6. Effects of gonadectomy. The incorporation of [^{25}S] methionine into protein (cpm/µg protein ± S.E.M.) over 24 hrs in the thalamus of male (—▲—) and of female (—●—) rats 3 weeks after gonadectomy. The rhythms observed for intact animals (.....), the numbers of animals used, and the dark periods are indicated.

degree of amplification can be correlated with the concentration of high affinity oestrogen binding sites detected in these various areas (Thomas, 1973). Hence a greater amplification is observed in the median eminence area than in the preoptic area than in the thalamus, etc.

The adrenal is an important source of sex steroids in the rat, and therefore the possible influence of this gland and its secretions was investigated. Following adrenalectomy, the circadian rhythms remain sexually differentiated (Fig. 8) though there is a phase shift towards that observed in the 30-day old prepubertal animals. Adrenalectomized females continue to show cyclicity though there is greater irregularity observed than in sham-operated groups of animals; thus the adrenal may play a minor regulatory role, possibly through its progesterone secretion. In connection with the phase shift of the adrenalectomized animals towards the patterns displayed by the prepubertal animals, the possible importance of the adrenal in the onset of puberty may be relevant (Gorski and Lawton, 1973; Weisz and Gunsalus, 1973).

CONCLUSIONS

From the results of the experiments described it is clear that the pattern of protein metabolism shown by any animal is under the direction of a number of influences, both internal and external. These many influences, some of which have been discussed in this chapter, are summarized in Figure 9.

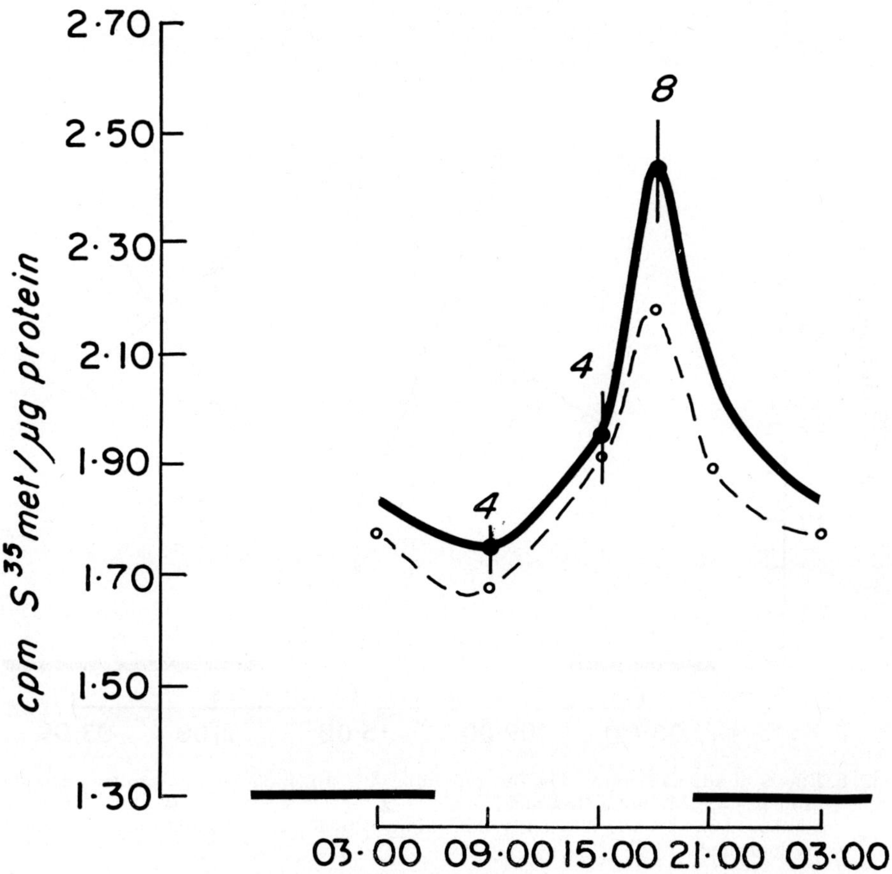

Fig. 7. Effect of oestrogen. The incorporation of [^{35}S]methionine into protein (cpm/μg protein $\pm$S.E.M.) in the thalamus of intact female rats following 3 daily injections (s.c.) of 20 μg of oestradiol benzoate. The rhythm observed for normal females (...○...), the numbers of animals used, and the dark periods are indicated.

From the physiological correlates of certain patterns of methionine incorporation, we have put forward the tentative hypothesis that a female circadian pattern of methionine incorporation is required in the rat for the animal to exhibit a positive feedback (or facilitatory) response to oestrogen in the form of a surge of LH release which leads to ovulation (Ter Haar et al., 1974). Thus the potential for oestrous cyclicity is not shown by males, neonatally-androgenized females, old females, animals kept under constant illumination, or in gonadectomized animals: none of these categories of animals displays a female pattern of rhythm of methionine incorporation into cerebral protein. On the other hand, normal females and neonatally-castrated males are both capable of showing adult oestrous cyclicity and, at the same time, show a female phase of protein metabolism.

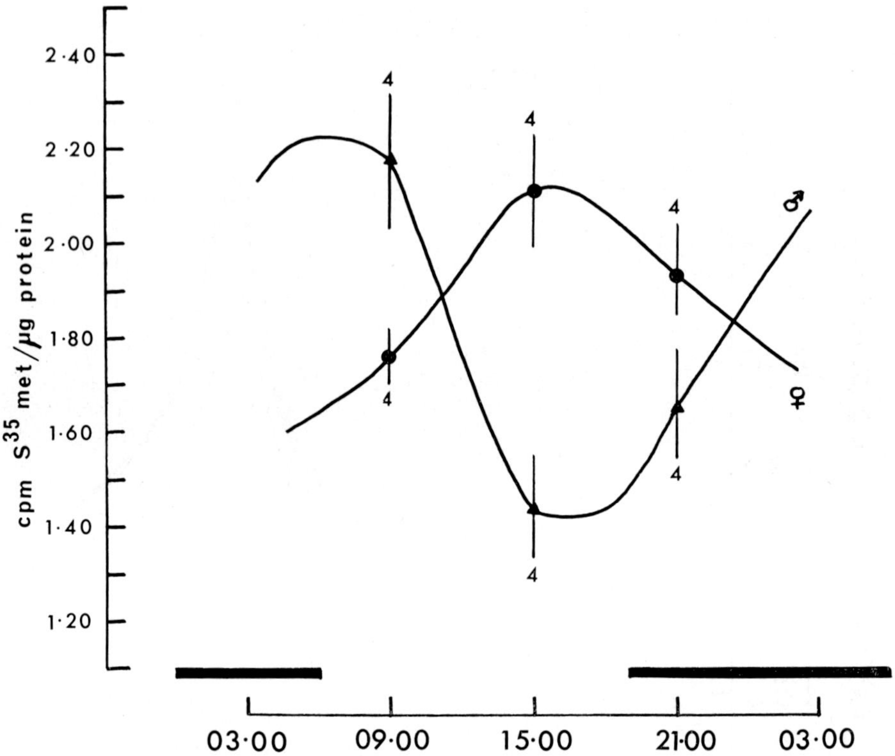

Fig. 8. Effects of adrenalectomy. The incorporation of [^{35}S]methionine into protein (cpm/µg protein ±S.E.M.) in the thalamus of male (—▲—) and of female (—●—) rats following adrenalectomy 3 weeks previously. The numbers of animals used and the dark periods are indicated.

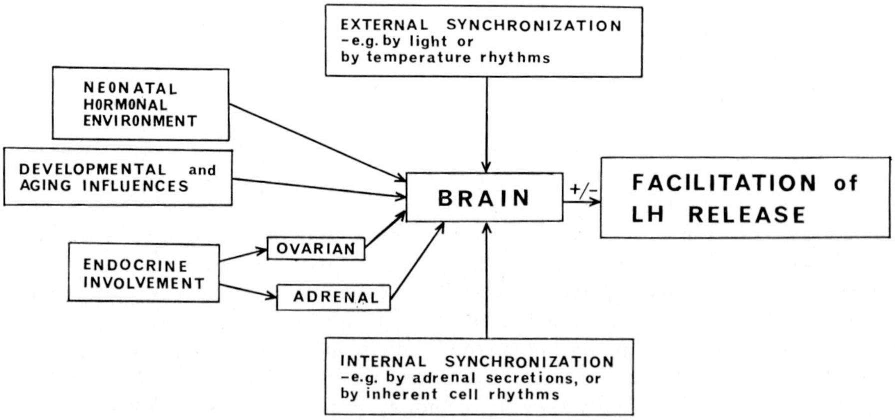

Fig. 9. Influences affecting the neural facilitation of LH release.

There is strong evidence that only a small area of the brain, the area incorporating the preoptic area and the median eminence area, is required for the maintenance of oestrous cyclicity and ovulation (Halasz, 1969). Thus in this part of the brain the functional relevance of a sex difference in biochemical systems responsive to hormonal and external environments is obvious. The functional significance, if such exists, of the sex difference in phase of cerebral protein metabolism in other brain areas is yet to be established, though the roles of other brain areas as pleasure centres etc., may be of relevance.

However, it would be premature to speculate at too great a length on the causes and the effects of rhythms described in this Section. Suffice it to say that we may be observing a biochemical mechanism by which external and internal milieux are synchronized to ensure continued reproductive activity under suitable environmental conditions.

ACKNOWLEDGEMENT: I should like to express my gratitude to Dr Pamela C. B. MacKinnon for the happy collaboration that brought the results described, and to Professor B. T. Donovan for his constructive criticism of this section.

REFERENCES: see pp. 259-260.

Circadian Rhythms in Food Intake and Control Processes Thereof in the Rat

by

M. B. ter Haar

INTRODUCTION

It is well known that under normal conditions the rat eats more at certain times of day than at others. However, the motivation behind this eating activity is not so clear. In this Section the pattern of food intake in the rat will be described, followed by a discussion of various processes involved in the control of food intake together with any possible correlations between known circadian rhythms and that of food intake.

PATTERNS OF FEEDING ACTIVITY

Our knowledge of rhythmic functions in the rat owes much to the extensive studies of Curt Richter who has shown that the rat exhibits rhythmic patterns of running activity with periods varying in length from a couple of hours to several months, with the rhythms of shorter period length sumperimposed upon those with a longer period (Richter, 1927, 1965). In an extension of these studies Richter also examined rhythms of various other parameters including those of feeding and drinking, and established that feeding shows a circadian rhythm which runs parallel with the daily activity pattern.

Most of these earlier investigations made on feeding patterns were qualitative in that they only measured activity towards feeding, and not the amount of food eaten. However, Siegel and Stuckey (1947) provided one of the earlier quantitative reports of food intake. Their measurements over 6-hourly intervals indicated that maximal food intake occurred in the male rat in the 6-hr period immediately after the onset of darkness. Measurements of food intake over 2-hourly intervals in the male and female rat showed that food intake was in fact maximal immediately after the onset of darkness, with a second though smaller peak later in the dark period (Fig. 1) (Ter Haar, 1972). These observations are in general agreement with those of Balagura and Devenport (1970).

In the female a departure from this pattern is observed on the day of pro-oestrus when the peak of feeding activity immediately after the onset of darkness is apparently suppressed (Fig. 1b). This suppression is possibly as the result of the elevated oestrogen levels observed earlier on the day of pro-oestrus (Brown-Grant et al., 1970), an effect which will be considered later.

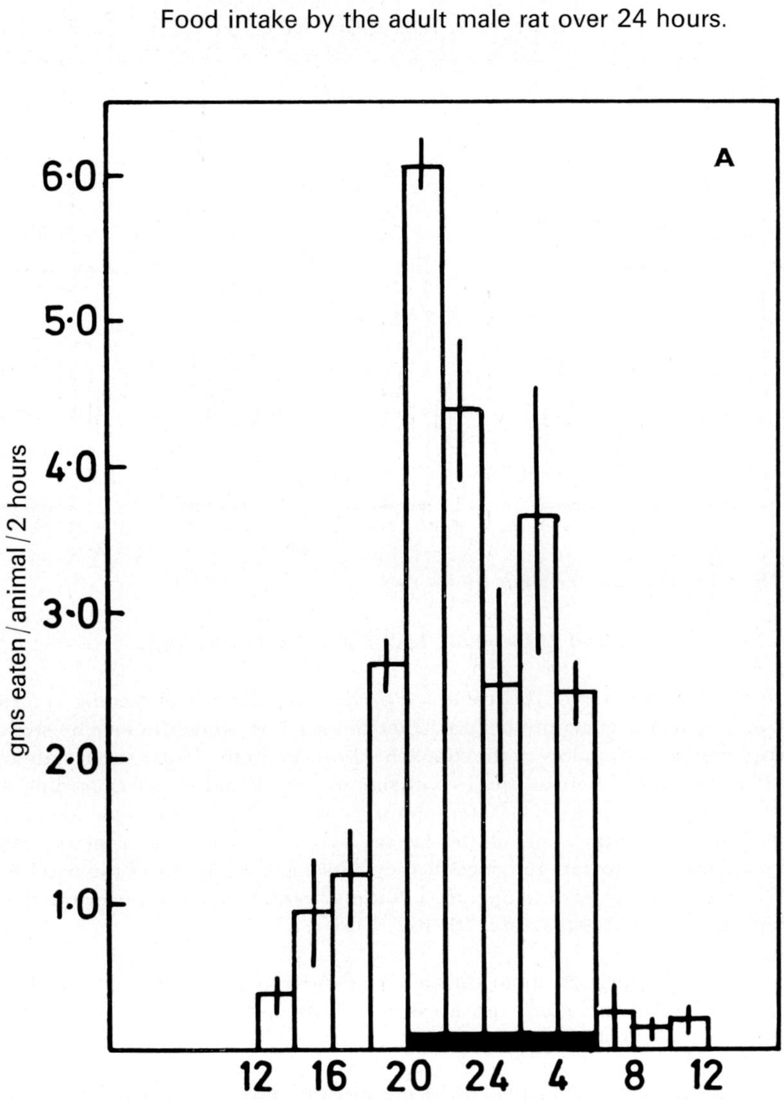

Fig. 1A. Food intake (±S.E.M.) by (A) adult male rats and (B) adult female rats as measured at 2-hourly intervals over 24 hrs and a 4-day oestrous cycle respectively. The insets in (B) indicate the timing and patterns of 17β-oestradiol and luteinizing hormone (LH) secretion as well as the time of ovulation in the same colony of rats. The hours of darkness are indicated. P = pro-oestrus; O = oestrus; M = metoestrus; D = dioestrus (From Ter Haar, 1972).

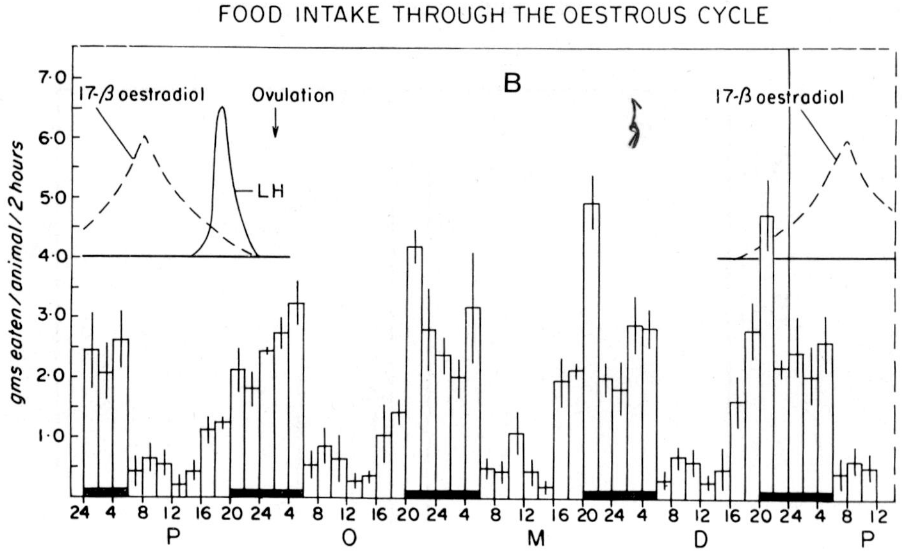

Fig. 1B. Food intake by the adult female rat.

SYSTEMS REGULATING FOOD INTAKE

Up to the first quarter of this century "peripheral" theories of feeding control held sway in that it was generally believed that hunger was brought about by stimulation of peripheral nerves, mainly in the stomach. However, in the 1940's surgical destruction of part of the brain, namely the hypothalamus, was found to affect feeding activity and give rise to hyperphagia. Greater sophistication of stimulation and lesion techniques resulted in the identification of the lateral hypothalamus as a primary centre for producing the urge to eat. In contrast, the ventromedial nucleus of the hypothalamus (VMN) has been suggested to operate a "satiety" centre which can overrule the eating urges of the lateral hypothalamus (Stellar, 1954).

It should also be borne in mind that other cerebral areas certainly have their effects upon food intake, including the areas involved in, for example, taste preference (Grossman, 1968; Rolls and Rolls, 1973).

Acting upon these neural centres are a number of influences, of which some are summarized in Figure 2. Due to restrictions of space, brief attention only can be paid to each of them; however, the references mentioned will provide much greater detail of each system.

The majority of these control systems are humoral, in that the active feedback components are carried in the bloodstream. However, before these humoral systems take effect and well before the products of digestion are absorbed, a state of satiety is normally reached: it appears that there is a sensory system which monitors the passage of food through, and the presence of food in, the digestive system. This sensory system, which is sensitive to the fat/carbohydrate/amino acid composition of the food in the digestive system is suggested to operate through a neural pathway to the VMN.

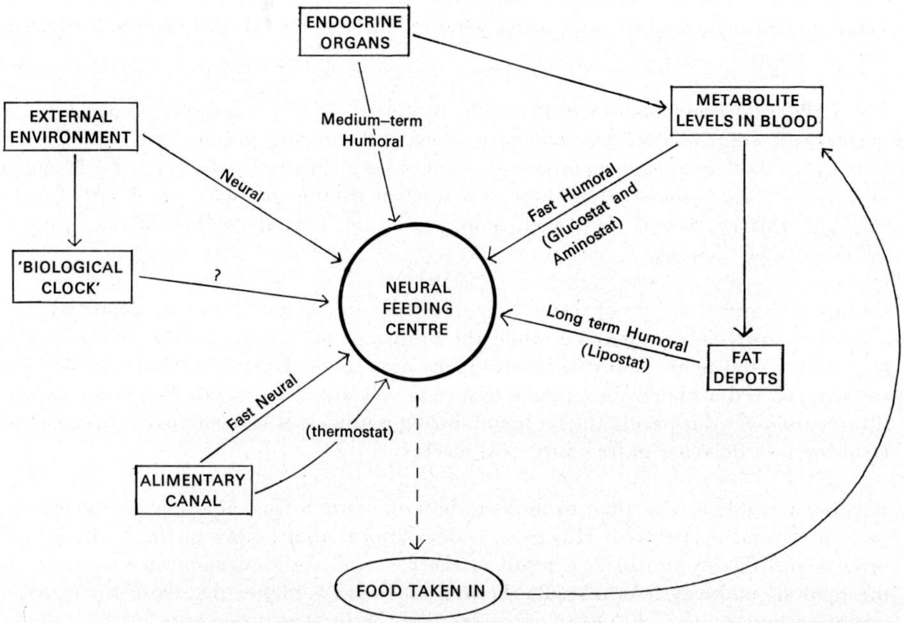

Fig. 2. A diagrammatic summary of the various direct and indirect influences of food intakes.

For a longer-term control system of food intake, Brobeck (1948) and Strominger and Brobeck (1953) proposed a THERMOSTAT mechanism of control of food intake through which the body balances energy intake with expenditure. They suggested that the hypothalamus detects changes in blood temperature, or some similar parameter related to the body's energy balance; however, it appears that though hypothalamic temperature changes generally run parallel with changes in food intake, these two functions are only related indirectly (Grossman, 1968).

Three CHEMOSTAT mechanisms of feeding control have been postulated. These are the "glucostat", the "lipostat", and the "aminostat" mechanisms, acting respectively through detection of blood levels of glucose, fats, and amino acids.

The GLUCOSTAT mechanism was first postulated by Mayer (1955) who showed in mice that gold thioglucose was concentrated in the VMN, and that this accumulation gave rise to damage in this area, and, as a result, obesity. On the other hand, Colin-Jones and Himsworth (1970) have shown with hypothalamic injections of 2-deoxy-D-glucose that there are also glucose-sensitive chemoreceptors in the lateral hypothalamus. Furthermore, the cells containing these receptors appear to respond to hypoglycaemia by the secretion of adrenaline (Himsworth, 1970). It would appear, then, that both the lateral and ventro-medial areas of the hypothalamus are involved in the "glucostat" mechanism of feeding control. This is further supported by the findings of Anand et al. (1964) who found that the unit activity of neurones in the satiety centre (VMN) was slower than in the lateral hypothalamus. When glucose was administered intravenously, unit activity in the VMN increased while that of the lateral area decreased.

The LIPOSTAT mechanism proposed by Kennedy (1953) is possibly responsible for longer-term control of energy balance through monitoring plasma fatty acid levels. It is in this system that a metabolic diurnal rhythm has been observed. Le Magnen et al. (1973) have shown that there is a diurnal rhythm of plama free fatty acids, and that this rhythm is destroyed, along with the diurnal rhythm of food intake, following VMN lesions.

From their measurements of respiratory quotients and of food intake it appeared that a lipolytic mechanism operated while the animals were resting and which gave rise to a release and hence elevated levels of free fatty acids. During the hours of activity the rats showed a higher food intake than was warranted by metabolic energy expenditure, and over this period the fat metabolizing pathways were switched to lipogenesis, resulting in a decrease in free fatty acid levels.

It is not possible at this time to indicate how the switch from lipolysis to lipogenesis, and vice versa, is operated. However, it does appear that lesions of the VMN which bring about obesity, do so as a result of the removal of the mechanism switching on the lipolysis pathways. As a result the animal shows a higher degree of lipogenesis: a state of hyperphagia ensues to accommodate the increased demands of this system.

The third "chemostat" mechanism, the AMINOSTAT mechanism, proposed by Mellinkoff et al. (1956) may also be of importance for short-term control of food intake. This mechanism was suggested as a result of the positive correlation between serum amino acid levels and the sensation of satiety. A fall in plasma levels of non-essential gluconeogenic amino acids may bring about an increase in food intake. On the other hand, a reduction in plasma levels of certain essential amino acids brings about a reduction in food intake, an effect possibly mediated through influences on brain protein metabolism (Leung and Rogers, 1969).

However, the quantitative importance relative to each other of these various postulated mechanisms in determining normal food intake has yet to be established.

CIRCADIAN INFLUENCES

The control processes described so far, apart from the lipolysis/lipogenesis diurnal switching system described by Le Magnen et al. (1973), do not appear to exert, or to be under, any direct circadian influence. Therefore it is expedient to discuss further sources of control that may bear a causal relationship to the circadian rhythm of food intake. For instance, it is now well established that various hormones have a major influence on the control of food intake, both directly and indirectly.

A number of hormones, such as insulin, the glucocorticoids, growth hormone, thyroxine, and possibly glucagon and adrenaline, have an indirect effect on the cerebral processes controlling food intake in that they influence metabolic processes and hence the energy balance. Thus a diurnal rhythm in the secretion of some of these hormones may affect the pattern of food intake: a cyclical secretion rate will have a differential effect on metabolic rates and hence could give rise to a diurnal rhythm in the concentrations of the humoral agents operating the control mechanisms proposed above.

Certain of the steroid hormones (e.g., 17ß-oestradiol), for which specific binding proteins have been demonstrated in the hypothalamus and other areas of the brain, have also been shown to have an affect on food intake. For example, Kakolewski et al. (1968) showed that ovariectomy of adult female rats gave rise to an increased body weight together with a somewhat increased food intake, while Tartellin and Gorski (1971) demonstrated that the administration of oestrogens to ovariectomized adult animals depressed the increased levels of food intake. Other detailed investigations have further emphasized that increased oestrogen levels give rise to lowered food intake (e.g., Zucker, 1969; Wade and Zucker, 1969, 1970; Bell and Zucker, 1971). Evidence that oestrogen is acting directly on the feeding centres comes from the findings of Wade and Zucker (1970) that VMN implants of oestrogen caused a decrease in food intake after a period of 12 hrs, while the effect took somewhat longer after systemic injection. (However, in implantation studies the question arises as to whether the implanted substances act at the implantation site or whether it diffuses a certain distance to act elsewhere.)

Hence, it is possible that the reduction of food intake observed on the evening of pro-oestrus in the female rat (Fig. 1b) is brought about by the oestrogen surge observed earlier on pro-oestrus (Brown-Grant et al., 1970). This reduction might on the other hand also be the result of elevated levels of prostaglandin E_1 since this hormone has also been shown to exert an inhibitory influence on the cerebral control of food intake (Baile et al., 1973).

It is unclear as to whether there is any direct correlation (as compared with the possible indirect relationship described above) between hormonal levels and the circadian rhythm of food intake. There is evidence that there are circadian rhythms in levels of plasma oestrogen (Yoshinaga et al., 1969), of plasma testosterone (e.g., Kinson and Chung-Ching, 1973), and of corticosterone (e.g., Hiroshige et al., 1973), all of which may have a circadian effect on food intake either directly or indirectly.

Only influences from the internal environment of the animal on the neural control of food intake have been considered so far. Of these influences, those showing circadian rhythms are those which can also be shown to be under a neural control initially, and thus consideration should be made of the cerebral circadian processes possibly bringing about these circadian rhythms.

These rhythms may be directed by one of various transmitter circadian rhythms that have been shown to occur in the brain e.g., for 5-hydroxytryptamine (Albrecht et al., 1956; Quay, 1965), noradrenaline (Reis and Wurtman, 1968), acetyl choline (Hanin et al., 1970), and general biogenic amines (Friedman and Walker, 1968). On the other hand, circadian rhythms exhibited in other biochemical systems e.g., RNA metabolism (Merritt and Sulkowski, 1970) or protein metabolism (Ter Haar et al., 1974) may be of greater importance.

As to the drive behind these circadian rhythms, knowledge is still very limited. There are indications (e.g., Richter, 1968, 1971) that a "biological clock" driving these circadian rhythms may be inherent in that adult rats will show an activity rhythm of about 24 hrs even if they have been blinded within the first few days of life, or if they have been kept in constant darkness. However, it is difficult to exclude possible influences on these animals from many other circadian cues, such as gravitational or pressure changes.

Furthermore, many of the circadian rhythms of the brain may be under the influence of the circadian rhythms exhibited by the pineal gland and its secretory products e.g., melatonin and serotonin. The anatomical pathways by which environmental light is transmitted to the pineal appear to be separate from the primary visual pathways: ablation of the primary optic tracts does not appear to delete activity rhythms, while ablation of the inferior accessory optic tracts does not interfere with normal vision, but affects pineal and gonadal responses to light (Moore et al., 1968).

Thus the effects of environmental illumination may be relayed by one or both of these visual systems to localized central structures with inherent circadian periodicities entrainable to light-dark fluctuations which in turn regulate the entrainment of a large number of circadian processes. Taking this concept further, Menaker (1971) suggests that the response of an organism to light is "not by measuring the length of the light period, nor by measuring the length of the dark period, nor again by measuring the ratio of the light period to the dark period. Rather, organisms distinguish between inductive and non-inductive daylengths by assessing whether or not light is present at a particular phase point on an endogenous circadian rhythm of sensitivity to its inductive effects."

CONCLUSIONS

Having very briefly considered the various influences controlling food intake described in Figure 2, the conclusion must be reached that though it is indisputable that the rodent displays a circadian pattern of food intake, it is not possible at this time to postulate how this rhythm is derived. One can only be aware of the controls exerted on this system, and hope that future research will show which of these control systems are of primary, and which are of secondary importance.

REFERENCES: see pp. 260-261.

Chapter 2, Section 1c

Photoperiodic Effects on Neuro-Endocrine Functions

by

G. L. Jackson

INTRODUCTION

Under natural conditions the duration of an animal's exposure to light varies with two cycles: a 24-hr light-dark cycle of day and night and an annual cycle of changing day length (absent near the equator). Many physiological functions change in relation to these cycles. Also, the periodicity of some functions is synchronized by these cycles. This Section examines briefly the effects of daily and annual light cycles on selected neuroendocrine functions, particularly functions of the brain, pineal gland, and anterior pituitary gland. Some examples of how light interacts with hormonal and neural factors to regulate specific physiological events are also discussed. Description of the various qualities of light which effect neuroendocrine function is found elsewhere (Van Tienhoven, 1968) and was purposely omitted from this Section.

The effects of light on the neuroendocrine system are mediated through a specialized and complicated system of photoreceptors, endocrine transducers, and neural analyzers. The primary, and probably only photoreceptor in the adult mammal is the retina. However, there are functional extraretinal photoreceptors in some immature mammals, birds, and in certain reptiles and amphibia. The primary endocrine transducer is the hypothalamic median eminence, which secretes its releasing hormones into the hypophysial-portal circulation. Another transducer of importance in some vertebrates is the pineal gland.

Neural analyzers relay messages from the photoreceptors to the hypothalamic median eminence or to the pineal gland (Moore and Lenn, 1972; Moore and Eichler, 1972; Wurtman, Axelrod and Kelley, 1968).

Relatively little is known about the neural analyzers; however, despite lack of detailed knowledge of the neural analyzer systems, the general mechanisms through which a light signal is received by a photoreceptor and transformed into an endocrine response can be described. The initial step is stimulation of the retinal photoreceptor. The stimulation generates nerve impulses which pass along specific neural pathways and induce, in sequence, alteration in secretion of neurotransmitters in the brain, alteration of secretion of hormone-releasing factors by the hypothalamus, corresponding changes in secretion of pituitary hormones, and finally changes in rates of function of endocrine target organs such as the adrenal cortex or gonads. By understanding these basic relationships one can recognize that a change in function of an endocrine target organ may reflect a preceding alteration in function within the anterior pituitary gland or the brain (Schwartz and McCormack, 1972; Everett, 1969).

Several experimental approaches have been used for studying physiological responses to photoperiod. A common one is to determine effects of various lighting regimens on the content either of neurotransmitters in the brain or of tropic hormones in the anterior pituitary gland. Interpreting such experiments is often difficult, for a change in content of a particular substance may reflect a change in rate of synthesis, degradation, release, or a combination thereof. Determining which processes are responsible for observed changes requires knowledge either of turnover or of release rate of that substance. Although the majority of reports fail to provide complete data, they remain of value for showing whether levels of a substance change in a rhythmic manner or are affected by changes in the photoperiod.

DAILY PHOTOPERIOD AND NEUROENDOCRINE EVENTS

RHYTHMS IN BRAIN NEUROTRANSMITTERS

The neurotransmitters serotonin, norepinephrine, dopamine, and acetycholine, play an important role in modulating secretion of hypothalamic hormone-releasing factors. Thus, it is probable that neurons secreting one or more of the transmitters relay neural impulses from photoreceptors to the hypothalamus. However, there has been only limited success in relating changes in metabolism of neurotransmitters and hormone releasing factors in the hypothalamus to changes in the photoperiod. Diurnal fluctuations occur in both serotonin and norepinephrine levels in several regions of the brain of the cat and rat; however, the phases of the rhythms vary among different regions of the brain. In male rats exposed to light from 04.00 to 18.00 hr, serotonin levels in the hypothalamus and pineal gland reached a peak at 14.00 hr and then dropped rapidly. However, a slightly different pattern occurred in the amygdala and hippocampus of rats exposed to light from 06.00 to 20.00 hr. Serotonin content rose steadily throughout the light period, and dropped only after the onset of darkness (Quay, 1968; Scapagnini et al., 1971).

Levels of norepinephrine follow circadian rhythms in the spinal cord, pons, lateral tegmentum, anterior hypothalamus, tuber cinereum, and pineal gland of the cat. There are also rhythms in levels of norepinephrine in the anterior and posterior hypothalamus and pineal gland of the female rat. When rats are kept on L/D: 12/12, norepinephrine levels in the anterior hypothalamus are low during the light period, but rise during the dark period. In contrast, levels in the posterior hypothalamus fall during the dark period. The rhythm in the pineal gland roughly parallels that in the anterior hypothalamus, i.e., rises during darkness and falls during the light period.

Constant exposure to light abolishes the rhythms of norepinephrine in the pineal gland of both the rat and cat; however, rhythms in the spinal cord and anterior hypothalamus of the cat are not affected. The rhythm of norepinephrine in the pineal of the rat is also abolished in blinded animals. Thus, the daily photoperiod may partially regulate norepinephrine levels in the pineal of the cat and rat; but it has no demonstrated effect on norepinephrine levels in other parts of the brain (Wurtman et al., 1968; Reis et al., 1968; Manshardt and Wurtman, 1968).

An apparent causal relationship exists between daily changes in metabolism of neurotransmitters in the brain and levels of circulating adrenal glucocorticoids. Specifically, in rats the diurnal rhythm in corticosterone levels in the blood roughly parallels the rhythm of serotonin levels in the amygdala and hippocampus. Treatment of rats with p-chlorophenylalanine, a drug which inhibits serotonin synthesis, abolishes both rhythms. Similarly, drugs which alter serotonin metabolism also upset the rhythm of 17-hydroxy corticosteroid levels in the blood of cats (Krieger and Rizzo, 1969).

DAILY PHOTOPERIOD AND THE PINEAL GLAND

Daily photoperiod regulates certain components of pineal function in the rat. The activity of hydroxyindole-O-methyltransferase (HIOMT), an enzyme localized to the pineal and that regulates the rate of synthesis of melatonin, clearly shows diurnal variations. Parallel changes are seen in levels of NE in the pineal. Both rhythms are abolished when rats are blinded or exposed to constant light.

The role of the pineal gland in regulating the function of the anterior pituitary gland, particularly its secretion of gonadotropin, has been reviewed in detail (Wurtman et al., 1968; Reiter, 1973). The pineal partially regulates gonadotropin secretion in rats and hamsters by synthesizing compounds, including melatonin and serotonin, which then inhibit the secretion of gonadotropic hormone-releasing factors by the hypothalamus. Although evidence for this relationship is strong in the rat and hamster, it is equivocal in birds and absent in the ewe.

ANTERIOR PITUITARY GLAND

Attempts to relate diurnal changes in levels of pituitary hormones to daily photoperiod have yielded variable results. No diurnal rhythm was detected in levels of FSH or LH in pituitary glands of either intact or orchidectomized male rats, or ovariectomized rats kept under standard conditions of L/D: 14/10. Similarly, there was no rhythm in levels of serum LH (Yamamoto et al., 1970). In contrast, levels of total gonadotropins (GTH) in the pituitary glands of birds kept under various lighting regimens show rhythmic fluctuations. However, the relationship between the fluctuating levels of hormone and the lighting regimens is so variable that it is difficult to assign a specific regulatory role to the daily photoperiod.

Examination of a few selected studies illustrates the apparent complexity of the relationship between the daily photoperiod and GTH secretion in birds. In male white-crowned sparrows (*Zonotrichia leucophrys*) kept under L/D: 20/4, peak levels of GTH occurred approximately every 4 hrs, but the highest level was reached 2 hrs after onset of darkness (Stetson and Erickson, 1970). In contrast, the GTH levels in pituitaries of Coturnix kept under a similar regimen followed a much different pattern: GTH rose gradually during the first 16 hrs of the light period, but then dropped suddenly before the onset of darkness, and levels remained low until the darkness ended. Levels of luteinizing hormone-releasing activity dropped coincidentally with pituitary GTH (Follett and Sharp, 1969).

The GTH levels in pituitaries of Coturnix kept under L/D: 16/8 L/D: 14/10 or L/D: 6/6 followed patterns different from those of Coturnix kept on L/D: 20/4. In all three groups GTH rose gradually during the light period, then dropped significantly beginning 2 hr after the onset of darkenss.

INTERACTION OF DAILY PHOTOPERIOD, NEURAL, AND HORMONAL FACTORS IN REGULATION OF ANTERIOR PITUITARY FUNCTION

A daily photoperiod of sufficient duration stimulates GTH secretion, as evidenced by elevation of pituitary GTH levels and stimulation of the gonads. However, it is not clear whether the photoperiod regulates daily fluctuations of GTH in the pituitary. Light may accelerate or entrain GTH secretion, but other factors may regulate daily secretion patterns. Furthermore, the relative importance of these factors may vary among species.

A classic example of this interplay is found in the regulation of timing of ovulation in the domestic hen. Although some hens ovulate and lay when kept in complete darkness, all require a daily photoperiod of 12 to 14 hrs for optimal laying efficiency. Under optimal conditions the hen lays one egg on each of two or more consecutive days, skips one day, and then lays again on two or more consecutive days. Eggs laid on two or more consecutive days constitute a sequence. In sequences of normal length, the first egg is laid early in the lighted period and subsequent eggs are laid progressively later until the sequence is terminated with the laying of an egg late in the lighted period. Oviposition does not occur during darkness. Various hypotheses have been presented to explain the cause of the skip, i.e., the failure of ovulation and oviposition. The intervals between LH release and ovulation, and between ovulation and oviposition, do not vary with different eggs of the sequence. Hence, a probable cause of the skip is failure to release the surge of LH required to induce ovulation. Apparently LH release, or the signal for LH release, occurs only during one phase of the light-dark cycle.

Thus, in the hen, photoperiod has the dual actions of stimulating the neuroendocrine reproductive system to function and of determining, to some extent, whether the preovulatory release of LH occurs each day. However, the precise timing of LH release, when it occurs, is also regulated by neural and hormonal stimuli. In fact, during the laying sequence, time of the preovulatory peak(s) of LH is related constantly to time of the last oviposition and not to hour of the day (Fraps, 1961).

In the laboratory rat, in contrast to the hen, the daily photoperiod precisely entrains the hour of ovulation. In this species, a multicomponent system regulates the preovulatory release of LH. Two of the essential components are a neural clock mechanism, probably located in the anterior hypothalamus, and a relatively high level of estrogen in the blood. Although the precise location and nature of the clock is unknown, it may be defined as a circadian pattern of neural sensitivity to the presence of estrogen. In animals kept under standard laboratory conditions (light 05.00 to 19.00 hr), this period of sensitivity, e.g., the "critical period", occurs each day between 14.00 and 16.00 hr. In intact animals optimal levels of estrogen are present only every 4th or 5th day. Thus, a preovulatory "surge" of LH occurs in the afternoon of every 4th or 5th day.

That this "critical period" is entrained by the daily light cycle has been demonstrated by changing the lighting schedule. An abrupt resetting of the lights to come on 3 hrs earlier triggers a gradual shift of the "critical period" that is completed within 2 weeks. Apparently both the onset and termination of lighting serve as signals for entrainment. Ovulation occurs between 02.00 and 04.00 hr whether rats are exposed to light from 06.00 to 18.00 hr (L/D: 12/12) or from 04.00 to 20.00 hr (L/D: 16/8). Time of ovulation is constantly related to the midpoints of the light and dark periods under both regimens (Everett, 1969; Alleva et al., 1970).

Daily photoperiod may have little or no role in regulating LH release in mammals other than small rodents. This species difference is demonstrated by comparing the timing of the LH release which can occur in response to an injection of estrogen. In steroid pre-treated ovariectomized rats, a properly timed injection of estrogen induces a significant elevation of serum LH. However, this elevation occurs only in the afternoon because timing of the LH release is controlled by a neural clock. In contrast, in the ewe an injection of estrogen induces LH release approximately 12-18 hrs later, but at any hour of the day. Ewes respond in this manner because they apparently lack a light-entrained neural clock for precisely timing the LH surge (Jackson, 1974).

ANNUAL CYCLES

Annual changes in photoperiod length have distinct effects on neuroendocrine function. Of those physiological events affected by annual rhythms in the photoperiod, the best described are reproductive cycles of seasonally breeding animals. There are two basic groups of seasonal breeders. The long-day breeders, represented by the cottontail rabbit (*Sylvilagus floridanus*) and most birds, show increased gonadal size and display reproductive activity coincident with increasing day length in the spring. The gonads regress in the fall. The other group, the short-day breeders, represented by the ewe and the white-tailed deer (*Odocoileus virginianus*), attain breeding condition coincident with declining day length in the fall.

There are apparent differences between the basic neuroendocrine responses of long- and short-day breeders to increasing or decreasing photoperiod; however, the specific neural pathways and mechanism involved are not known in either group. Existence of a direct neural pathway between the retina and hypothalamus has been demonstrated in only a few species (Moore and Lenn, 1972). In species with extraretinal photoreceptors, neither the nature of the photoreceptors nor the pathways to the hypothalamus have been identified (Menaker, 1971). There is little information relating changes in metabolism of neurotransmitters to annual changes in the photoperiod. Secretory activity in the pineal gland of some species changes markedly in response to seasonal shifts in photoperiod, but these changes are not related to changes in gonadal function (Barfuss and Ellis, 1971; Reiter, 1973).

Since growth of the gonads depends on stimulation by gonadotropins it is not surprising that some species show distinct annual rhythms in the levels of GTH in the pituitary glands. In long-day breeders, changes in pituitary gonadotropin levels precede or parallel changes in gonadal size and activity. This is seen clearly in birds such as the ring-necked pheasant (*Phasianus colchicus*) in which total GTH levels are nearly undetectable in January or February, rise quickly to plateau levels by April, then decline sharply in June (Greely and Meyer, 1953). Both testicular and ovarian size follow similar patterns, indicating that changes in GTH levels in the pituitary reflect parallel changes in secretory activity.

The pronounced changes in levels of GTH in the pituitary of long-day breeders may not occur in short-day breeders. The ewe has no distinct annual pattern in pituitary gonadotropin levels. Ewes in the northern hemisphere, if not bred, exhibit estrous cycles from September through April. They enter anestrus during May and remain in this state until August. Levels of total gonadotropins and LH in the pituitary during anestrus are nearly as high as average levels during the breeding season, but levels of LH in the circulation are very low or undetectable. Thus, in this species, anestrus is accompanied by a partial damping of the synthesis and release of GTH rather than by a total cessation of synthesis.

The neural mechanisms by which long photoperiods induce seasonal anestrus in short-day breeders remain a mystery. In the anestrous ewe the ovary is sensitive to exogenous gonadotropins. The anterior pituitary contains both LH and FSH which can be released by exogenous estrogen or releasing factor (Jonas et al., 1973), and the hypothalamus contains significant levels of LH-releasing activity (Jackson et al., 1971). Thus, while the organs which operate the reproductive cycle appear capable of functioning at this time, they are quiescent owing to the absence of some triggering impulse from the brain.

Perhaps there is temporary inactivation of a stimulatory center or activation of a peripheral inhibitory center. The idea of an inhibitory center is supported by studies of the ferret, also a seasonal breeder. When lesions are placed in the anterior hypothalamus of anestrous ferrets, the ferrets resume estrous behavior significantly earlier than non-lesioned animals (Donovan and Van der Werff ten Bosch, 1959).

SUMMARY

Photoperiod can regulate both daily and annual neuroendocrine events. Light is detected in mammals by retinal photoreceptors, but in non-mammalian vertebrates, particularly some birds, extra-retinal structures also function as photoreceptors. The signals are analyzed by neural centers or pathways not yet known. They are then passed to neurons in the hypothalamus. These neurons, which secrete neurotransmitter compounds, alter the release of hypothalamic hormone-releasing factors that in turn modulate the secretion of tropic hormones of the anterior pituitary gland. The final step is a change in the rate of function of endocrine organs such as the gonads and adrenals.

That photoperiod can exert its effect through different pathways in different species is shown by comparing the changes occurring in the endocrine system of long-day breeders with those in short-day breeders. In the former, gonadotropin secretion fluctuates between full activity under increasing day length and complete cessation under decreasing day length. In the latter, gonadotropin secretion appears to remain static during anestrus, neither functioning fully nor ceasing.

The literature is replete with studies concerning rhythms in levels of compounds in various compartments of the body or the effect of different light regimens on levels of compounds in single compartments. However, few studies have been designed to determine if the rhythms in question are regulated by the photoperiod or if rhythms in different compartments are dependent on or correlated with each other. Thus, with the exception of the pineal gland in the rat, data concerning the pathways and mechanisms by which changes in the photoperiod alter neuroendocrine functions are largely incomplete.

Major problems to be solved include determining (a) the neural pathways and centers which transmit and modulate impulses from the photoreceptors to the endocrine transducers, (b) the relationship between specific neurotransmitters and hormone-releasing factors, and (c) the effects of photoperiod on the rate of metabolism of the neurotransmitters, hormone-releasing hormones, and tropic hormones of the anterior pituitary gland.

REFERENCES: see pp. 261-262.

Photoperiodic Effects on Mammalian Reproduction

by

R. J. Reiter

INTRODUCTION

Photoperiod as a determining factor in controlling the sexual physiology of mammals is often overlooked by scientists who use common laboratory rodents, e.g., rats, mice and hamsters, as their experimental animals. This is due primarily to the fact that in the majority of animal quarters the experimental subjects are virtually always maintained under rigidly controlled photoperiodic conditions, usually with the animals being exposed to 12-16 hrs of light per day. The effect of such constant conditions often is to make continuous breeders out of animals that are seasonal breeders in their natural habitat. In other words, the effects of light and darkness on sexual physiology are minimized.

This is somewhat unfortunate since such controlled conditions represent an unnatural situation for the animal. The rodents most frequently utilized for these studies are nocturnal fossorial species and in the wild they are exposed to only short daily photoperiods of irregular length. They certainly do not witness anything reminiscent of 12-16 hrs of uninterrupted lighting per day. Such carefully regulated photoperiodic conditions has in fact severely stymied research for many years on one important endocrine organ, the pineal gland. It is now well documented that photoperiodic length, pineal activity and reproductive physiology are inextricably linked (Reiter, 1972).

In addition to the rodents, many other scientists use larger domesticated species which also exhibit seasonal reproductive phenomena. Partly due to the size of these mammals and partly because these investigators were more concerned with the important role of light in regulating the pituitary-gonadal axis, such species were often kept under less photoperiodically controlled environments. This allows the sexual physiology of the animals to fluctuate with the changes in the environmental conditions (Hafez, 1964).

Both extremes of photoperiod, i.e., long days or continuous light exposure and short days or total light deprivation dramatically influence the functioning of the hypothalamo-pituitary axis. In terms of the common laboratory species, long days are generally considered to be stimulatory to reproduction whereas short days are often inhibitory. There may be, however, many exceptions to this generalization. The mechanisms whereby extremes in photoperiod influence sexual functions are undoubtedly different. Long daily periods of light seem to have a direct effect via retinohypothalamic fibers on the neuroendocrine-gonadal axis whereas darkness involves the pineal gland as an intermediate (Reiter and Fraschini, 1969).

STIMULATORY PHOTOPERIODS

What exactly constitutes a stimulatory photoperiod is difficult to define. Generally, anything over 12 hrs of light per 24-hr period may well fall into this category. On the other hand, progressively increasing photoperiodic lengths are not uncommonly stimulatory. Finally, some animals are sexually most active during short day periods and, thus, such photoperiods even though short or decreasing in length may be considered to be stimulatory to the reproductive system.

One of the earliest demonstrations of the stimulatory influence of extra light on sexual photoperiodicity was provided by the work of Bissonette (1932). Using the ferret (*Putorius vulgaris*), a seasonal breeder and a highly photosensitive species, he observed that exposure of females to 6 hrs of extra light daily during the November and December advanced estrous onset from March, when it normally occurred, to January. This stimulatory effect of increased photoperiodic length on various aspects of the sexual cycle has been repeatedly confirmed in other carnivors, in lagomorphs, in rodents, and in ungulates. In addition to initiating genital activity in adult animals during their reproductively dormant season, exposure of immature animals to long daily photoperiods or to constant light may cause precocious sexual development (Critchlow, 1963). This is especially noticeable in the common laboratory rodents.

The phasing of the estrous cycle in rodents is closely linked with environmental illumination. This is readily demonstrable by reversing the day-night lighting conditions. After a short period of adjustment, the phases of the estrous cycle also shift accordingly. Likewise, changing the onset of the light period, in an alternating light-dark situation, is accompanied by a shift in the critical period of proestrous in rats. The critical period, defined by Everett and Sawyer (1950), is the interval during which the luteinizing hormone (LH) required for ovulation is released from the anterior pituitary. LH release invariably occurs during the light phase of the light-dark cycle. The circadian mechanisms triggering LH release and ovulation presumably are essentially endogenous but the normal "zeitgeber" (the timing device which ensures precise circadian rhythms of various functions) for the ovulatory responses seems to be alternating periods of light and darkness (Jöchle, 1967).

Exposure of several common rodents to constant light induces a condition of persistent vaginal cornification. The sequence of events preceding continuous estrous in rats under such conditions are rather complex. After short term exposure, the ovaries and uteri exhibit a mild hypertrophic response perhaps because of the direct stimulatory effect of light on the neuroendocrine-gonadal axis. After several weeks of continuous lighting, however, the ovaries begin to regress in size and actually become smaller than normal. This response represents the loss of corpora lutea from the ovaries in that these animals are incapable of a spontaneous LH surge and ovulation every 4 to 5 days. The ovaries of such animals consist primarily of large vesicular follicles which secrete estrogen continuously; the continual presence of high titers of this steroid in the blood accounts for the persistent cornification of the vaginal epithelium. Although in a sense, long term continual illumination seems to be inhibitory to reproduction in female rats in that spontaneous ovulation is suppressed, the animals are in persistent estrus, are

sexually receptive and if they mate the copulatory activity induces ovulation. Thus, the reproductive system may be considered to be hyperfunctional (Jöchle, 1967).
The mechanisms of the persistent estrous syndrome induced by continual exposure to light are not completely known. Presumably the light is detected by the retina, the specific retinal element involved still remains in doubt, and the neural information is transmitted across the optic nerve to the hypothalamus (via the retinohypothalamic fibers). Within the hypothalamus this neural information probably inhibits the cyclic LH release center which resides in the anterior hypothalamus-preoptic area. With the activity of this center quelled, the cyclic phenomena associated with the normal estrous cycle disappears and the animals enter the persistent estrous state. As noted, however the cyclic LH release center can still be activated by copulatory behavior. There seems to be no direct effect of light on the brain or on the neuroendocrine axis in mammals such as has been shown in some birds.

INHIBITORY PHOTOPERIODS

Definition of what determines an inhibitory photoperiod is shrouded with the same ambiguities as is that which describes the stimulatory photoperiod. For the purpose of this discussion an inhibitory photoperiod will be defined simply as one that induces gonadal regression or one that suppresses reproductive physiology. This obviously varies considerably from one species to another.

It has been shown that completely depriving various species of photic stimuli or merely limiting them to short daily photoperiods retards sexual maturation and generally lowers sexual capability. After light deprivation, the rat estrous cycle, gauged by vaginal smears, is often disrupted or lengthened (Hoffmann, 1967) and the gonads and secondary sex organs hypotrophy (Reiter, 1972). Unquestionably, these effects of darkness are a consequence of a stimulated pineal gland since its removal negates any influence of light deprivation on reproduction (Reiter, 1973a).

One of the most sensitive species to the absence of photic stimulation is the hamster (*Mesocricetus auratus*). Restricting the amount of light (1 to 2 hrs of light per day) either males or females of this species sees results in severely atrophic reproductive organs within 4 to 8 weeks (Hoffman and Reiter, 1965; Reiter and Fraschini, 1969). Accompanying the involution of the reproductive organs is the disappearance of cyclic vaginal phenomena and infertility (Sorrentino and Reiter, 1970). The pituitary levels of LH and prolactin are also significantly altered by light deprivation due to blinding. Eight weeks after blinding the pituitary content and concentration of LH and prolactin are depressed in adult male hamsters (Fig. 1) while in females treated similarly pituitary prolactin levels are depressed but the quantity of LH within the pituitary is actually elevated (Fig. 2). As in the rat, all or nearly all of the sexual consequences of light deprivation in the hamster are obviated by removal of the pineal gland or by removal of the superior cervical ganglia which provide the pineal gland with its functional innervation (Reiter, 1973a).

The mechanisms of gonadal suppression by darkness are quite obviously different from

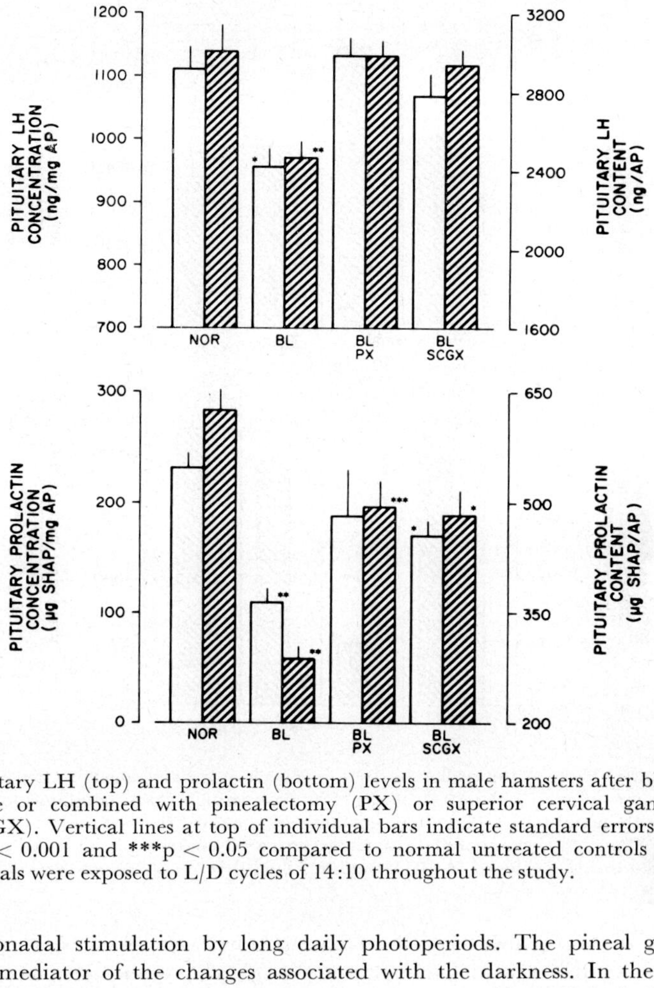

Fig. 1. Pituitary LH (top) and prolactin (bottom) levels in male hamsters after blinding (BL) alone or combined with pinealectomy (PX) or superior cervical ganglionectomy (SCGX). Vertical lines at top of individual bars indicate standard errors. *$p < 0.01$, **$p < 0.001$ and ***$p < 0.05$ compared to normal untreated controls (NOR). All animals were exposed to L/D cycles of 14:10 throughout the study.

those of gonadal stimulation by long daily photoperiods. The pineal gland is the important mediator of the changes associated with the darkness. In the absence of environmental photoperiodic stimuli this organ is known to be physiologically activated and strongly antigonadotropic. Without this organ, such as following pinealectomy, darkness is incapable of appreciably altering the functioning of the neuroendocrine-gonadal axis (Reiter, 1973a). Although light effectively inhibits pineal activity the sexual changes associated with exposure of experimental animals to continual lighting seem to be virtually independent of the pineal gland.

PHOTOPERIODS, PINEAL AND ANNUAL REPRODUCTION

The ability of mammals to respond to different photoperiods must provide them with a physiological advantage. One fact that essentially all animals which inhabit the

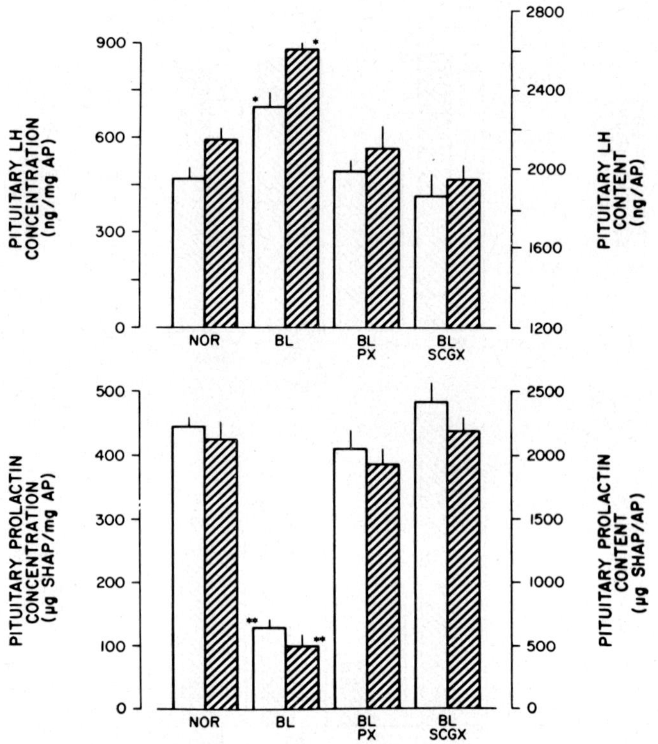

Fig. 2. Pituitary LH (top) and prolactin (bottom) levels in female hamsters after blinding (BL) alone or combined with pinealectomy (PX) or superior cervical ganglionectomy (SCGX). Vertical lines at top of individual bars indicate standard errors. *$p < 0.01$, **$p < 0.001$ and ***$p < 0.05$ compared to normal untreated controls (NOR). All animals were exposed to L/D cycles of 14:10 throughout the study.

temporate or the boreal areas of the earth have in common is that they breed seasonally. In these species, it is essential that the young be born at the time most propitious for their survival, i.e., the spring or summer. To synchronize birth of the young with these seasons it is important for the animals not only to recognize but to anticipate the season. For this they must depend on external environmental stimuli. One of the most consistent annual changes in the environment is, of course, the photoperiod. Hence, it is natural that annual reproductive cycles would be linked with seasonal changes in daylength. Knowing and, more importantly, being able to respond to daylength and changing (either increasing or decreasing in length) photoperiods is essential to the survival of all species. The organ that keeps them attuned or in phase with the changing photoperiod is the pineal gland. Without this organ a number of species are known to lose the ability to synchronize breeding activity, and thus the time of birth, with the spring of the year (Reiter, 1973a). Hence, environmental lighting acting via the pineal gland is essential for normal reproductive rhythms and, in essence, ensures the survival of the species in their natural habitat. Such relationships have been demonstrated in several species including the golden hamster (Reiter, 1973b), djungarian hamster (Hoffmann, 1973) and in the ferret (Herbert, 1971).

CONCLUDING REMARKS

The studies discussed above provide us with only a brief view of the problems of environmental lighting and reproduction. Many investigators still question the existence of the retinohypothalamic fibers which would probably be important in mediating the effects of light on the neuroendocrine axis. Even the photoreceptor element involved in detecting the light which will ultimately change reproductive physiology is unknown. The problem of extra-retinal photoreception in mammals has not been completely resolved. All of these questions must be answered before we can understand the stimulatory effects of light on the functioning of the hypothalamo-pituitary system.

Whereas we know that the consequences of the absence of light on reproduction are mediated by the pineal gland, we are still ignorant as to the pineal hormones which are responsible for the effects. Definitive information is lacking concerning the mechanisms whereby the pineal hormones alter sexual physiology. Surely, changing photoperiods and pineal activity are somehow linked with annual reproductive rhythms but much needs to be done in defining more closely these interrelationships. It seems safe to predict that we will identify one of the physiological niches of the pineal gland as work continues along these lines.

REFERENCES: see pp. 262-263.

Chapter 2, Section 1e

Seasonal Rhythms in Animals

by

J. L. Cloudsley-Thompson

See Progress in Human Biometeorology, Part IB, Chapter 6, Section 22b (1), pp. 578-580.

Seasonal Molting Cycles in Wild Mammals

by

Elizabeth Johnson

INTRODUCTION

A seasonal molt in wild mammals is most strikingly observed in those which have a seasonal colour change. A change from a brown or grey summer coat to a white winter coat has been described for many Arctic and northern species, such as the stoat, *Mustela erminea* (Rothschild, 1942), the weasel, *Mustela frenata* (Wright, 1942) and the varying hare, *Lepus americanus* (Lyman, 1943). The change to a white winter coat is linked to the presence of snow; the mountain hare, *Lepus timidus*, has a range from Northern Scandinavia, where change to winter whiteness is complete, to Ireland where a winter colour change seldom occurs. In the Scottish mountain hare it has been shown that the rate of colour change varies with the climatic conditions (Hewson, 1958; Watson, 1963; Flux, 1970).

It seems likely that one adaptive function of such a change in pelage is that of camouflage (Cott, 1940). A similar adaptive function is ascribed to the white spotted summer coat of the European fallow deer, *Dama dama*, which provides effective camouflage amongst the trees in summer. A colour change of coat may have other adaptive functions. Severaid (1945) has suggested that a white winter coat is an aid to thermoregulation because it reduces the radiation of heat. The change from a dark, dull winter coat to a bright russet, glossy summer coat in the roe deer, *Capreolus capreolus*, is certainly no aid to concealment but may aid the reflection of sunlight from the body as seems to occur in some tropical breeds of cattle with similarly coloured coats (Bonsma and Pretorius, 1943).

Many wild mammals have less obvious but equally important seasonal coat changes which are of thermoregulatory value. The changes which occur in the density and type of hair produced at different seasons are possible because the molt is a cyclic phenomenon.

THE HAIR CYCLE

Primary hair follicles develop as downgrowths from the epidermis and are usually arranged in groups of three (De Meijere, 1894). In many mammals secondary follicles develop as branches from the primaries; and this group forms the adult complement of hair follicles (Lyne, 1966; Khateeb and Johnson, 1971a). No further development of follicles occurs under normal conditions, but the number of hairs retained by each follicle may vary.

The cyclic events which occur in individual hair follicles were first described by Dry in 1926 and are illustrated in Figure 1. During the growing stage or ANAGEN cells surrounding the dermal papilla divide actively, stream upwards and become keratinised

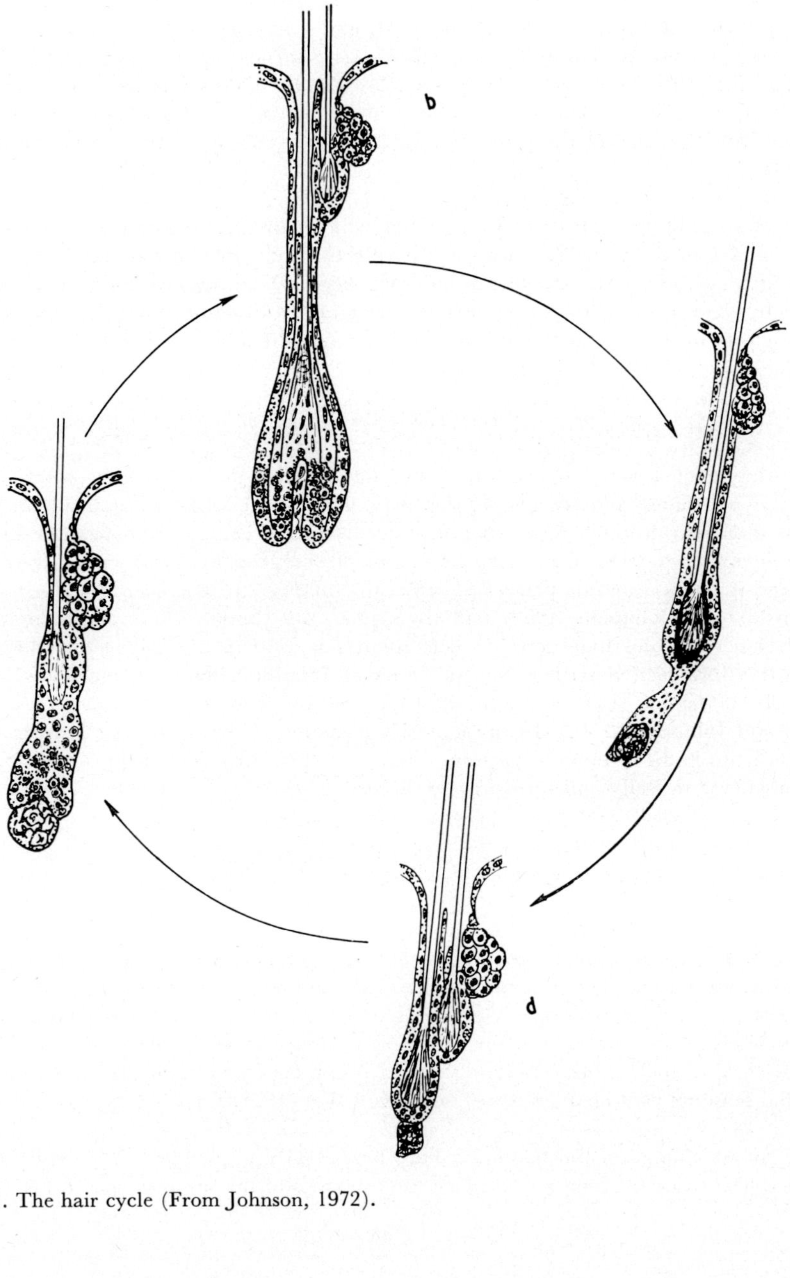

Fig. 1. The hair cycle (From Johnson, 1972).

to form the growing hair (Fig. 1a). At the end of the growing phase mitoses cease and the hair develops a characteristic brush-like anchoring "club" at its base (Fig. 1b). This transition or CATAGEN follicle migrates upwards through the skin, leaving behind an epithelial strand of cells which subsequently shortens to form the hair germ and expose the dermal papilla (Fig. 1c). At this resting or TELOGEN stage the hair follicles will not fall out if the skin is cured, and a fur skin with all its follicles at this stage is described as a prime pelt. At the start of a new cycle cells of the hair germ divide (Fig. 1d), extend downwards to enclose the dermal papilla and thence upwards to produce a new hair alongside the old. During active growth pigment may be introduced into the cells of the growing hair from melanocytes contained in the dermal papilla.

At each stage of the cycle there are variables which will affect the type of coat produced. The length of the growing phase or the rate of production of hair may vary, either of which will affect the amount of hair produced. Once the new hair has grown the old club hair may be shed immediately or retained for some time, thereby affecting the density of the coat. The initiation of activity in the follicle will determine the time of the molt.

MOLTING PATTERNS

The timing of activity in individual hair follicles will determine the pattern of the molt. Synchronous activity all over the body, as illustrated diagrammatically in Figure 2a is unusual, although near-synchrony occurs in the change from juvenile to adult coat in many rodents. Conversely, each follicle may behave independently of those around it. Such a mosaic pattern of replacement (Fig. 2b) has been described for the scalp of man (Kligman, 1959) and the guinea pig (Bosse, 1965). A similar diffuse replacement of hair may occur in wild mammals towards the end of their life span; and this has been described for the muskrat (Shanks, 1948), the meadow mouse, *Microtus californicus* (Ecke and Kinney, 1956) and the field vole, *Microtus agrestis* (Khateeb and Johnson, 1971a). In the majority of mammals waves of hair growth spread around the body in definite patterns (Fig. 2c), either from the belly to the back or commencing dorsally and spreading to the belly.

SEASONAL CHANGES OF COAT

Typically hair follicle groups produce two types of hair: coarse overhairs or guard hairs, and fine underfur or wool. Such an arrangement provides an insulating layer of warm air trapped between the fine hairs at the body surface. The efficiency of this insulation as a thermoregulatory device is shown by the fact that the Alaskan red fox relies solely on an increase in fur thickness to adapt it to the Arctic winter (Irving et al., 1955). Such a body covering would prove an embarrassment in the summer, so that seasonal coat changes are of adaptive value.

The colour change of the mountain hare has already been noted. The white winter coat is also twice the length and twice the thickness of the summer coat (Flux, 1970).

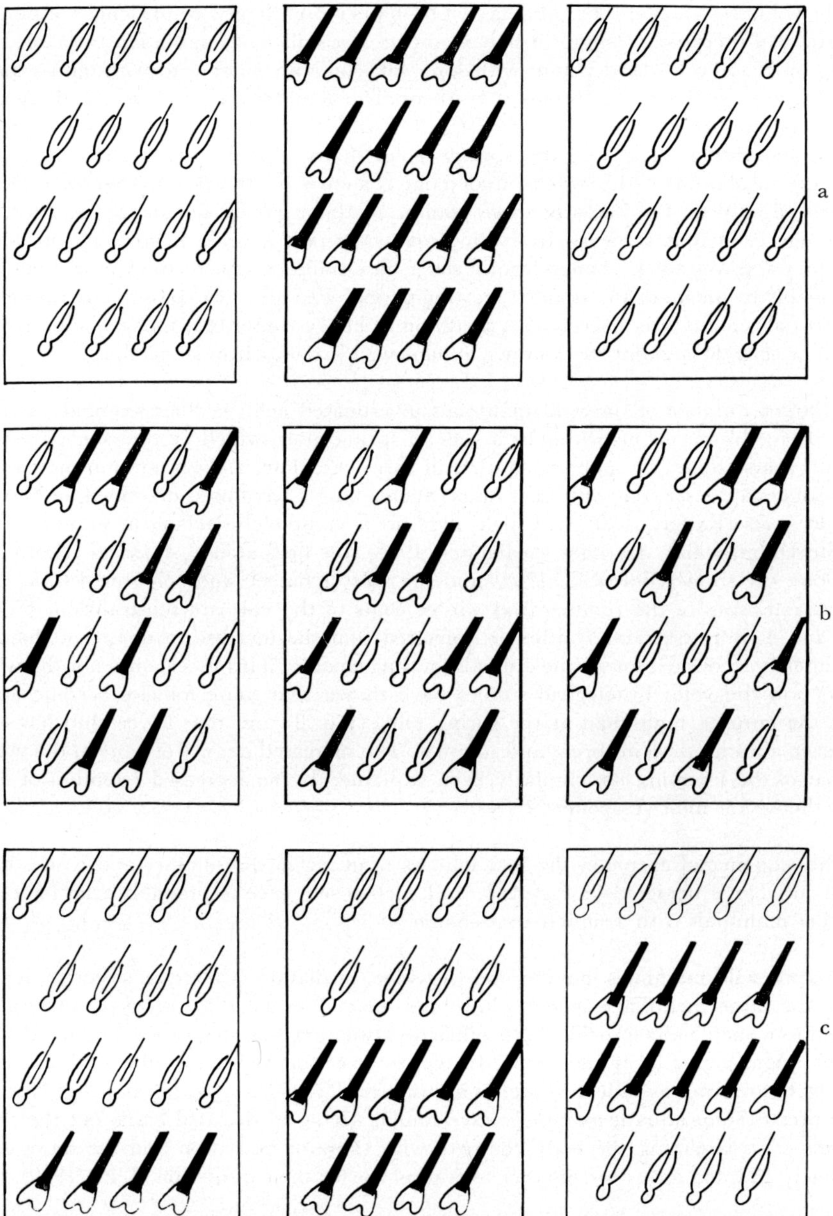

Fig. 2. Molting patterns. a: synchronous; b: mosaic; c: wave (From Johnson, 1972).

The wild ass (Mazak, 1962) also has winter hairs more than twice the length of summer hairs. An increase in length of hair is not itself a sufficient adaptation to season. For example, some lowland ponies with long silky hair are unable to withstand exposed conditions. In harsh conditions it is essential to maintain a dry layer of air next to the body surface and moorland ponies achieve this by having a winter coat comprising fine underfur overlain by wiry, springy guard hairs (Speed, 1960). The effectiveness of the insulation of this woolly undercoat is shown by the fact that snow does not melt rapidly on the backs of these ponies. In these ponies and in the wild ass, the change in coat is achieved by having only one molt a year; a shedding of hair in spring and a gradual change from a short, fine summer coat to the winter coat, presumably by means of an extended growing period. The silver fox (Bassett and Llewellyn, 1948) also molts only once a year, shedding its heavy winter coat in April and growing a new coat throughout the summer to achieve a prime pelt in November.

A larger number of the wild mammals investigated achieve their seasonal variation of pelage by means of two molts a year; a spring molt, which produces a sparse coat with coarse hairs, through which air will easily circulate, and an autumn molt which produces a denser coat with finer hairs. Such changes have been described for Scottish wild goats (Ryder, 1970), in which underfur is completely lacking in summer, mink (Stevenson, 1962), deermice (Sealander, 1951), the field mouse, *Apodemus* (Haitlinger, 1968), the shrew (Borowski, 1958) and the vole (Khateeb and Johnson, 1971a). The characteristics of the summer and winter coats of the vole are illustrated in Figures 3 and 4. Borowski and Haitlinger suggested that the increase in density of hairs in winter is accounted for by the animals' getting smaller. This is certainly not the whole story in the vole; histological studies have shown that more follicles become active at the autumn molt than at the spring molt (Fig. 3) and that fewer club hairs are lost in autumn than in spring and summer. The increased density of hairs in the winter coat of the lemming has similarly been explained by an increased retention of hairs at the winter molt (Koponen, 1964).

The sequence of events in the hair follicles resulting in the summer and winter coats of the vole is depicted in Figure 5, and such a sequence is probably true for many other mammals with a similar coat change.

In some wild mammals, notably the Cervidae, insulation of the coat seems to depend on the large medullary spaces within the hairs. Flerov (1952) suggests that in the reindeer such hairs provide both efficient insulation and buoyancy in the seasonal migrations across lakes and rivers. In the roe deer the highly medullated guard hairs (Fig. 6) are twice as wide in winter as in summer (Pavlovic, 1965; Johnson and Hornby, in press). Some aquatic mammals have similar highly medullated hairs, but the problems of streamlining the body compete with those of insulation and the more completely adapted forms use blubber instead of fur for their insulation (Ling, 1970).

ENVIRONMENTAL INFLUENCES ON HAIR GROWTH

Bissonnette (1935) was the first to show that increased daylength in winter would both

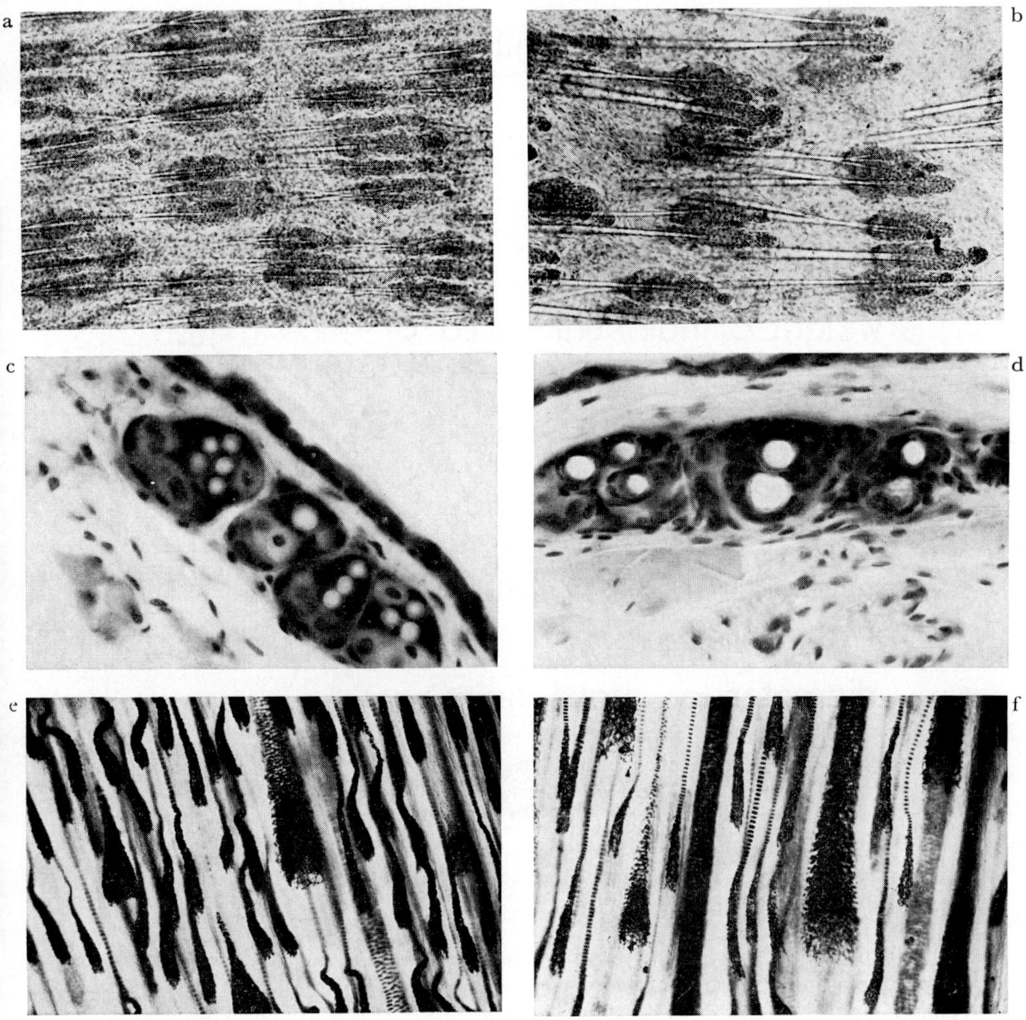

Fig. 3. Summer and winter coats of the field vole (*Microtus agrestis*). a: Whole mount of skin with resting hair follicles: winter. b: Whole mount of skin with resting hair follicles: summer. c: Hair follicle group cut transversely: winter. d: Hair follicle group cut transversely: summer. e: Whole mount of skin with growing hair follicles: winter. f: Whole mount of skin with growing hair follicles: spring. a and b; c and d; e and f; at same magnification. (From Khateeb and Johnson, 1971a).

induce oestrus in ferrets and alter the timing of the molt. These, and further studies by Hammond (1952) and Harvey and MacFarlane (1958) have shown that with decreasing daylength ferrets become reproductively quiescent and grow a dense winter coat. With increasing daylength, on the other hand, ferrets become reproductively active

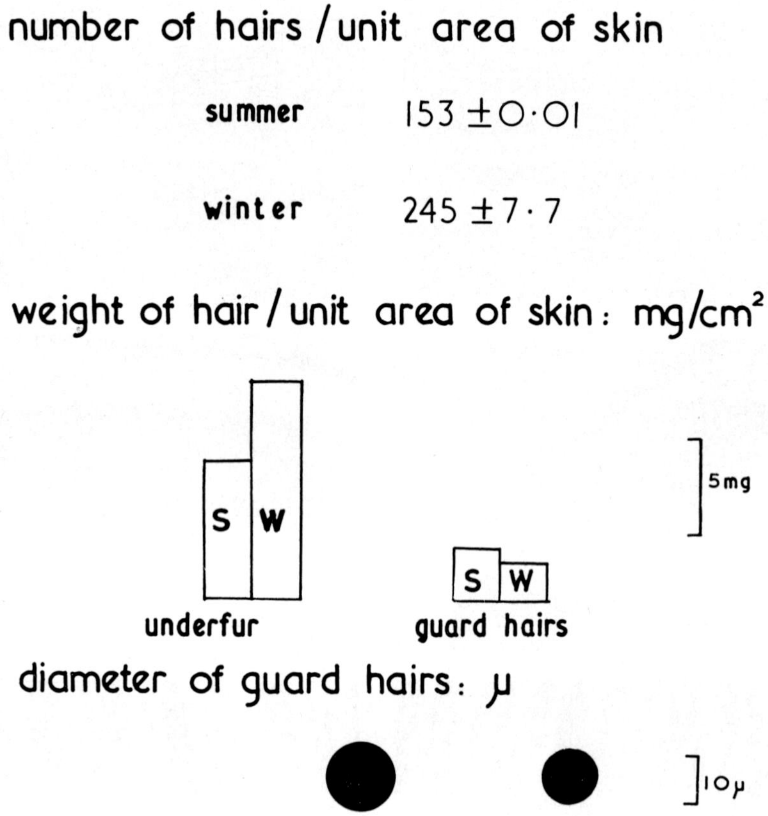

Fig. 4. Characteristics of summer and winter coats in the field vole (*Microtus agrestis*). (Data from Khateeb and Johnson, 1971a).

and change to a sparse, coarse summer coat. A reversal of daylength by six months resulted in a similar reversal of the reproductive cycle and the molt. Similarly, artificial shortening of the hours of daylight brought forward the autumn molt of mink (Bassett and Llewellyn, 1949), whereas extended daylength through the spring and summer encouraged the single annual molt of the silver fox to proceed faster (Bassett et al., 1944). Increased illumination in winter accelerated the onset of oestrus and the shedding of the coat in the mare (Burkhardt, 1947) and brought forward the spring pelage change in the white-tailed buck as well as the subsequent rut and fall pelage change (French et al., 1960). In *Microtus agrestis* (Khateeb and Johnson, 1971b) animals kept on long days in autumn did not regress sexually and produced a summer type coat with low density of hairs and coarse guard hairs. Conversely, short days in summer advanced the onset of anoestrus and the autumn molt of the voles. Extended light in winter caused the varying hare to remain brown, whereas reduced illumination in summer induced a change to winter whiteness (Lyman, 1943).

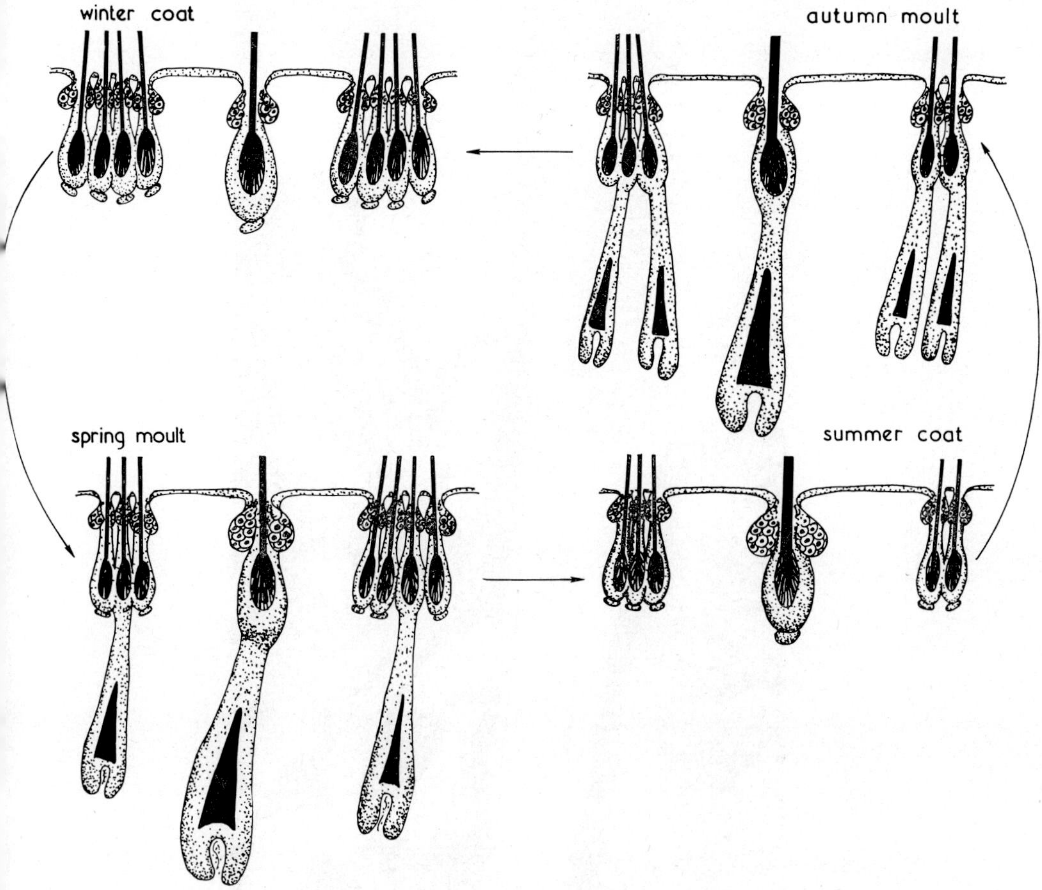

Fig. 5. The sequence of events in the hair follicles resulting in the summer and winter coats of the vole (*Microtus agrestis*). (From Johnson, 1972.)

In experiments with animals in captivity it becomes obvious that temperature and food supply also affect both the reproductive cycle and the pelage. The role of nutrition has been investigated in *Microtus montanus* reared in conditions of long or short days and on two planes of nutrition. The voles on long days molted earlier than those on short days, but the well fed groups molted earlier than the low feed groups (Pinter, 1968). *Peromyscus* showed an increase in pelage density when kept at low temperatures in the summer and a decrease in density at high temperatures in winter (Sealander, 1951). The increased density and finer hair of species of chinchilla living at increasingly higher altitudes (Seele, 1968) may also be related to the temperature difference at different heights, and Jackes and Watson (1975) suggest that temperature is important in the seasonal colour change of the mountain hare.

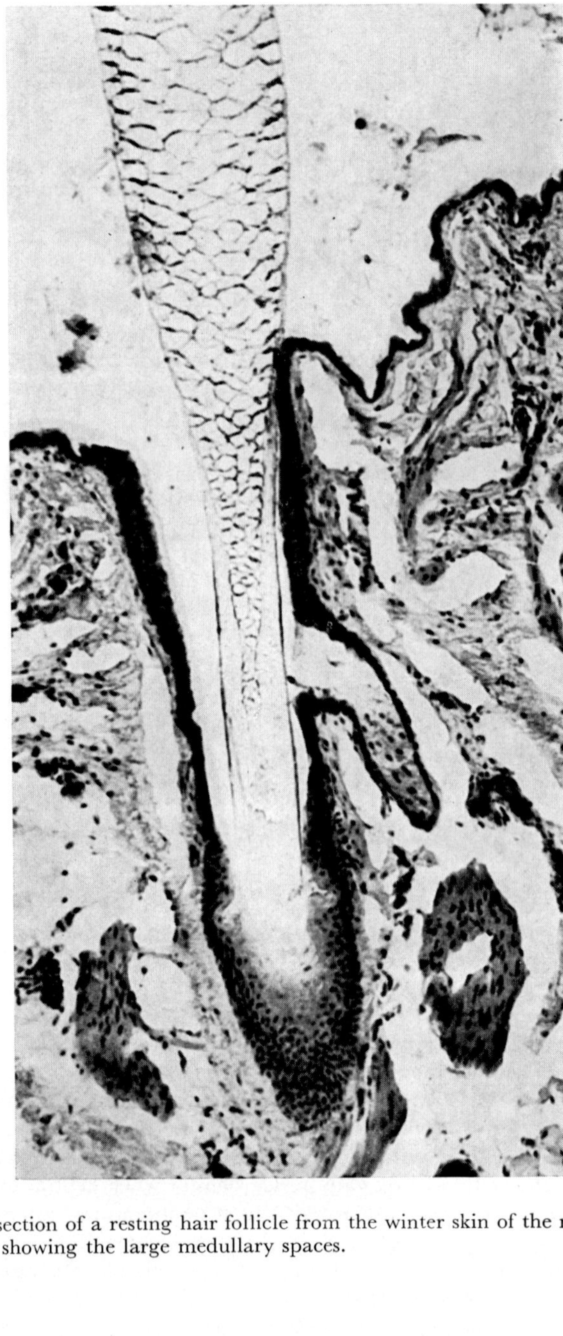

Fig. 6. Saggital section of a resting hair follicle from the winter skin of the roe deer (*Capreolus capreolus*) showing the large medullary spaces.

PHYSIOLOGICAL CONTROL OF THE MOLT

Experiments with laboratory rodents provide evidence that the hair follicle has both an inherent rhythm of activity as well as being subject to some systemic control (Ebling and Johnson, 1961). Such an experiment is illustrated in Figure 7. In the laboratory rat the resting phase between each molt is longer as the animals get older. It is possible, therefore, to exchange skin between a young and older rat so that their molt cycles are out of phase. Skin removed from a 28 day old rat and grafted into one aged 49 days continues to grow hair for at least two cycles as if it had not been removed from its original site, thus demonstrating an inherent rhythm of activity. On the other hand, the graft from the older to the younger rat is stimulated to grow hair at the same time as its recipient and much earlier than any growth would have occurred on its own body. This indicates that systemic factors are influencing the timing of the molt.

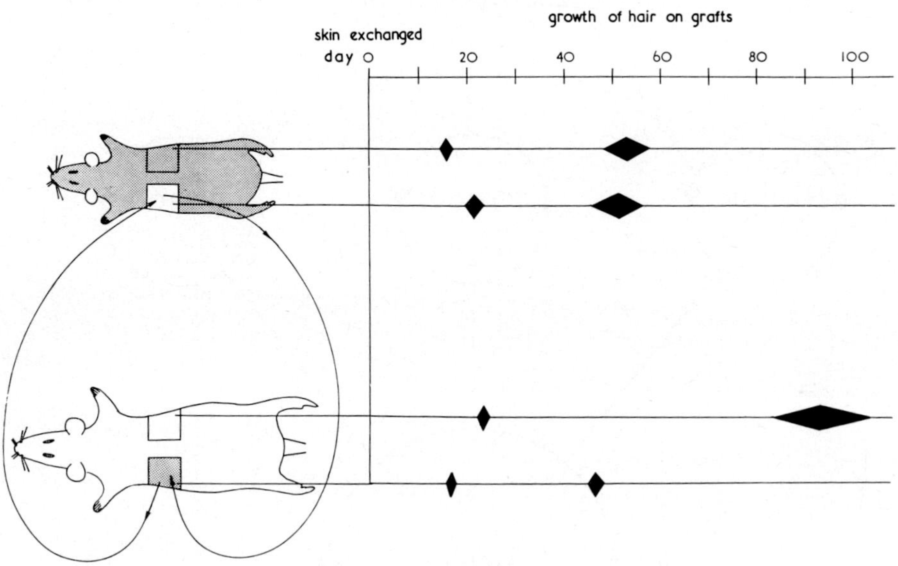

Fig. 7. Growth of hair on skin grafts exchanged between young rat aged 28 days (stippled) and older rat aged 49 days. (Data from Ebling and Johnson, 1961).

The most likely route of this systemic control is by way of the endocrine system, since many hormones have been shown to affect the timing of follicular activity. In laboratory rodents follicular activity is inhibited by sex hormones (Hooker and Pfeiffer, 1943; Houssay and Higgins, 1949; Johnson, 1958a; Mohn, 1958) and by hormones of the adrenal cortex (Baker and Whitaker, 1949; Johnson, 1958b), and stimulated by thyroid hormone (Butcher, 1940; Dieke, 1948; Ebling and Johnson, 1964). Hormones also affect other aspects of hair growth. For example the rate of hair growth is inhibited by oestrogen (Johnson, 1958a) and stimulated by thyroxine (Ebling and Johnson, 1964), and the loss of club hairs is delayed by the female sex hormone (Johnson, 1958a).

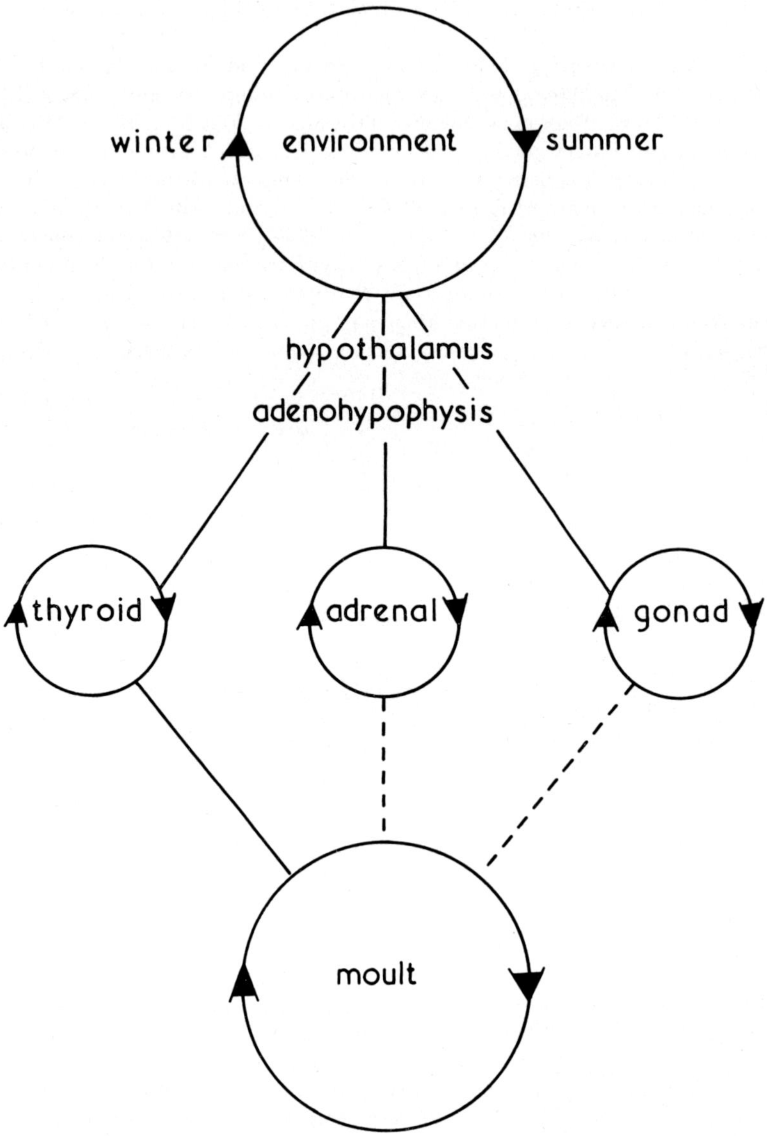

Fig. 8. Relationship between the environment, endocrine system and the molt. Broken lines indicate inhibiting factors; solid lines indicate stimulating factors. (From Johnson, 1972).

In laboratory rodents there is no evidence of a seasonal variation in the production of hormones and no seasonal molt. In wild mammals with a seasonal molt it is reasonable to look for the control pathways in the endocrine system. The synchrony between the sexual cycles and the molt cycles is well documented and has already been discussed, but there is less information available on the relation between the molt and other endocrine glands.

In *Microtus agrestis* there is evidence that the pituitary, thyroid and adrenal glands, as well as the gonads, show a seasonal variation in activity, with high activity in spring and summer and low activity in autumn and winter (Khateeb and Johnson, 1971a). The spring hair replacement occurs when the glands are increasing their activity, whilst the autumn replacement coincides with regression of these glands.

Experiments with voles in captivity indicate that in spring thyroid hormone may be necessary for the initiation of hair growth, whilst the increasing levels of sex hormone inhibit the follicles, so that fewer hairs grow. Thyroxine stimulates the amount of hair produced by individual follicles and adrenal cortical hormones may encourage the loss of club hairs. The result is a sparse pelage with coarse hairs. In autumn, the lowered amount of sex hormone allows more hairs to grow, whilst less adrenal cortical hormones encourage retention of club hairs; lowered secretion of thyroxine results in finer hairs. Thus, the voles grow a dense coat with fine hairs (Khateeb and Johnson, 1971c).

In the mink it has similarly been shown that the thyroid gland is essential for growth of the spring pelage (Reineke et al., 1962), and the effect of thyroxine in promoting wool growth in sheep is well known (Labban, 1957; Coop and Clark, 1958; Kirton et al., 1959).

Lyman (1943) showed that the pituitary gland was involved in the change from white to brown pelage of the varying hare in spring, and more recently it has been shown that melanocyte-stimulating hormone is responsible for the similar colour change in the weasel (Rust, 1965). Rust and Meyer (1969) have also shown that implantation of melatonin into weasels nearing their spring molt resulted in a change to a white coat instead of brown, whilst at the same time the animals became reproductively quiescent.

CONCLUSIONS, PROBLEMS AND PROSPECTS

The evidence which has been presented supports the view that the seasonal changes in pelage in wild mammals are adjusted to environmental changes by way of the endocrine system. The molt cycle appears to be linked with cycles of activity in the gonads, adrenals and thyroid gland. Such a scheme is illustrated in Figure 8 in which it is suggested that the adjustments to environmental conditions occur by way of the hypothalamus and adenohypophysis in the same way that control of the sexual cycle is mediated.

As a result of natural selection, the environmental cycle is engaged with both the endocrine cycle and the molt cycle. Domestication, on the other hand, tends to disengage the cycles and in such animals as laboratory rodents there are no seasonal changes in the endocrine system and no seasonal molt, but hair growth is still influenced by hormonal factors.

There are as yet very few studies which relate seasonal changes in pelage with a complete analysis of the seasonal changes in the endocrine glands. It is important that any such studies take care to relate the experimental situation to the field conditions. There is always a danger of turning wild mammals into laboratory mammals which respond in quite a different way to their counterparts in natural conditions. For example, it is interesting that the often quoted experiments on the effect of cold in stimulating the thyroid gland of laboratory rats (Rand et al., 1952) and guinea pigs (Stevens et al., 1955) give a misleading view of the role of the thyroid in thermoregulation. There is increasing evidence to suggest that many mammals in the wild have more active thyroids in summer than in winter (Hart, 1964), and it has been shown that white rats acclimatized to outdoor conditions do not increase their thyroid activity in winter as laboratory rats do when subject to cold stress (Héroux, 1960). It is obviously more efficient for a mammal to adapt to cold by changing its insulation than by increasing its metabolism when food is scarce. Scholander et al. (1950) have shown that for many mammals adaptation to cold rests entirely on fur insulation.

The control of seasonal molting may be more complicated than has been presented here. The story will certainly differ in detail for those mammals that are reproductively active in the autumn rather than the spring.

REFERENCES: see pp. 263-265.

Photoperiodic Effects on Hair and Wool Growth of Domestic Animals

by

J. C. D. Hutchinson

1. INTRODUCTION

One of the most important factors adapting animals to the changing seasons of cold and temperate climates is the control of coat character which provides a deep coat for winter and a shallow coat for summer (Hart, 1956). Some effects of coat change in adaptation to hot environments are discussed in Chapter 1, Section 13a and Chapter 2, Section 2a. One might suppose that the seasonal changes of temperature would cause the animal to develop a suitable coat. However, shedding and regrowth of fibres are involved and this is a relatively slow operation and the animal requires advanced warning of a change of season. Changes in temperature from month to month could give signals of approaching hot or cold seasons, but these are erratic in comparison with daylength. Therefore daylength is more suitable as a signal and it is not surprising that photoperiodic effects play an important part in control of coat character.

A simple general account of the many photoperiodic phenomena in animals is given by Lofts (1970). Early work established that daylength regulated shedding of the coat of the silver fox, ferret, mink and horse (Bisonnette and Wilson, 1939; Bassett et al., 1944; Burkhardt, 1947; Hammond, 1951, 1952, 1953). Subsequently a more detailed study on the ferret has been made by Harvey and Macfarlane (1958). They established that there were two distinct moults during the year, a major shedding in late summer and autumn and a minor spring shedding (Fig. 1). In addition, there were seasonal variations in hair density, coarseness of the coat, hair growth rate and weight of hair per unit area. They succeeded in entraining the annual cycle of coat change to accentuated reversed photoperiodic seasons and this entrainment was complete within two months of the start of the experimental treatment. It is noticeable that the minor shedding was somewhat heavier from the animals experiencing reversed photoperiodic cycles than from the control animals.
Photoperiodic effects on the hair and wool growth of cattle and sheep have been studied in more detail than those of other domestic animals and these are discussed below.

2. CATTLE

Bos taurus, *Bos indicus* and their crosses have two definite seasons of shedding and regrowth of hair, one in spring and one in autumn (Dowling and Nay, 1960; Hayman and Nay, 1961). Many details of the coat changes have been studied on about 70 calves of various breeding which lived on pasture near Rockhampton, a coastal district

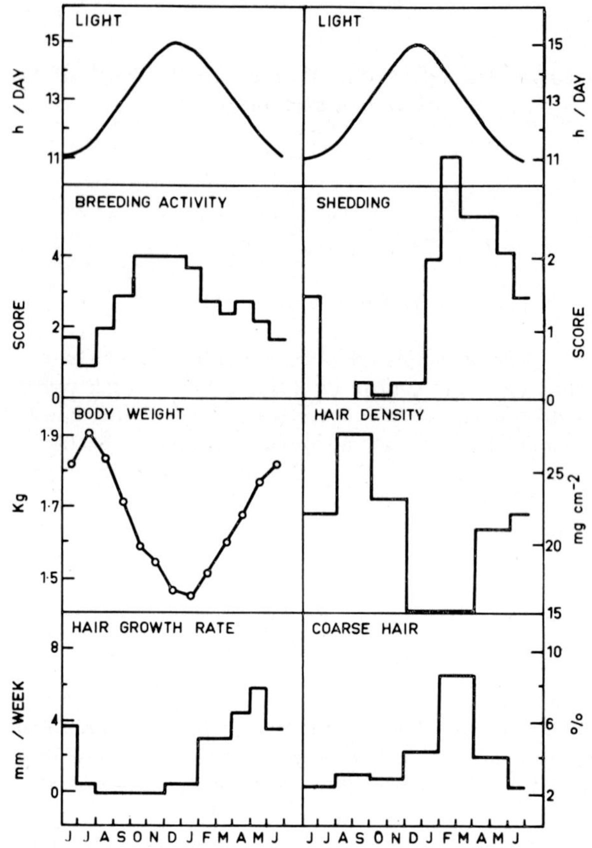

Fig. 1. Correlation of light changes in the Southern Hemisphere with breeding season, body weight and coat characters of male ferrets. There were small differences between the hair cycles of the males and females. All the characters were reversed when the photoperiodic seasons were reversed. (From Harvey and Macfarlane, 1958).

of Queensland (latitude 23.5° S). Some idea of the complexity of the changes may be gained from Figures 2 and 3 and these illustrate only part of the information gathered. It was found that although there are two fairly well defined seasons of shedding and regrowth, each follicle on an average produced a little more than two hairs per year.

Yeates (1955), working at Brisbane, Queensland (latitude 27.5° S), exposed polled Shorthorn calves to reversed exaggerated photoperiodic seasons for nearly a year, starting at the beginning of February. The animals shed their winter coats in May, had summer coats by the end of June and had winter coats in December. Yeates (1957) next tried to find out whether cattle experiencing an equatorial daylength would exhibit an endogenous rhythm of coat change, whether the changes would be disrupted,

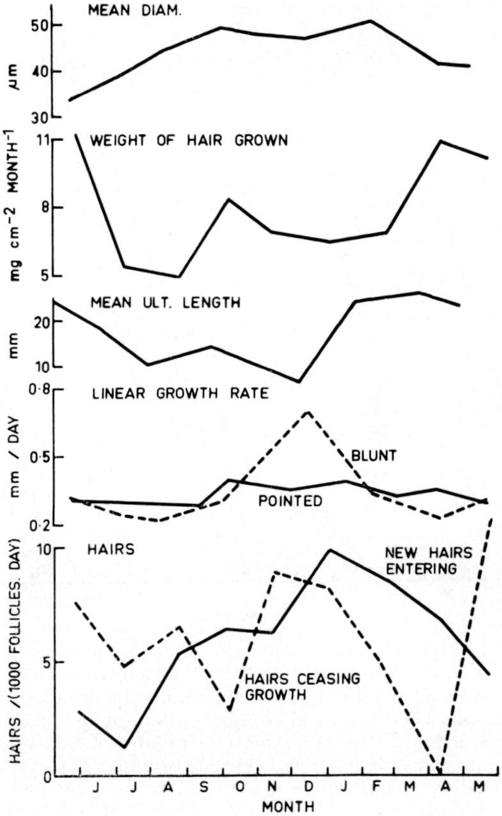

Fig. 2. Seasonal changes in various parameters of hair growth in Hereford and beef Shorthorn male calves. "Mean diameter" refers to growing hairs. "Mean ult. length" describes the ultimate length attained by hairs beginning growth at the dates marked on the abscissa. "Blunt" means a hair that had been present at the last examination and therefore clipped. "Pointed" means a hair that appeared since the last examination. (From Turner and Schleger, 1970).

or whether the coat would reach a constant condition. He found that regardless of the time of year he began his treatment, the coat reached a constant condition. The depth was intermediate between that of summer and winter, and there seemed to be a steady slow turnover of hair fibres. Calves born in late winter or early spring and placed immediately in the experimental environment, failed to shed their coats either in the spring of their birth or the next spring. These experiments showed that the annual rhythm of coat change did not become endogenous when the normal photoperiodic signals that control it are absent. In this it resembles the annual rhythm of moulting and egg-laying in the domestic fowl (Greenwood, 1962; Hutchinson, 1962). In the fowl an equatorial daylength with a constant temperature results in cessation of laying within about three years and ultimately premature death from adeno-

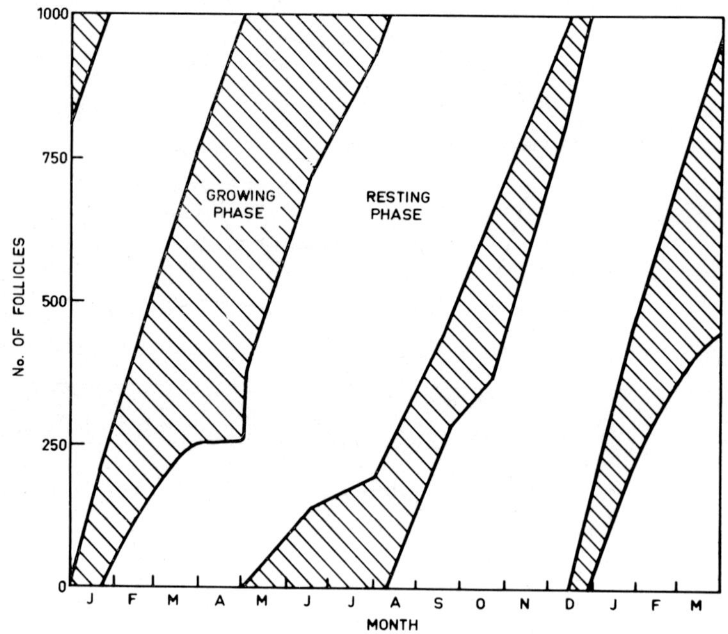

Fig. 3. Illustration of the cyclic behaviour of follicles, which is the main determinant of seasonal variation in coat type. The left hand boundary of each shaded section is a plot of cumulative number of new hairs and its slope represents the number of new hairs per 1000 follicles per month. The right hand boundary is a plot of cumulative number of hairs completing growth, and its slope represents the number of hairs ceasing growth per 1000 follicles per month. Any horizontal line represents presumed successive phases of growth, quiescence, shedding and regrowth undergone by an individual follicle; any vertical line indicates proportions of follicles which are active or resting at a given date. (From Turner and Schleger, 1970).

carcinomata of the reproductive system, which is a common cause of death in old age in untreated fowls. However, no ill effects have been noticed when cattle from breeds native to high latitudes are kept in equatorial environments that are not hot enough to overstress them. Experiments in long or short constant daylengths have not yet been made.

In the experiments discussed, only spring shedding and the general appearance of the coat were examined. It is presumed from Harvey and Macfarlane's (1958) study of ferrets that when shedding is entrained to daylength, the other changes of the follicular cycle are likewise entrained.

Yeates (1965) considered that the coats of equatorial breeds of cattle were not influenced by photoperiodic changes. He gave as an example an Ankole animal born in Bedfordshire, England, which as an adult had a glossy, shallow coat in March. The Ankole breed is native to equatorial east Africa. Both *Bos taurus* and *Bos indicus* breeds that are not native to equatorial latitudes have distinct summer and winter coats.

Effect of nutrition. Daylength is not the only factor controlling the annual cycle of coat change. Yeates (1958) obtained preliminary evidence that malnutrition could delay the change from winter to summer coats. Hayman (1965) examined the effect of undernutrition on coat change in adult F1 Brahman × Friesian cows. When undernutrition was imposed during summer, shedding in spring was delayed. Fibres of the winter coat were longer and less medulated than those of well nourished animals. When undernutrition was continued into the next summer, a portion of the preceding winter coat remained unshed. In Hereford heifers undernutrition imposed in winter can postpone shedding of the winter coat till the middle of summer (Allen et al., 1970). *Lantana camara* poisoning, like malnutrition, can delay shedding of the winter coat (Turner and Schleger, 1970). It has been observed that well fed heifers calving in spring retain their winter coats even when in top condition. This suggests that the stress of late pregnancy and lactation can delay shedding (Hayman, pers. comm.).

Although the coats of animals living close to the equator tend to have a constant coat character throughout the year, there is preliminary evidence of an effect of nutrition. On the Athi plains of Kenya grazing is good in December following the lesser rains which come after a period of relative scarcity of pasture. At the end of December three out of four Ayrshire and Hereford animals were found to be shedding but only one out of five nearly purebred Boran steers, a breed native to equatorial latitudes (Finch and Hutchinson, unpubl.). The remainder had a few loose hairs as described by Yeates (1957). These casual observations suggest that systematic study of the influence of nutrition on coat change in an equatorial environment might give interesting results.

(2) *Effect of temperature.* Although photoperiodic signals, with some influence from nutrition and other factors, control the timing of the cycle of coat change, they do not adjust the coat for particular climates. However, it has been known from Hagenbeck's time (1911) that some animals native to a warm environment and transported to a cool environment grow deeper winter coats than in their natural habitat. Recent observations show that in cattle the influence of winter temperature can be large (Webster et al., 1970). The subject is discussed in detail in Chapter 1, Sections 4a, 5a, 8a, b and 13a, b.

3. SHEEP

A. ANNUAL RHYTHM OF WOOL GROWTH

It is well known that wool sheep on a constant ration have an annual rhythm of wool growth. The growth is faster in summer than in winter except for the Bikaneri breed in which it is the other way about (Nagarcenkar and Bhattacharya, 1964a). Fibre diameter as well as length is affected (Morris, 1961; Hutchinson et al., unpubl.). In the Merino, the rhythm is negligible in amplitude in those commercial strains studied; but in the Camden Park Merino, a non-commercial strain, the rhythm is considerable (Ferguson et al., 1949). There are several general accounts of the subject (Hutchinson and Wodzicka-Tomaszewska, 1961; Hutchinson, 1965; Ryder and Stephenson, 1968). The cycles of individual follicles are complex, as in cattle, and are described in detail by Ryder and Stephenson (1968). On the whole the picture is of an archaic pattern of a spring shedding and regrowth of secondary follicles and two sheddings per annum of the primary follicles, and it has been suggested that the difference in wool growth between summer and winter in modern wool sheep is a modification of this archaic pattern (Hutchinson, 1965; Ryder and Stephenson, 1968). In these modern sheep the fibres continue growing for longer than a year. In those that have a pronounced annual rhythm and live on the hills in cool climates, the low wool growth in winter may conserve nitrogen for other purposes, such as gestation in ewes (Doney, 1964).

Morris (1961) showed that the annual rhythm in Romney Marsh ewes could be reversed by reversed accentuated photoperiods (Fig. 4), starting in the autumn. The reversal was slow and took two years to complete. Reversing seasonal temperatures had no influence on the rhythm exhibited in the natural daylength of Brisbane, Queensland. The mean winter temperatures in both groups do not seem to have

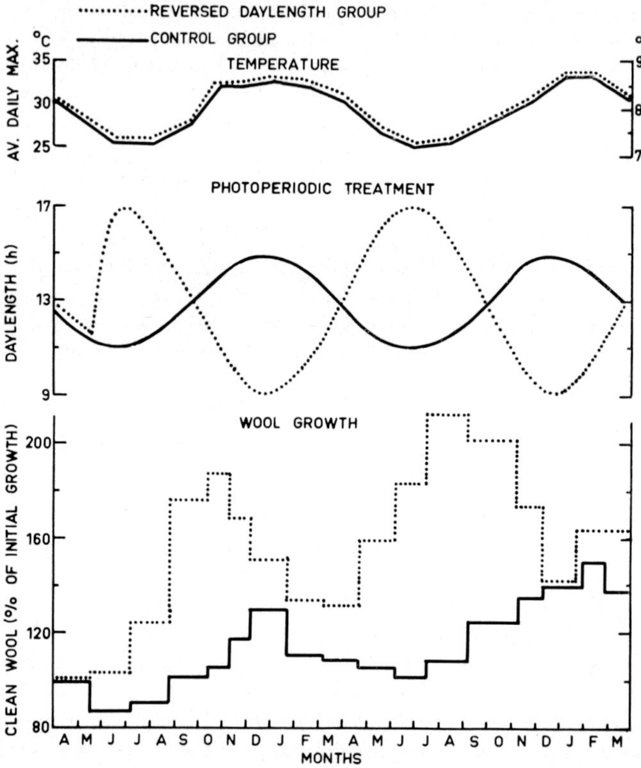

Fig. 4. Effect of a reversed accentuated photoperiodic cycle on the annual rhythm of wool growth in Romney Marsh ewes. Wool growth was measured on a mid-side patch of skin. The control group had 4 ewes and the reversed daylength group 7 ewes. (From Morris, 1961).

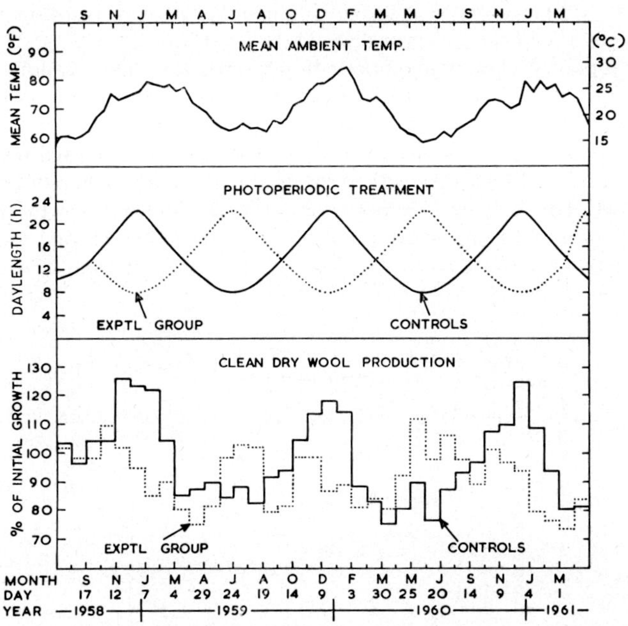

Fig. 5. Effect of reversed photoperiodic cycles on the annual rhythm of wool growth of Southdown ewes. Wool growth was measured on mid-side and thigh patches of skin. Both the normal photoperiodic cycle (controls) and the reversed cycle (exptl. group) correspond to a 60° including civil twilight. Lighting was artificial. There were 6 ewes in each group. (From Hart, et al., 1963).

dropped below 20° C in the coldest month. Hart et al. (1963) also tried to reverse the rhythm in two localities by accentuated reversed photoperiodic treatment. In Christchurch, New Zealand, they used Corriedale ewes and near Sydney, N.S.W., Southdowns and Peppin Merinos. The Christchurch trial started in autumn and the Sydney experiment in spring. In both experiments the rhythm became bimodal with small peaks in winter or late autumn, and in spring (Fig. 5). The effect of the light treatment was immediate and not spread over two years as in Morris' (1961) experiment. The split rhythm bears some resemblance to the increase of shedding in the minor shedding season of ferrets when their annual cycle was reversed by reversed photoperiods (Harvey and Macfarlane, 1958). The difference between Morris' results and those of Hart and his colleagues may represent a breed difference. When the Sydney experiment was continued over two seven-month photoperiodic cycles, there was an immediate change in the rhythm to one with a single peak per photoperiodic cycle, 2-3 months after the longest day (Hutchinson, 1965). This suggests that the rhythm was now completely entrained to daylength. In normal daylengths hair and wool on the legs of both Merino and Southdown animals was observed to become loose in the spring. This rhythm was reversed when the photoperiodic seasons were reversed (Hutchinson, 1965).

Rougeot (1957, 1961) showed that the shedding and growth cycles of the kemp follicles of the Limousine sheep can be controlled by daylength.

It might be supposed that if photoperiodic changes were not available the rhythm would disappear, as in Yeates' (1957) experiment with cattle. Alternatively it might be maintained by the seasonal variations of temperature or it might run endogenously. To elucidate this point, Southdown ewes were exposed to accentuated reversed seasons of air temperature and humidity with a constant equatorial daylength (Bennett et al., 1962). They maintained a nearly normal annual rhythm of wool growth on skin patches protected by covers with about the same thermal resistance as that of their fleeces, which were kept at a constant depth (Fig. 6). The peak of wool growth was about

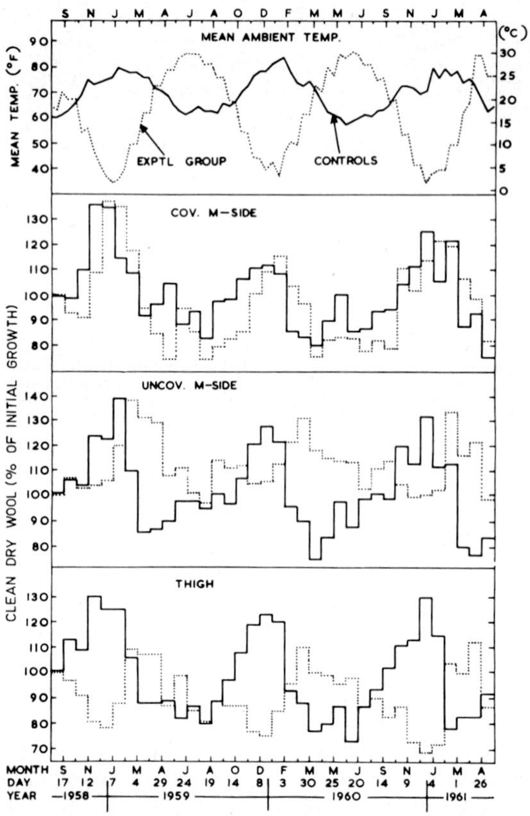

Fig. 6. Effect of an accentuated reversed cycle of ambient temperature and humidity on the annual rhythm of wool growth in Southdown ewes. Lighting of "controls" (——) as in Fig. 5. Daylength of "exptl. group" (....), 12 hrs 50 min, corresponding to an equatorial day including civil twilight. M-SIDE, mid-side patch of skin; COV., clipped patch of skin protected by an insulating cover. There were 7 ewes in each group, 3 of which had protected mid-side patches. No thigh patches were protected. (From Bennett, et al., 1962).

a month after that for the control sheep. On unprotected mid-side and thigh skin patches on the same animals, the peak was delayed by about 3 months so that the rhythm was distorted but certainly not reversed. This suggests that the rhythm has endogenous properties. However, there was no suggestion that the rhythm of any individual animals had a period significantly different from a year. Therefore, one cannot from this experiment exclude entrainment to other environmental variables that have a period of a year, such as atmospheric pressure, electromagnetic radiation in the radio spectrum, cosmic rays and magnetic fields (Chapman, 1951; Ratcliffe, 1960; Hines et al., 1966).

Examination of other experiments in which daylength has been maintained constant shows several cases in which the rhythm of wool growth continued (Table 1). The authors did not discuss the possibility of endogenous properties, probably because at that time little was known about circannual rhythms. In two of the experiments listed, the rhythm was disrupted. Hart (1961) found that hooding the eyes of Corriedale ewes, so that they were in darkness, abolished the rhythm after about a year. In Slee's (1965) investigation on newborn Wiltshire Horn lambs in a continuous dim light, shedding of the birth coat was greatly delayed and it was incomplete. These experiments, in which the rhythm was disrupted, suggest that neither temperature nor other seasonal influences maintained it in the other experiments listed and therefore that it was truly endogenous. As mentioned above, the experiments of Morris (1961) and of Bennett and his colleagues (1962) likewise suggest that temperature does not influence the rhythm except on unprotected patches of skin at environmental temperatures below about 5° C. In addition, Symington (1959) found that Persian Blackhead rams, housed for eight months, failed to shed their coats which continued to grow. He ascribed the results to the decrease in light intensity on housing.

The information in Table 1 raises a problem in the definition of annual rhythms. These are usually considered to be true biological circannual rhythms if they continue endogenously when the photoperiodic signals that normally entrain them are removed. However, if a rhythm can continue with one constant daylength regime and fails to continue with another, it is rather difficult to define it. There is a parallel case in another ungulate, the Sika deer (*Cervus nippon*). In this animal the annual rhythm of antler shedding and regrowth continues endogenously in a number of different fixed daylengths, but fails in a 12-hr daylength (Goss, 1969a, b). There is a somewhat similar situation in the physiology of circadian rhythms. Constant light can extinguish them although they run endogenously in constant darkness (Pittendrigh, 1960; Wurtman et al., 1968).

(1) *Effect of nutrition.* In field conditions the considerable seasonal variations in wool growth of Merino sheep are caused by variations of nutrition (Brown and Williams, 1970). On the other hand in the Cheviot breed, which has a pronounced annual rhythm, nutrition has little influence on wool growth in winter but a considerable effect in summer (Doney, 1966). The Scottish Blackface and the Soay may be similar (Doney, 1964; Doney quoted by Ryder and Stephenson, 1968). In the Wiltshire Horn breed undernutrition causes less complete moulting than normal but does not seem to affect the time at which moulting starts (Slee, 1965).

Table 1. Effect of constant photoperiods on maintenance of an endogenous annual rhythm of wool growth or shedding and regrowth

Experiment	Treatment		Duration (months)	Breed and Sex	Effect on Rhythm		Author
	Light (h. min)	Dark (h. min)			Timing	Amplitude	
1	0	24*	29	Unmated Corriedale ewes	Lost within a year	Lost within a year	Hart, 1961
2	8	16	23	Mated Corriedale ewes	Unchanged	Slightly*** reduced	Hart, 1961
3	8	16	13	Unmat d Corriedale ewes	Unchanged	Slightly*** reduced	Hart, 1955, 1961
4	7	17	11	Columbia-Southdown Merino and Hamsphire ewes	Unchanged	Unchanged	Wodzicka, 1960
5	4	8	16	English Romney rams and wethers	Unchanged	Unchanged	Wildman, 1957
6	2	4	13	Unmated Corriedale ewes	Unchanged	Slightly*** reduced	Hart, 1955, 1961
7	12.50	11.10	29	Unmated Peppin Merino ewes	Unchanged	Unchanged	Bennett et al., 1962
8	12.50	11.10	29	Unmated Southdown ewes	Unchanged	Unchanged	Bennett et al., 1962
9	24**	0	10	Unmated Corriedale ewes	Unchanged	Unchanged	Hart, 1961
10	24****, dim light			Wiltshire Horn lambs	Shedding delayed or incomplete		Slee, 1965

* Hooded to exclude light from eyes. Kept with control animals in natural indoor daylight of Christchurch, New Zealand.

** Natural indoor daylight supplemented with artificial light at night, at about one tenth of the intensity of the natural daylight.

*** Wool production was increased during the winter and early spring so that amplitude of the rhythm was somewhat less than normal.

**** Animals born in experimental environment.

(2) *Effect of temperature*. This is discussed in Chapter 1, Section 13a and b. The effect of temperature on clipped patches of skin illustrated in Figure 6 is now known to be a local effect as would appear from the figure (Downes and Hutchinson, 1969, and other authors). Figure 6 suggests that the effect depends on change of environmental temperature rather than temperature per se, and this has been confirmed in a 7-month temperature cycle (Hutchinson and Bennett, unpubl.). There is no evidence that a cold winter stimulates sheep to produce a warmer winter coat (Webster et al., 1969). In this they differ from cattle.

B. CONSTANT DAYLENGTHS AND ANNUAL WOOL GROWTH

In several experiments in which constant short daylengths have been used, wool growth over the year has been greater than for animals exposed to daylengths of the local environment (Table 1, Exp. 2, 3, 5, 6; Fig. 7). Food intake was approximately constant between seasons. The light periods were half the length of the dark periods, but in experiments 5 and 6 there were more than one light and one dark period per 24 hrs. The increased wool growth was mainly in winter and early spring, the time of year

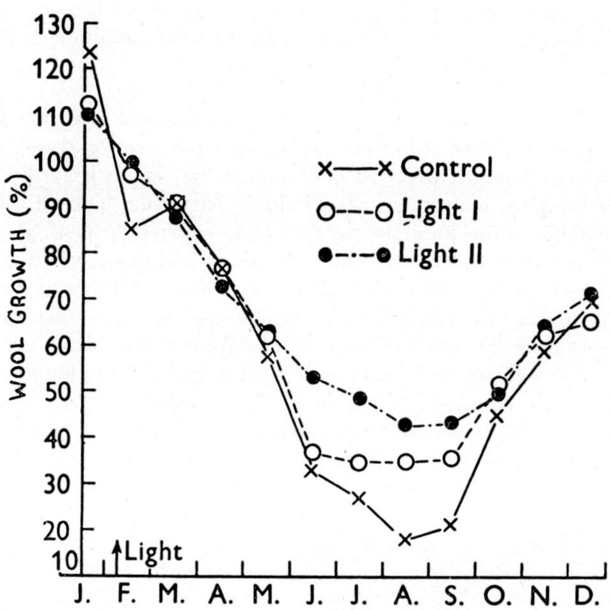

Fig. 7. Effect of two constant daylength treatments on wool growth of Corriedale ewes. Wool growth is expressed as a percentage of average wool grown on a mid-side patch of skin from Dec. 1951 to Feb. 1952 inclusive. "Control" means normal daylength of Christchurch, New Zealand; "Light I", L/D:8/16; "Light II", L/D:2/4. There were 4 control ewes and 3 ewes in each experimental group. (From Hart, 1961).

at which wool growth is lowest in natural daylengths (Fig. 7). In one experiment with a constant short daylength, wool growth was not increased (Table 1, Exp. 4).

In an experiment, not listed in Table 1, unmated Bikaneri ewes were exposed to 8 hrs light and 16 hrs dark periods from the summer to the winter solstice (Nagarcenkar and Bhattacharya, 1964b). They grew more wool than animals exposed to normal daylengths. Another group receiving the long daylength of 16 hrs light and 8 hrs dark likewise grew more wool than the controls, but only a little more.
At present no clear theoretical explanation has been advanced to explain these interesting effects of fixed daylengths.

4. MECHANISM OF PHOTOPERIODIC EFFECTS

The central nervous system is supposed to receive the photoperiodic signals via the retina and the optic nerve. However, it should be mentioned that the skull of the sheep can transmit a very small amount of sunlight directly to some regions of the brain (Ganong et al., 1963), and that in the duck, which has a much thinner skull, direct transmission is an important pathway for photoperiodic stimuli (Benoit, 1964). In the domestic fowl change of photoperiod is the most important function of daylength governing the moulting and laying cycles (Hutchinson, 1962; Morris, 1962). On the other hand, absolute daylength is still considered to be the photoperiodic signal that controls breeding in wild birds (Farner, 1970; Lofts, 1970).

Animals such as the ox and sheep, have had to accompany man on his migrations through considerable changes of latitude. At a particular season, daylength can vary widely with latitude; but the direction of change in daylength is constant within the northern or the southern hemisphere. It would be surprising if natural and artificial selection had not given the animals the ability to respond to change of daylength. It is true that the activation of the hair follicles in spring to produce the summer coat takes place close to the equinox (Hayman and Nay, 1961; Turner and Schleger, 1970), and that at this time daylength (including civil twilight) does not vary more than about 45 min over the latitudes inhabited by these species. However, the autumn shedding and the growth of the winter coat start soon after the longest day, though they do extend to the equinox. At the time when shedding starts, daylength varies widely with latitude. Moreover, the most obvious difference in photo-periodic conditions between the two seasons at which the hair follicles have to make their preparations for the two different types of coats is that daylengths are increasing in the former, and decreasing in the latter season. Calves can be born at any season of the year, and it would seem that they need to respond to changes of daylength in order to discriminate between spring and autumn in their first year of life.

The sheep, having an endogenous rhythm of wool growth, should not require more than one photoperiodic signal per year to maintain the rhythm properly synchronized. In some breeds, such as the Merino, the lambs can be born, like the calves of cattle, over many months of the year. For these it would be advantageous to be able to respond to change of daylength as well as absolute daylength, so that the coat cycle can be synchronized with the correct half of the year.

No systematic experiments have yet been published to discriminate between the effects of absolute daylength and change in daylength on hair and wool growth.
There is preliminary evidence that reducing the light intensity with polaroid spectacles delays spring shedding considerably (Y. S. Pan, unpubl.).

Bünning (1960; see also Pittendrigh and Minis, 1964) suggested that living organisms measured daylength by a diurnal rhythm that runs endogenously from the time of day at which it is synchronized, for example from dawn. He envisaged this rhythm as consisting of two half-cycles (the photophil followed by the scotophil), each of approximately twelve hours duration. He considered that photoperiodic induction of a physiological process requiring long days occurred only when the duration of the natural day extended into the scotophil part of the endogenous diurnal cycle. Evidence for such a system in the breeding of wild birds is summarized by Lofts (1970). The system could presumably react to changes in daylength as well as fixed daylengths. When the light and dark periods add up to 12 or 6 hrs instead of 24 hrs, as in experiments 5 and 6 of Table 1, the system worked out by practical experiments for Japanese quail (*Coturnix coturnix japonica*) would classify the treatments as long absolute daylengths although the dark periods in experiments 5 and 6 were in fact twice the length of the light periods (Follet and Sharp, 1969). The Japanese quail require long daylengths for testicular growth, which was the parameter of reproduction studied. If the sheep were sensitive to long nights rather than long days and the system were therefore, sensitive to the photophil period only, experiments 5 and 6 would be classified as long-night experiments.
One might suppose that the photoperiodic signals, perhaps interpreted by Bünning's sytem, are conveyed by the neuroendocrine system to their target organs as in reproduction (Thibault et al., 1966; Farner, 1970). At the present time, the evidence is that neither cortisol nor thyroxine plays an important part in seasonal variation of wool growth (Ferguson et al., 1965). However, there are other hormones that could be involved, for example hormones of the anterior pituitary other than thyroid stimulating hormone, or activity of the pineal gland, which is supposed to influence reproduction (Wurtman et al., 1968).

There is no information on transmission of photoperiodic information by direct neural pathways to the hair or wool follicles. Likewise there is no information on any part that the follicles may play as a delay factor in the annual rhythm of wool growth nor on any part they may play as direct recipienıs of light.

This discussion of mechanisms does not indicate why the annual rhythm of wool growth in sheep developed such a strong endogenous property. It is easy to understand why a Long-tailed Shearwater (*Puffinus tenuirostris*), travelling through many daylengths in a short time, needs an endogenous rhythm to assist regulation of its breeding season (Marshall, 1960). It is not clear why a primitive sheep should need such an apparatus to produce a winter coat at the right season and the reason for an endogenous rhythm in modern wool sheep is no clearer.

5. SUMMARY AND FUTURE DEVELOPMENTS

It is now clear that coat change in the domestic animals studied and the annual rhythm of wool growth in sheep on a constant ration are mainly controlled by daylength.
In the sheep the annual rhythm can be endogenous in a number of fixed daylengths but not in darkness. In cattle the cycle of coat change does not continue in a constant equatorial daylength. Malnutrition and other stress factors are important in delaying spring shedding in cattle. In sheep that have a small annual rhythm, such as the Merino, wool growth in different seasons depends mainly on nutrition. Nutrition can also affect shedding in the Wiltshire Horn breed. Very little is known about the mechanism of the photoperiodic effects. Even the elementary question whether absolute or change of daylength is the more important has not been definitely answered.

The early studies of photoperiodic effects on fur changes and the reproductive cycle of fur bearing domestic animals have led to attempts to control quality of fur and the season at which it is obtained (Beljaev et al., 1964; Ignatovich, 1970; Stout and Adair, 1972; and other authors). It is to be expected that these attempts will be intensified with more sophisticated photoperiodic treatments.

In cattle husbandry a problem in heat tolerance is delay in coat shedding in spring, particularly in animals inhabiting low latitudes. As mentioned above, it is thought that such delay is caused by poor nutrition. However, the latitude at which photoperiodic entrainment of follicular cycles becomes weak in good nutritional conditions has not been investigated, nor how this latitude varies with genotype. Further work is required on this problem. In addition, photoperiodic effects need to be examined on parts of the coat cycle other than spring shedding.

In recent years interest in the chemical shearing of sheep has increased. It may well be that the effects of particular agents depend upon the point in the annual photoperiodic cycle of wool growth at which the treatment is applied. Fortunately, the photoperiodic effect on wool growth of Merino sheep is so small that such complications may not arise, so that only effects of previous nutrition need be considered. However, in breeds that have a large annual rhythm of wool growth on a constant ration, such as the Cheviot, season may need to be considered apart from nutrition. This may lead to further work on photoperiodic effects.

Much more work on the mechanism of the photoperiodic cycle is required to make it intelligible. Such studies may help in elucidating the general theoretical properties of annual rhythms, both those that have and those that do not have endogenous properties.

ACKNOWLEDGEMENT: I am most grateful to Mr F. B. Spence for help in calculations and for the preparation of other material.

REFERENCES: see pp. 265-268.

Effects of Solar Radiation on the Sensible Heat Exchange of Mammals

by

J. C. D. Hutchinson, G. D. Brown and T. E. Allen

1. INTRODUCTION

The thermal effect of solar radiation on animals, as opposed to plants, has excited relatively little interest despite its importance in the thermoregulation of mammals This chapter deals with mammals other than man except where a comparison with man is relevant. The thermal effect of solar radiation on the sensible heat exchange between an animal and its environment is not defined physiologically by the radiant energy absorbed, as was measured by Riemerschmid (1943) and Riemerschmid and Elder (1945). It is the heat exchange at the base of the coat that interests physiologists because heat exchange at this level can then be assessed in terms of the effect of insolation on heat production and its dissipation by sweating, respiration and by sensible exchanges.

2. PHYSICS OF HEAT FLOW AT THE BASE OF THE COAT

An equation of heat balance for an animal in the sun can be written as follows:

$$M + H_{abs} = A_r \varepsilon \sigma T_r^4 + A_c h_c (T_r - T_a) + C + H_{ex} + H_{sw} + H_{st} \quad (1)$$

where M is the metabolic rate, H_{abs} is the amount of radiation absorbed by the animal surface, A_r and A_c are the coat surface areas for emission of infrared radiation and convective heat exchange respectively, ε is the emissivity of the coat and skin, σ is the Stefan-Boltzmann constant, T_r is the radiant temperature (°K) of the coat surface, h_c is the convective heat transfer coefficient which is dependant on wind speed and posture of the animal, T_a is the air temperature, H_{ex} is the respiratory heat loss, H_{sw} is the cutaneous evaporative heat loss, C is the heat conducted to the ground, for instance, by an animal lying on the ground, and H_{st} is the amount of heat stored by the animal.

The heat absorbed by the surface of the animal (H_{abs}) is derived from both the incident solar radiation (R) i.e., the radiation with wave-lengths between 0.3 and 2.2 μm, sometimes referred to as "shortwave" radiation and incident infrared or "longwave" radiation emitted from the atmosphere and surroundings with wavelengths greater than about 3 μm. H_{abs} in equation 1 is defined as:

$$H_{abs} = R\alpha + A_g R_g \alpha_l + A_a R_a \alpha_l \quad (2)$$

where α is the average absorptance of the coat for shortwave radiation, R_g is the incident longwave radiation emitted from the ground and surrounding objects, R_a is the incident

longwave radiation emitted by the atmosphere, A_g and A_a are the surface areas exposed to the diffuse radiation from the ground and atmosphere respectively, α_l is absorptance of the coat for longwave radiation from the ground and atmosphere.
R in equation 2 is defined as:

$$R = A_pS + A_as + A_gr_gF_g(S \sin \theta + s) \quad (3)$$

where S is the flux of the direct solar beam measured on a surface normal to the sun, s is the diffuse solar radiation from the sky, A_p is the area of the profile of the animal subtended to the direct beam, θ is the altitude of the sun above the horizon, r_g is the reflectance of the ground for direct and diffuse shortwave radiation and F_g is a factor representing the correction for the effect of the animal's shadow on the radiation reflected from the ground. The above equations deal with energy exchanges at or near the surface of the coat and are similar to those used by Porter and Gates (1969). As mentioned before physiologists are more interested in the heat flow at the base of the coat. By rearrangement of equation 1 the sensible heat flow into the animal from the environment (H_s) is:

$$H_s = H_{abs} - A_r\varepsilon\sigma T_r^4 - A_ch_c(T_r - T_a) - C \quad (4)$$

and:

$$H_s = H_{sw} + H_{ex} + H_{st} - M \quad (5)$$

The areas involved in these equations, A_p, A_r, A_g, A_a are functions of the coat surface area; but biologists require H_s to be based on skin surface area. For large mammals with shallow coats, such as most ruminants living in hot climates, these areas are nearly identical. For animals such as the sheep, with deep coats adjustments may have to be made. The heating effect of solar radiation is H_s in the sun minus H_s in the shade.

The accuracy of prediction of the effect of solar radiation on sensible heat exchange depends on the accuracy with which the quantities in these equations are known. In this Section we shall discuss them individually and practical attempts will be examined to predict and measure the effect of solar radiation. A fuller treatment of longwave exchanges will be found in Chapter 1, Section 13a. There is also a recent review of the physics of radiative loss (Cena, 1974).
Instruments used for measuring radiation and other meteorological quantities are described by Drummond et al. (1958) and Monteith (1972).

A. ABSORPTANCE

Reflection from coats or skin is a complex phenomenon which can be influenced by structural factors such as interference, diffraction and scattering of light as well as pigments. These subjects are discussed by Fox (1953), Fox and Vevers (1960) and Land (1972).
The absorptance of the coat and skin (1 − reflectance) is clearly an important quantity in the heat exchange of animals exposed to solar radiation. Riemerschmid and Elder (1945) devised a simple method for measuring it on cattle coats. Unfortunately, it can only be performed on a hide spread on the ground thus involving the sacrifice of the animal. They found that absorptance decreased abruptly at oblique angles of incidence of the direct solar beam (Fig. 1), and this is confirmed by Finch (1973).

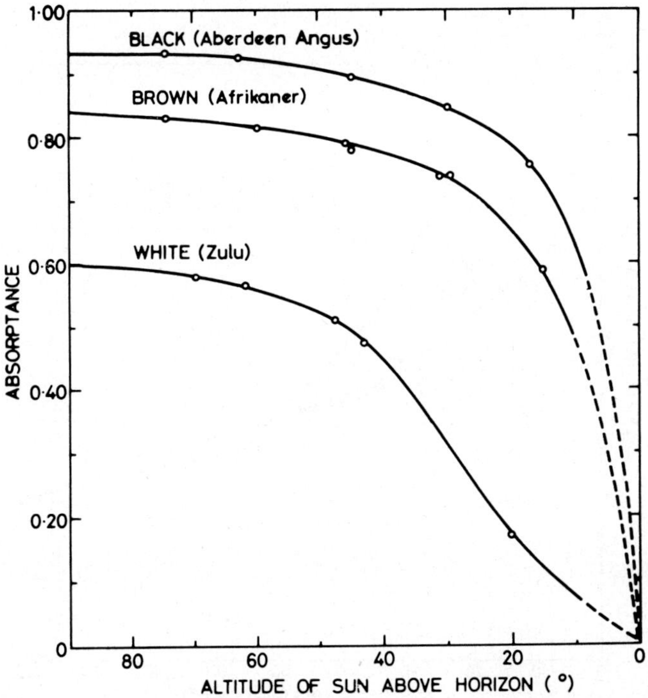

Fig. 1. Absorptance of the direct solar beam at different sun altitudes by coats of different colours laid on the ground. (From Riemerschmid and Elder, 1945).

The most obvious reason for this would be an increase in specular as opposed to diffuse reflection for at oblique angles of insolation the reflected flux increases with a decreasing angle of viewing (Fig. 2). This increased reflected radiation is seen as an increase in gloss and has the colour of sunlight rather than that of the coat. Whether gloss has an important effect on absorptance is not known. Riemerschmid and Elder (1945) showed how to calculate the absorptance of the coat of a live animal from observations made on a pelt at various angles of solar incidence.

Cena (1966) succeeded in measuring the absorptance of a portion of the coat of live animals at a position on the coat normal to the solar beam. Under these conditions the amount of reflected light is independent of viewing angle; and he was, therefore, able to restrict the solid angle of acceptance of his radiometer so that it embraced the coat only, not the surroundings. Hutchinson et al. (in press), using a microsolarimeter, devised a method of measuring the absorptance of the entire coat of a live animal determined from the average of measurements of the absorptance of many areas on the animal.

Absorptance of animal coats varies widely with the wavelengths of the solar spectrum (Fig. 3). It is greatest at the short end and decreases over the visual spectrum, reaching

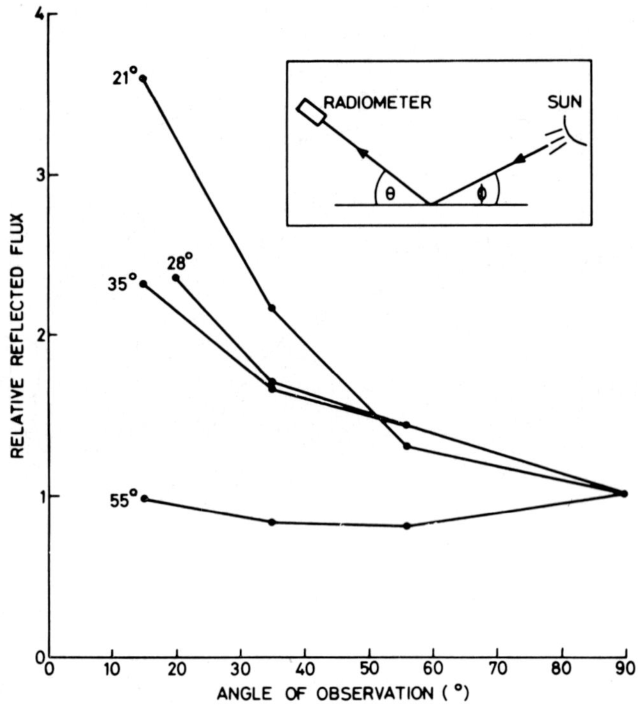

Fig. 2. Relative flux reflected from a black winter cattle coat laid on the ground at variou angles of observation. Abscissa, angle of observation in the plane of the sun's altitude. Ordinate, flux relative to that at 90° angle of observation. Angles written besides the lines represent the sun's altitude. On the small diagram Φ is altitude of the sun and Θ is the angle of observation. (From Hutchinson, Allen and Spencer, 1975. Comp. Biochem. Physiol., 50A: 343-349).

its minimum in the near infrared, after which it increases. Much of the reflected radiation is invisible. In his measurements of absorptance on live animals Cena (1966) found a lower absorptance at short wavelengths than that for hides shown in Figure 3 and in other similar experiments (Stewart, 1953). He suggested that coats of live animals might be more glossy than hides.

The direct solar beam, diffuse light from the sky, and light reflected from the ground have different spectra. Thus, absorptance of the same piece of coat mounted on the animal will be different according to its site on the animal (Fig. 4). Such variations are in nature combined with variations in colour of the animal; and the combined effect is illustrated in Figure 4 for a Jersey cow, which was light brown on the back and flanks, and off-white on the belly. In calculating a mean absorptance for use in studies of sensible heat exchanges, the individual values for absorptance must be weighted for variations in incident shortwave radiation over different areas.

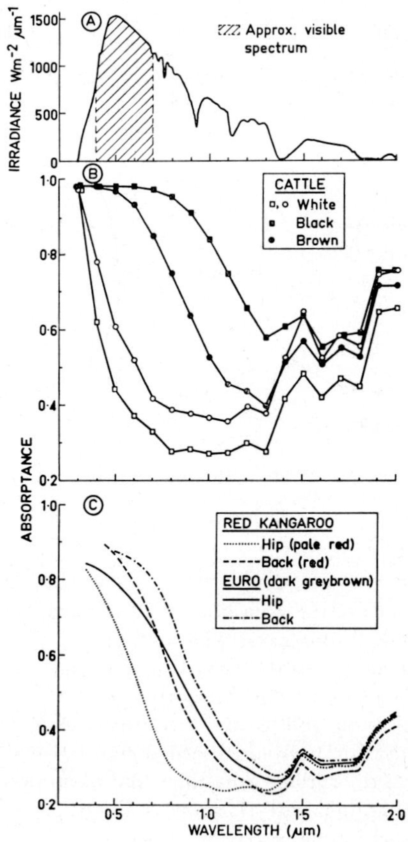

Fig. 3. A, spectral energy distribution of direct solar radiation at 90° sun altitude above the horizon. B, C, absorptance at different wavelengths of the coats of cattle and two species of kangaroo. (From Hutchinson and Brown, 1969; Dawson and Brown, 1970).

The absorptance of solar radiation by animal coats varies from about 0.9 to 0.5 (Stewart, 1953; Birkebak, 1966 and other references cited in this Section). The absorptance of longwave radiation is close to 1.0 (Hammel, 1956; Birkebak, 1966).

B. SURFACE AREA EXPOSED TO RADIATION

1. *Direct solar beam.* The incident radiation from the direct solar beam is calculated from the profile area (A_p) and the intensity of the beam (S). This profile area is the area of the shadow of the animal cast on a surface normal to the solar beam. Because the direct solar radiation is the largest proportion of the shortwave radiation, it would be expected that measurements of profile area would have been made on a number of species. In fact, a profile based on an open ended cylinder is still a favoured model and the cylinder is always placed sideways on to the sun. Only two measurements have been made on animals. Riemerschmid (1943) determined the profile for a single

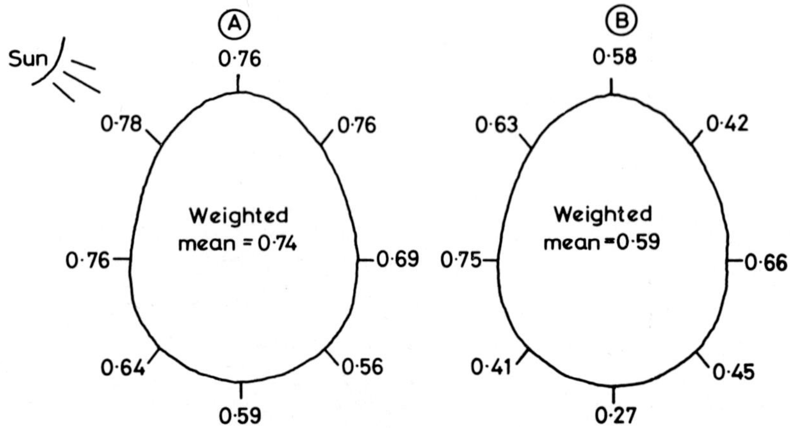

Fig. 4. A, absorptance of a piece of brown cattle coat attached at different positions round the trunk of a Jersey cow (shown in cross-section) standing sideways on to the sun. B, absorptance of the cow's own coat in the same positions. (From Hutchinson, Allen and Spence, 1975: Comp. Biochem. Physiol., 50A: 343-349).

Sussex bull by measurements of the area of its shadow throughout the sunlight hours, a simple and exact method. Profile areas were obtained when the animal's spine was in line with and at right angles to the sun. Clapperton et al. (1965) measured the solar profiles of a number of shorn and unshorn British sheep using a photographic method to obtain profiles applicable to the sun altitudes for any country. The solar profile areas of both the Sussex bull and shorn sheep, when sideways on to the sun, are a smaller proportion of the surface area than that of an open ended cylinder (Fig. 5). The profile area of an animal with its spine in line with the sun is smaller at low sun altitudes than that of an animal sideways on to the sun.

2. *Diffuse radiation.* We know of only two estimates of the surface area available for diffuse radiation exchange (long and short wave). Clapperton et al. (1965) reckoned that it averaged 85% of the surface area of the skin for a shorn sheep and 119% for a fleeced sheep. Mount (1968), with a sophisticated technique, found values for the newborn pig increasing from 67 to 76% of the surface area as air temperature rose from 20 to 30° C. For calculating the incident diffuse radiation, these percentages are applied to half the skin surface area which is then used separately for sky and ground radiation. In reality, because the short and longwave fluxes from the sky and ground are often widely different, there should be a separate fraction for each flux as in equations 2 and 3. Unfortunately, there is no information on the separate exposed areas. There appears to be no calculation of the effect of the animal's shadow on the amount of shortwave radiation received by reflection from the ground (F_g in equation 3).

C. THERMAL RESISTANCE OF THE COAT AND CONVECTIVE HEAT TRANSFER

Riemerschmid (1943) calculated that during a 15-hr day in midsummer at Johannes-

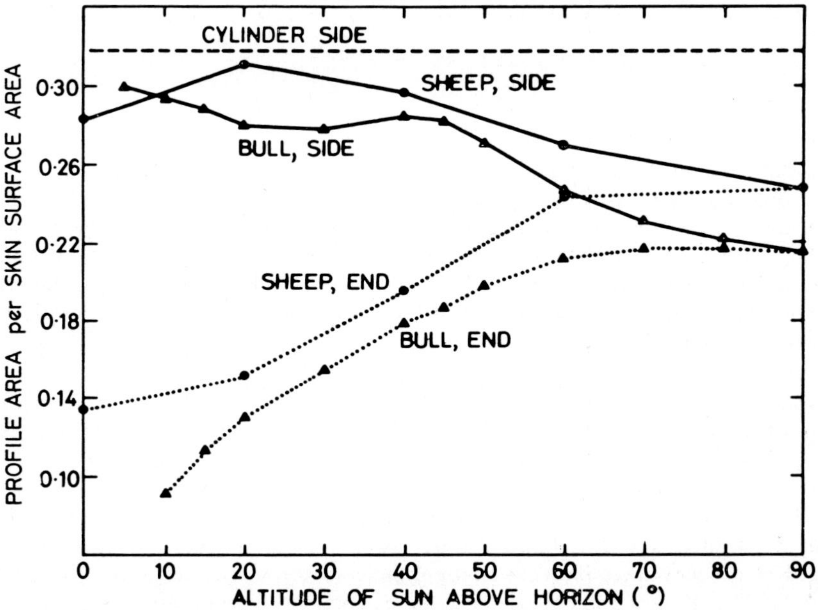

Fig. 5. Profile area to the direct solar beam as a proportion of skin surface area for a Sussex bull, shorn British sheep and an open ended cylinder. Sideways presentation to the sun is indicated by "side"; end presentation by "end". For the sheep, values for end presentation are averages for animals facing towards and away from the sun; for the bull, only facing the sun. (From data of Riemerschmid, 1943; Clapperton, Joyce and Blaxter, 1965).

burg, a brown ox would absorb an amount of solar radiation on the surface of the coat which was about 2.5 times its heat production. However, the effect on the sensible heat flow at the base of the coat (H_s) is less than this because H_s depends on the thermal resistance of the coat and its boundary air layer.

The effect of these thermal resistances is more easily understood by using the notation of Burton and Edholm (1955) than that used in equations 1-5. For an animal which is not exposed to solar radiation:

$$H_s = -\frac{(T_s - T_a)}{(I_a + I_c)} \qquad (6)$$

where T_s is the skin temperature, I_c is the thermal resistance of the coat; and I_a is the thermal resistance from the surface of the coat to the surrounding air. In fact, I_a is a composite term in which both convective and longwave radiative exchanges are involved and strictly speaking, is only applicable if the mean radiation flux from the atmosphere and the surroundings corresponds to that from a black body at a temperature equal to T_a. In these conditions, the longwave radiation exchange is $h_r(T_r - T_a)$ where h_r is the radiant heat transfer coefficient and the convective heat flow is $h_c(T_r - T_a)$. I_a is defined as $1/(h_r + h_c)$.

When the animal is exposed to solar radiation:

$$H_s = \frac{R\alpha(I_a + I_p)}{(I_a + I_c)} - \frac{(T_s - T_a)}{(I_a + I_c)} \qquad (7)$$

where I_p is the average thermal resistance of the coat penetrated by solar radiation before absorption. Penetration will be discussed later. The first term on the right of equation 7 represents the effect of solar radiation on the heat flow at the base of the coat, and this is obviously less than the incident radiation absorbed at the surface of the coat ($R\alpha$). Figure 6 shows diagrammatically the direction of heat flow within the coat and the temperature gradients for three levels of solar radiation. An alternative equation for expressing the combined effect of radiation and wind has recently been devised (Øritsland, in press).

1. *Thermal resistance of the coat.* The thermal resistance of pelt samples from many species has been measured on the "hot plate" apparatus (Scholander et al., 1950). For live animals only the resistances of the coats of oxen and sheep have been published (Blaxter and Wainman, 1964; Joyce et al., 1966). The resistances per unit depth obtained are a little more than half those for these species measured by the hot plate method (Bennett and Hutchinson, 1964; Bennett, 1964). The difference could be caused by the movement of the animal pumping air through the coat. Recently, Clark et al., 1973), using a radiometric technique, obtained a value for sheep fleece that is about 1/3 of that determined by Blaxter and his colleagues. Thus there is still some confusion as to the value for thermal resistance which should be used in calculating the thermal effect of solar radiation.

2. *The convective heat transfer coefficient* (h_c). This has been measured for a few species only (Joyce et al., 1966; Nishi and Gagge, 1970). In calculations for other mammals, values of h_c derived for cylinders sideways on to the wind have often been used despite the lack of resemblance between the shape of cylinders and mammals. The convective coefficient is dependent on wind speed. Because wind velocity falls near the ground, air movement should be measured at several points between the ground and the top of the animal if the energy budget of the animal is investigated. A fuller account of thermal resistance of the coat and the convective heat transfer coefficient will be found in Chapter 1, Sections 13a and 13b.

3. *Penetration and transmission of solar radiation.* The thermal effect of penetration of radiation into a coat was first recognised by Burton and Edholm (1955) when discussing the warming effect of "see through" clothing in sunshine. More recently, Kovarik (1964) predicted by mathematical reasoning that some solar radiation would penetrate the coat by multiple diffuse reflection before absorption and that penetrance would be deeper with light coloured than with dark coats. He calculated that the deeper the penetration, the greater is the heat load per unit of solar radiation absorbed. However, the heat load per unit of incident solar radiation would still be greater for a black than for a white coat. Kovarik considered that a coat of some undefined intermediate colour between black and white could have a greater heat load than either per unit of incident solar radiation.

Hutchinson and Brown (1969) made measurements of penetration of radiation into samples of cattle coats mounted on a hot plate apparatus. The average thermal resistance of the coat penetrated by radiation (I_p, Fig. 6) was greater in white than in black coats. The effect of penetration on the heat flow at the base of the coat is twofold. If the radiation is absorbed below the surface of the coat, rather than at the surface, there is less thermal resistance between the site of absorption and the skin and more between the site of absorption and the air, resulting in a greater H_s (equation 7). It was also shown (Hutchinson et al., 1973) that the amount of thermal resistance penetrated by solar radiation is influenced by the bulk density of the coat (weight of hair per unit volume), penetration being less for coats of higher bulk density. The heat load of solar radiation of black coats was greater than that of white coats of equal thermal resistance. However, when the differences in bulk density and total thermal insulation between coats of different colour were taken into account, Hutchinson et al. (1973) calculated that the heat load for a brown coat was slightly higher than for a black coat, as was predicted by Kovarik (1964).

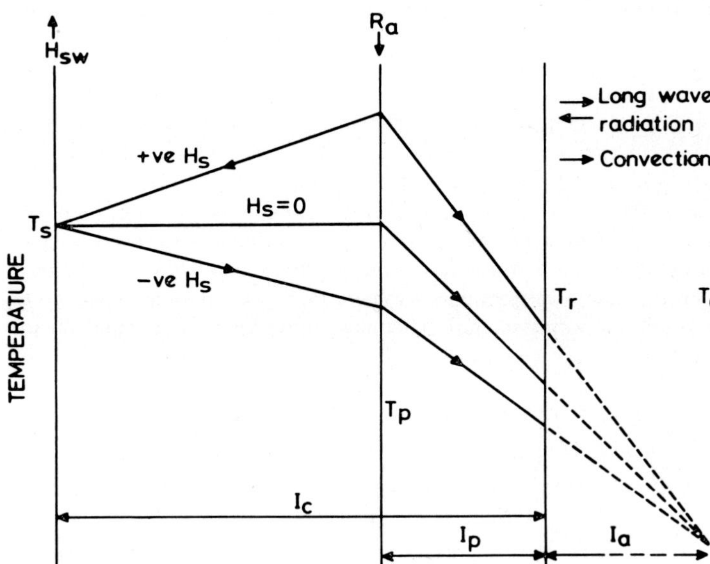

Fig. 6. Diagram to illustrate sensible heat flow within an irradiated coat for different values of absorbed radiation ($R\alpha$). A notional average site of absorption within the coat is taken and the thermal resistance between this site and the surface is I_p. In practice there would be absorption at all levels and a little radiation would be transmitted to the skin. Therefore the lines representing temperature gradients would in practice be curved. The upper line (+ ve H_s) illustrates the unexpected situation that can occur with a large I_p where there is a sensible heat flow to the skin. The coat surface temperature is lower than the skin temperature and the highest temperature is within the coat (T_p). The direction of heat flow is indicated by arrows in each case shown. H_{sw} is the latent heat of evaporation of moisture from the skin; T_r is the coat surface temperature. The remaining terms are explained under equations 6 and 7 in the text.

Øritsland (1971), examining the effect of solar radiation on the coats of harp seals (*Pagophilus groenlandicus*), found that parts of the coat with "bright" coloured hair of a silver-grey colour transmitted more heat to the skin when exposed to solar radiation than the "dark" brown coloured parts. Hairs of the bright pelt are coarse but have no medulla, appearing transparent under a microscope. This suggests that scattering of incident light rays may be mainly toward the skin rather than toward the coat surface, as in the case of "see through" clothing described by Burton and Edholm (1955). Such "non-isotropic" scattering would increase H_s as compared with "isotropic" scattering, e.g., scattering that is equal in both directions. It is interesting to speculate that the medulla, especially when it contains cell detritus, may tend to make the scattering isotropic; and it is an interesting fact that heat tolerant cattle tend to have coats that are more medullated than those of heat intolerant animals (Dowling, 1959).

Recently, Cena and Monteith (1975) developed a system for calculating the transmission of solar radiation through a coat based on measurements of coat structure and the absorptance of the coat. The calculated transmittance through cleaned white and grey-black sheep's fleeces agreed with experimental measurements. Presumably, an analogue to I_p could be calculated from the transmission coefficient. This approach gives a more rational assessment of the effect of coat structure on penetration than bulk density as used empirically, but successfully, by Hutchinson et al. (1973).

Both Hutchinson and Brown (1969) and Øritsland (1971) made their measurements with pelt samples. There are no published measurements on live animals which might assess the thermal effect of penetration in live animals. However, Finch (unpubl.) made measurements on a white and a black Boran steer (*Bos indicus*) out of doors in sunlight. From these measurements we have made calculations based on equation 4, using the profile of a Sussex bull (Riemerschmid, 1943) and found an inward heat

Table 1. Comparison of H_s between a white and a black zebu (Boran)

	WHITE ($\alpha = 0.53$)				BLACK ($\alpha = 0.87$)			
Time (h)	T_a (°C)	T_r (°C)	T_s (°C)	H_s (Wm^{-2})	T_a (°C)	T_r (°C)	T_s (°C)	H_s (Wm^{-2})
0915	21.3	35.8	36.5	55.1	19.3	42.8	38.4	30.7
1015	23.7	36.1	37.7	71.2	22.8	40.1	38.1	118.6
1115	25.5	35.8	37.9	106.1	25.5	42.3	39.0	147.9
1220	27.0	35.4	38.2	94.9	27.0	43.1	39.2	152.1
1310	28.0	36.1	38.4	79.5	28.5	41.7	38.6	158.4
1445	29.2	36.0	38.1	99.1	28.5	39.2	38.3	134.0

Note: The animals were presented side to the sun on a clear day. Wind varied throughout the day from 0.54 ms^{-1} in the early morning to 2.3 ms^{-1} in the afternoon and was from the east. Coat depth 3—3.5 mm, bulk density 35 mg cm^{-3}, both measured at a later date. There were similar results on other experimental days (calculated from unpublished observations of Finch).

flow at the base of the coat in both animals (H_s positive). As would be expected, the average coat surface temperature was greater than skin temperature for the black animal, but, surprisingly, for the white animal T_r was less than T_s (Table 1). A simple explanation is that penetration was greater in the white coat than in the black coat; and thus in the white coat the highest temperature was not at the surface but within the coat; and this temperature (T_p) was greater than T_s (Fig. 6). For the black coat, penetration would be negligible; and T_p would be almost the same as T_r.

3. ENERGY BUDGET OF ANIMALS IN A NATURAL ENVIRONMENT

If all the quantities in equation 1 are estimated correctly, the energy budget of an animal in its natural environment can be predicted or measured. Using similar equations, Priestley (1957) calculated the heat balance of sheep standing in the sun in order to demonstrate the importance of the thermal effects of solar and longwave radiation. In an extension of this type of analysis, Porter and Gates (1969) calculated the energy budgets of a reptile, a bird and four species of mammal, including a sheep, in an attempt to delineate the "climate space" which a given animal must occupy in order to survive. The climate space was defined in terms of radiation, air temperature, wind and humidity. However, it seems likely that more exact estimates for several of the quantities in equation 1, such as h_c etc., are needed. Their calculations suggested that when sheep are standing in the sun, they could not endure for long air temperatures above 24° C unless wind speeds were greater than 1 m sec^{-1}. Nevertheless, on the shadeless Mitchell-grass plains of northern Queensland air temperatures during summer are commonly over 38° C for much of the day and often reach 46° C but sheep do not die of heat stroke. Rectal temperatures of sheep in pens were only about 40.5° C (Macfarlane et al., 1958). Indeed, studies in New South Wales using radio-telemetry have shown that the rectal temperatures of sheep grazing unshaded pasture at an air temperature of 30° C are lower during the day than at night (Brown, 1971a). Despite the differences between the calculations of Porter and Gates and what happens in nature, their work was important in stimulating others to attack this difficult field.

Adopting a slightly different approach, Joyce et al. (1966) tried to predict the thermal effects of a natural environment on the energy budget of the sheep from meteorological measurements. Their work was confined to experiments in climates colder than the thermoneutral and apparently with low levels of solar radiation. They expressed the thermal effect of solar radiation in terms of a notional change in thermal resistance of the coat and air ($I_c + I_a$). They obtained good agreement between predicted and measured values of the notional thermal resistance of the coat and air. However, the predicted values are not independent of their measurements of heat production in the field because the sensible heat flow, calculated from heat production, appears in the prediction equation.

Analysis of the energy budget of domestic mammals may have considerable practical importance. Ojo (1971), by making calculations of thermal exchange in the sun, tried to assess the suitability of Jersey and Shorthorn cattle for different regions of Nigeria but details of the information used in these calculations are not sufficient to enable the work to be properly evaluated.

In order to understand how the shallow summer coats of cattle are adaptive to hot environments, Hutchinson, Brown and Allen (1973) calculated the air temperature at which the sensible heat flow at the base of the coat (H_S) of an animal standing in the sun would be zero at certain specified wind speeds. The penetration of solar radiation into coats of known characteristics was taken into account. The surface areas exposed to direct and diffuse radiation were based on those for shorn British sheep. They estimated that in warm environments there would be a sensible heat flow from the environment to the skin for most of the hours of sunshine (Fig. 7) i.e., the air temperature would be greater than that for $H_S = 0$. Because of differences in solar

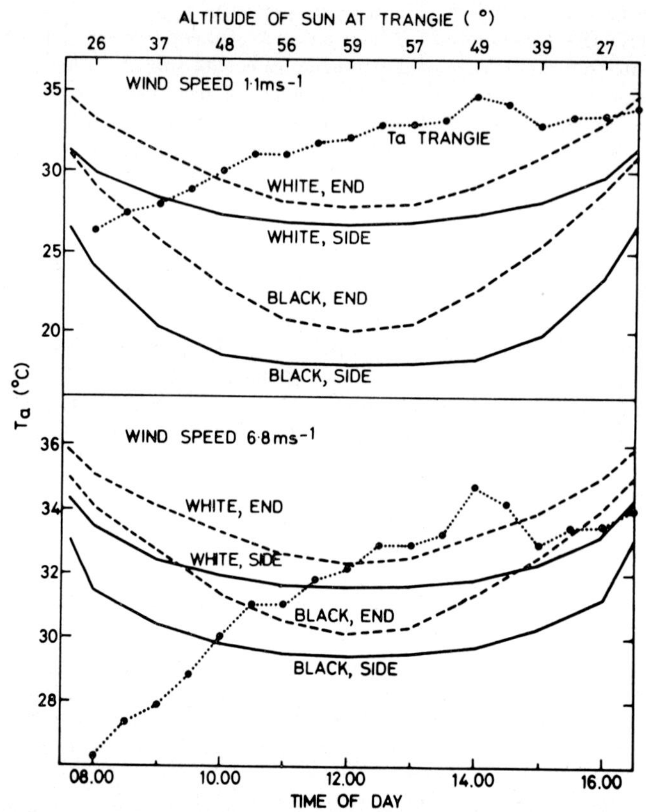

Fig. 7. Calculated air temperature for which sensible heat flow at the base of the coat (H_s) is nil in two different winds for black and white cattle with side and end presentation to the sun. It is compared with air temperature on an autumn day at Trangie, N.S.W. There is a heat flow through the base of the coat from the environment to the animal when the air temperature at Trangie exceeds the plot of air temperature for $H_s = 0$. (Redrawn from Hutchinson, Brown and Allen, 1973).

profile areas, H_S is also greater when the animal is side on to the sun than when end on. The effect of a ground temperature 10° C above the air temperature was to reduce the air temperature for $H_S = 0$ by 1-2° C according to the wind speed.

Finch (1972a, b; 1973) experimentally evaluated the terms of equation 1 in studies with the eland (*Taurotragus oryx*), Coke's hartebeest (*Alcelaphus buselaphus*) and Boran cattle (*Bos indicus*) in studies in the Kenyan highlands (alt. 1450 m). In these experiments the animals were placed sideways on to the sun. Some idea of the magnitude of the heat flow at the base of the coat (H_S) in these animals can be obtained from the results for each species at about 13.30 hr each day. Applying these results to equation 4 and using the profile area to the direct beam for a bull (Riemerschmid, 1943) and a value of 85% of the skin surface area as the area exposed to diffuse radiation, we have calculated values for H_S of the order of 90 Wm^{-2} for the cattle and 100 Wm^{-2} for the eland. Air temperatures were about 30° C and wind speeds about 4 ms^{-1}. In these conditions, H_S for the hartebeest was -17 Wm^{-2}. The values for H_S calculated from equation 5 for the cattle and the eland were about twice those given above and about 60 Wm^{-2} for the hartebeest. The resting metabolic rates for each species were 67, 97 and 83 Wm^{-2}, respectively. The difference in the results obtained from the two equations must reflect the present inadequacies of technique and illustrate the difficulty in measuring the quantities involved under field conditions.

Finch's results also demonstrated the value of the protection against the heat load of solar radiation given by thick coats. The heat load at the base of the coat for the hartebeest with its relatively thick coat (8 mm) was much less than for the thin coats of the cattle (2.5-4.0 mm) and eland (1-2 mm). In a graphical breakdown of the relative contributions of the terms of equation 1 to the heat gains and losses for each of these species (Fig. 8), it can be seen that whereas eland and cattle rely mainly on sweating for evaporative cooling, the hartebeest relies on panting. There are also marked differences in the proportion of heat dissipation attributed to longwave radiation from the coat due to differences in coat surface temperature which again reflect the effect of coat thickness of each species.

4. ECOLOGICAL IMPLICATIONS

The manner in which the thermal effect of solar radiation interacts with the temperature regulation of mammals is dependent on both the type of environment in which they live and the size of the animal. In cold environments, the warming effect of solar radiation will reduce the amount of metabolic heat production required to maintain a stable body temperature during the hours of sunlight. In warm and hot environments, the heat load of solar radiation will increase the amount of heat which must be dissipated by evaporative cooling if a stable temperature is to be maintained. The thermal effect of solar radiation is of particular importance to mammals in arid and semi-arid regions with a savannah grassland or open woodland vegetation where water is often scarce. It is in these habitats that the size of an animal is also important. Most small mammals are able, by virtue of their size, to avoid the heat load of solar radiation by sheltering under rocks, in caves or burrows or by building nests. For example, most

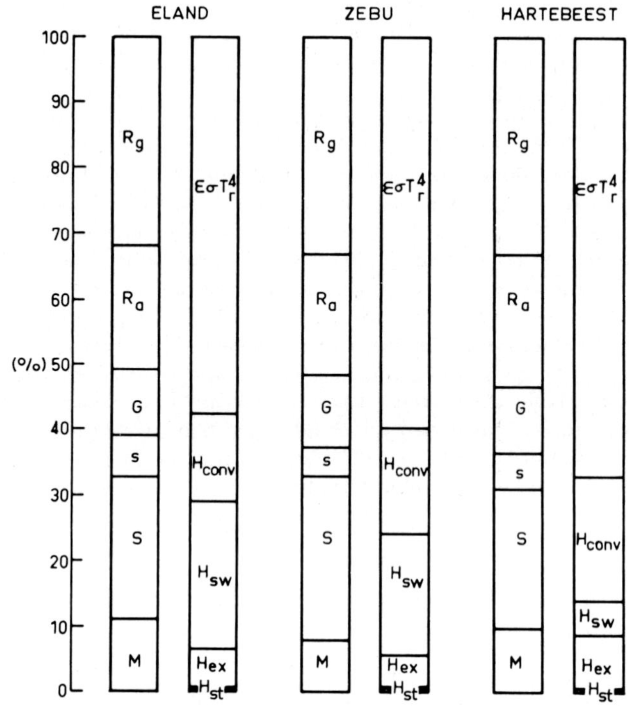

Fig. 8. Percentage gains of heat from different sources and losses by different pathways for eland, zebu and a hartebeest between 09.00 and 16.00 h on a clear day. For each species the left hand column represents ingoing heat and the right hand column outgoing heat. Radiation gains are in terms of what is absorbed on the coat and does not represent the heat flow at the base of the coat (H_s). Symbols are similar to those in the text (equations 1-5), but indicate only the source of heat and the pathways by which heat is dissipated. G is shortwave radiation reflected from the ground; H_{conv} is convective heat loss. Coat absorptance of eland, 0.75; zebu, 0.78-0.81; hartebeest, 0.65. The animals were sideways on to the sun and the profile area exposed to the direct beam was taken as that of an open ended cylinder. The area exposed to or emitting diffuse radiation was taken as the area of the cylinder. Wind was from the east and increased from about 1.8 to 4.3 m s^{-1} during the experiments. (Redrawn from Finch, 1973).

Australian arid-zone marsupials avoid solar radiation by taking advantage of the protection given by these types of micro-habitat (Brown, 1973). The larger mammals of the arid regions, such as the red kangaroo (*Megaleia rufa*), the African antelopes and other bovids, and domestic animals, can only avoid the heat of the sun by sheltering in the rather sparse shade typical of many of the trees in these areas. In these species, the coat will be important in providing some protection against solar radiation because the shade is not used all the time.

A. COAT CHARACTERISTICS

Considering those characteristics of the coat discussed above which would reduce the thermal effect of solar radiation, the coats of arid-zone animals should be very pale in colour, relatively deep, have a high bulk density and perhaps should be glossy. Although little information on the characteristics of the coats of arid-zone mammals is available, they do not seem to conform to all these requirements. The most common coat colours of wild animals are various shades of fawn, red and brown. Many appear relatively pale, and some species such as the impala (*Aepyceros melampus*) have very pale or even white hair on the underside of the body. Absorptance of the coats of a few wild species has been measured. In the red kangaroo it may vary from 0.7 for fur on the back to less than 0.5 for the palest hip fur (Dawson and Brown, 1970). Finch (1972a) found absorptances of 0.75 for the eland and 0.65 for the hartebeest. Because some brown coats may have a heat load as high as black coats due to the effect of penetration as described above, we are uncertain of the advantages afforded by the various shades of buff and brown.

It is obvious that the advantage of a low absorptance is not the only factor of evolutionary significance influencing the coat colour. In some populations of red kangaroos, most females have a relatively dark "blue" coat while that of the males is a pale red (Dawson and Brown, 1970). In African mammals, populations of the same species from different locations may vary in colour; and variations in colour due to protective camouflage are common, but few animals have very dark coats.

The coats of most mammals that cannot avoid solar radiation in warm or hot arid regions appear to be shallow. There are a few measurements showing that the coats of eland, zebu and European cattle on the equator in Kenya, and the summer coats of the red kangaroo in Australia are 4 mm or less in depth (Dawson and Brown, 1970; Finch, 1972a; Finch and Hutchinson, unpublished), but our ignorance of such an elementary coat character is emphasized by the case of a hartebeest that at different periods of its life on the equator had coat depths of 8, 4.3 and 3.2 mm (Finch, 1972a; Finch and Hutchinson, unpubl.). Shallow coats must surely aid temperature regulation in the heat because the summer coats are in general shallower than winter coats. Indeed there is a strong positive correlation between coat depth, and Turner and Schleger's coat score. This score is negatively correlated with growth of beef cattle in the tropical environment of coastal Queensland, Australia, (Turner and Schleger, 1960; Schleger and Turner, 1960). This suggests that a shallow coat is advantageous to animals that spend part of the day grazing in the sun.

Coats with a high bulk density have a relatively low penetration by radiation. Almost nothing is known about bulk density. There are differences between species. That of the hartebeest mentioned above had a density of 120 mg cm^{-3} in comparison with a mean of 38 mg cm^{-3} for two Hereford and 12 Boran steers measured on the same day (Finch, unpubl.).

Coats with white tips to the fibres as in the hartebeest (Finch, 1972a) and some Brahman cattle may be particularly advantageous because low absorptance could be combined with a low penetrance by virtue of the darker hair below the tips.

The summer coats of cattle and some African bovids such as impala are quite glossy but the importance of this is not known and measurements of gloss have not been made.

The deep coats of wool sheep, which can endure hot semi-arid climates, have quite different characteristics from those described above. The value of a deep coat in reducing the heat load from solar radiation in a hot climate has been vividly demonstrated in experiments showing the considerable increase in evaporative water loss when sheep and camels, exposed to the sun, are shorn (Macfarlane et al., 1958; Schmidt-Nielsen, 1964). The daily water turnover rate of shorn sheep is much higher than for unshorn sheep (Macfarlane et al., 1966), showing that H_s, integrated over the day and night, is greater in the shorn animals. These results suggest that shallow coats have not evolved to give the most protection against the heat load from solar radiation; and therefore, other factors such as evaporative cooling, bulk density of the coat, superior sensible heat dissipation at night or in the shade, and behaviour are involved. Sweating may be more effective with a shallow coat because evaporation of sweat from the skin is more effective thermally than from the coat and a deep coat would become wet more readily than a shallow coat. However, Allen et al. (1970) were unable to find free moisture in the coats of sweating cattle with coats varying widely in depth and in a variety of environments. Several species of arid-zone or tropical mammals such as the red kangaroo (Dawson, pers. comm.) and the hartebeest (Finch, 1972a) spend much of their time in the sun, can have shallow coats and apparently do not sweat much. Robertshaw and Taylor (1969) studied sweating rates of several species of African bovids and suggested that sweating ability was related to body size, the larger species having a greater sweating rate. However, coat depths were not given.

It is well known that cattle with a shallow summer coat are more heat tolerant in the sun than those which have failed to shed their winter coat (Bonsma, 1943; Yeates, 1965). This seems to imply that shallow coats give a better protection from the sun than deep coats. One explanation of this apparent anomaly is that the heat intolerant animals are intolerant for other reasons which result in delay in the shedding of the coat (Turner and Schleger, 1960). Secondly, Schleger and Bean (1971) have shown that sweating ability in cattle is greater when a high proportion of the hair follicles is active. Thus animals with shallow summer coats would be expected to have a greater sweating ability since more new hairs appear during summer than in winter (Turner and Schleger, 1970). A third reason for the higher heat tolerance of the animals with shallow summer coats is related to the structure of the coat. The bulk density is higher in summer coats than winter coats (Berman and Volcani, 1961; Allen et al., 1970), and thus the penetration of radiation into the shallow coat will be less. Hutchinson and Brown (1969) measured the heat load from radiation of a sample of red winter cattle coat and found that the heat load was slightly greater when the hairs were combed erect than when flattened. With the hairs erect, the increase in penetration (I_p) outweighed the protective effect of the greater thermal resistance of the coat.

Measurements of body temperature of arid-zone mammals over 24 hrs in a cool equatorial climate have shown that differences between species are to be found in the degree to which body temperature falls at night rather than differences in the maximum observed during the day (Bligh and Lampkin, 1965; Bligh and Harthoorn, 1965).

Although there is evidence that in a hot sunny environment, H_s over 24 hours is greater with a shallow than with a deep coat (Macfarlane et al., 1958; Schmidt-Nielsen, 1964; Macfarlane et al., 1966), shallow coats favour heat loss at night and conditions at night may be important. There is indeed evidence from laboratory experiments with pregnant ewes that high night-time air temperatures are particularly important in causing stunting of lambs in a hot environment (Shelton and Huston, 1968; Alexander and Williams, 1971).

In hot climates some large mammals that tend to avoid the sun, such as the hill kangaroo (*Macropus robustus*) which shelters under rock ledges or in shallow caves, have deeper and woollier coats than those such as the red kangaroo which only seek shelter under trees (Dawson and Brown, 1970). Other mammals such as various species of wild pig and water buffalo have sparse coats, but they utilize wallows which both cool the animals and provide protection against the sun.

Man, when he evolved as a hunter-gatherer, had to be capable of work at a high rate in a hot savannah environment over many daylight hours, as judged by the behaviour of modern hunter-gatherers (Abbie, 1969; Weiner, 1971). Yet surprisingly, man has no coat for protection from solar radiation; indeed, his sparse hair cover is similar to that of wallowing animals. Loss of the coat suggests that man is primarily adapted to a fast rate of work over many hours in warm or hot environments and not specifically for heat tolerance in the sun. At a fast rate of work, man's powerful sweating would probably wet out any coat, thus becoming relatively inefficient for cooling the body.

B. BEHAVIOUR

Certain aspects of the behaviour of arid-zone mammals may be as important as the characteristics of the coat in reducing the thermal effects of solar radiation. As has been mentioned above, many small species and a few larger mammals avoid the heat load of solar radiation by sheltering in caves and burrows. Those species that do not avoid the solar radiation may reduce the heat load by presenting the smallest profile area to the sun and adopting postures that will facilitate convective heat losses.

1. *Orientation to the sun.* The topi (*Damaliscus korrigum*) from Kenya is commonly found lying in the sun and orientated so that the spine is in line with the sun (Jarman, unpubl.). Dehydrated camels have also been observed to adopt this alignment to the sun, lying on one spot throughout the day, only moving to keep their spine in line with the sun. Hydrated camels, however, frequently stretch out in the sun (Schmidt-Nielsen, 1964). The surface of animals exposed to the sun can also be reduced if they huddle together in tightly packed groups during the hottest part of the day, a type of behaviour observed in camels (Schmidt-Nielsen, 1964) and Merino sheep (Brown, 1971b). However, Merino sheep usually seek shade if it is available.

2. *Thermal effects of lying down.* Although it is well known that many species lie in the sun, for example hartebeest and wildebeest defending their territories, information of the thermal conditions at the time or the length of time spent lying is meagre. However, Schmidt (1969) found that during summer on the shadeless Mitchell-grass plains

of northern Australia, all cattle spent the day standing or lying close to the water troughs when air temperatures were usually between 36 and 43° C (see also Yeates and Schmidt, 1974). The number lying down increased as the air temperature rose. Apart from the reduction in profile area mentioned above, it might be expected that lying down would reduce the amount of short and longwave radiation received from the ground. There will also be direct conduction of heat between the animal and the ground (C), but the net effect of this is not known. There will be some reduction in metabolic rate when lying down. Because the wind velocity is relatively low near the ground, convective losses of heat will be reduced by lying down. Lying in deep grass, sheltered from the wind, will accentuate the reduction in convective loss. There is evidence from northern Australia that newborn calves lying in deep grass may die from heat stress (Smith and Alexander, 1966).

3. *Shade seeking*. From the examples quoted above, most large mammals in hot environments appear to have relatively shallow coats, particularly in summer. It would seem, therefore, that they are adapted to environments in which some shade is available. In open shade with no direct solar radiation, H_s would be negative for as long as air temperature was less than skin temperature. The advantage given by the relatively sparse shade of many trees typical of savannah-woodland areas would be less but still important because a small tree can reduce the direct solar radiation by as much as 60% (Dawson, 1972). There are few well documented studies of shade-seeking behaviour in the larger arid-zone mammals. Red kangaroos spend much of the hottest part of the day standing in the shade of small trees (Dawson and Denny, 1969). Measurements of incident solar and longwave radiation indicate that this would be marginally lower for a kangaroo lying in the shade than for standing. The preference for standing may result from the greater convective heat loss because air movement would be greater than when lying down (Dawson, 1972). Detailed observations of the behaviour of cattle during hot weather (air temperatures of 36-43° C) in desert country with open woodland vegetation in northern Australia have been made by Schmidt (1969) (see also Yeates and Schmidt, 1974). Most animals camped during the day close to water troughs with shade nearby, but not all made use of the shade. Some cattle camped in the shade of trees well away from water, near where they grazed. These would graze at higher air temperatures than the others and had sleeker coats.

5. PRACTICAL CONSIDERATIONS

Alleviation from the heat stress of solar radiation in hot climates can improve the productivity of domestic animals. The introduction of shade into the Californian pen feeding system is a successful example (Elam, 1971). Sunshades have been used in experiments on the work capacity of bullocks in India (Mukherjee et al., 1961). It can be argued that such improvements in husbandry could be, and were made with limited knowledge of the effect of solar radiation on sensible heat exchange. Further advances in alleviation, such as the provision of a highly reflecting, artificial coat to reduce heat induced sterility in rams, or to protect high producing dairy cows, with their low heat tolerance, require more advanced knowledge of the biophysics of heat exchange. Such knowledge could also lead to the definition of better selection criteria in breeding for productivity in hot climates. For example, a desirable cattle coat might be one which combines high reflectance with low penetration of radiation.

Tourists are as important to developing countries as food production because the tourist industry provides them with foreign exchange needed for development. To maintain national parks for recreation and for conservation of wildlife, both the animals and their environment must be managed. A knowledge of the climate space required by each species, one of the eventual aims of the studies outlined above, will help to define the conditions which must be maintained by management if each species is to persist. We anticipate that there will be an increased effort to solve the problems of physics and behaviour that we have outlined, in order to acquire some of the expertise required for management. It will be clear to readers that knowledge of these problems is, at present, most inadequate.

ACKNOWLEDGEMENT: We are grateful to Mr. F. B. Spence for many calculations and for the preparation of other material.

REFERENCES: see pp. 268-270.

Effects of Light Intensity on Reproduction of Mammals

by

J. C. Hoffmann

INTRODUCTION

The effect of changes in light intensity appears to be one of the least-studied aspects of the influence of light on biological activities, particularly in mammals. If one limits the focus of interest to light intensity effects on reproduction, the amount of data available is decreased even further. The reason for this is probably that there are very few situations in nature in which light intensity is a limiting factor, at least for terrestrial forms, because the light from the sun that most animals encounter is well above threshold intensity. Nevertheless, studies have been done under laboratory conditions which suggest that light intensity must be considered as an important variable when examining other characteristics of light.

UNITS OF LIGHT INTENSITY

Light intensity may be expressed in footcandles, lumina, or lux. One footcandle (ft-c) is defined as the amount of light from a standard candle reaching a square foot oı surface placed one foot from the light source (Wurtman, 1967). One footcandle is also equivalent to 1 lumen/square foot or 10.76 lux (Hertig, 1968). One lux, on the other hand, is the amount of illumination reaching a one square meter surface from one standard candle located one meter away and is equal to 1 lumen/square meter or 0.0926 footcandle (Hertig, 1968). The sun reaching the earth's surface provides about 10,000 ft-c of light, reduced to about 500 ft-c in the shade; the average laboratory is provided with about 50-150 ft-c at eye level (Wurtman, 1967).

EFFECTS OF LIGHT INTENSITY ON REPRODUCTION

The earliest reports of studies on the effects of light intensity on reproduction came from Rowan (1938) who did not believe that intensity was a factor in reproductive system responses of birds to light once a low threshold is reached. However, when ducks were exposed to rather restricted periods of light, there was a correlation between testicular growth and light intensity (Benoit, 1961). In ferrets, Marshall (1940) found that females placed closer to a light source came into estrus sooner than those placed further away. In general in these early studies the intensity of light was varied by placing the animals at different distances from the same light source. When actual measurements of light intensity were reported in some later studies, it was shown that only 1 footcandle was necessary for normal egg production in poultry (Dobie, 1946), and that while a light intensity-response relationship existed for testicular weight of sparrows exposed to a range of 0.04 to 10 ft-c, there was no further increase in the response even up to 250 ft-c (Bartholomew, 1949). In a recent study of the quail, Oishi and Lauber (1973) used very low energy levels (one 7.5 w incandescent lamp reduced by means of filters to 1/10 and 1/100 of that amount) and also found a light intensity-response relationship for testicular weight. Farner (1961) has summarized the early work on light intensity in the ferret, stating that the threshold for a response is below 100 lux (about 9 ft-c), and that intensities between 100 and 1300 lux are equally effective.

Perhaps the best study on the effect of light intensity on reproduction is a very recent one by Piacsek and Hautzinger (1974). These workers used the day of vaginal opening as an indicator of sexual maturity in female rats. Pregnant females were exposed to the experimental light intensity from the day before expected delivery onward, and the females from the litters were kept in the same lighting conditions until autopsy which was done no later than 3 days after vaginal opening. The photoperiod was either 14 hours light, 10 hours dark (L/D) or 24 hours of light per day (L/L). In L/D, increasing the intensity from 30 lum/m^2 to 100 resulted in significantly earlier vaginal opening, but a further increase to 600 lum/m^2 resulted in a delay of 3 or 4 days in vaginal opening. Paradoxically, increasing intensity to 2400 lum/m^2 returned the time of vaginal opening to a day that was intermediate between those of the 30 and 100 lum/m^2 groups. In L/L, the effect of increasing intensity was more linear, with a steady decrease in the mean age of vaginal opening as light intensity was increased from 30 to 2400 lum/m^2. One must conclude that the effects of changes in light intensity depend to an extent on the pattern of light exposure i.e., alternating light and dark or continuous light.

In a number of studies in our laboratory, we have found that the length of the daily photoperiod can determine the length of the estrous cycle in the albino rat, with shorter photoperiods tending to result in more 4-day cycles and longer photoperiods in more 5-day cycles (Hoffmann, 1968, 1970, 1973). This effect has been seen in a variety of different housing conditions: rats in wire cages in a room with incandescent lights, in wire cages in a room with fluorescent lights, and in plastic cages in specially built cabinets where fluorescent bulbs are about 6 inches from the cages. The intensity of light surely was different in these three situations, though we never measured it, and yet the photoperiod length effect was always seen. In view of the work in birds (Bartholomew, 1949) and in ferrets (see Farner, 1961), it seems reasonable to conclude that in our experiments the light intensity was well above threshold and in the range where further effects are not expected.

EFFECTS OF LIGHT INTENSITY ON ACTIVITY RHYTHMS

Although it is somewhat outside the scope of this Section, it is interesting to look at effects of light intensity on activity rhythms where considerably more work has been done.

In most natural settings, animals are exposed to alternations of light and dark and for a given species there will be a concentration of physical activity during either the light or the dark period. When the light-dark signal, and therefore the timer or Zeitgeber, is removed by placing animals in either constant light or constant dark (or by blinding), animals still maintain a rhythm of activity and rest, but the period, or length of time between onsets of successive activity blocks, is usually no longer exactly 24 hrs but may be shorter or longer and hence the active phase will drift with respect to solar time. Through extensive experimentation and observation of the results from many laboratories, Aschoff (1961) proposed a general rule which states that animals which are light-active tend to show a shortening of the period between active phases as light intensity (with continuous exposure) increases, whereas dark-active animals tend to show a lengthening of the period with an increase in light intensity. This rule appears to apply across a wide range of animal species and even in plants.

Body temperature also shows a diurnal rhythm which is synchronized with the light-dark cycle. McGuire et al. (1973) recently tested the influence of variations in light intensity and spectrum on the speed with which the temperature rhythm in rats entrained to a lighting schedule shifted abruptly by 6 hrs. They found that both factors were important, and that the lowest intensity of light was required for those wave lengths (mainly green) which correspond to the highest region of sensitivity of rat rhodopsin. These results conflict with those of Piacsek and Hautzinger (1974) who studied the effects of various spectra on sexual maturation. They found that in L/D: 14/10, green light of 800 lum/m^2 did not advance vaginal opening compared to yellow light of 650 lum/m^2. In L/L, however, the same intensity of green light did appear to advance puberty in the majority of animals compared to the group in yellow light. The reason for these differing results may be that the spectrum of light does not have the same effect on rhythm shifts as it does on the tonic rate of secretion of gonadotropins which presumably determine the time of vaginal opening.

POSSIBLE CORRELATION BETWEEN ACTIVITY RHYTHMS AND REPRODUCTIVE CYCLES

While it is not possible to monitor all biological rhythms as easily as the rhythm of physical activity or temperature, it seems likely that in constant light or dark many functions normally showing a diurnal rhythm would adopt a circadian rhythm with the same period as that of the activity rhythm. Indirect evidence for this is provided by Browman (1937), who examined both the activity rhythm and the estrous cycle in rats in either constant light (L/L), constant dark (D/D), or L/D: 12/12. Wheel-running activity in number of revolutions was recorded every 12 hrs. In intact animals in L/D, there was an obvious concentration of running activity during the dark period, with a much higher peak of activity seen during the night of proestrus at a time when the vagina is cornified. For animals in constant dark, it is evident that the peak of activity coincides with the period of vaginal cornification as it does for animals in L/D. In constant light, the activity patterns were reported only after the animals were already showing persistent vaginal estrus (see below), so all that is evident is a gradual shift in activity peaks from solar day to solar night and vice versa.

Albino rats placed in constant light will eventually go into a state of persistent estrus in which ovulation does not occur and the ovary contains only follicles and no corpora lutea (Everett, 1961). Age, genetic makeup (Everett, 1961), and lighting history (Hoffmann, 1973) of the animals are factors that determine how soon this state is reached following exposure to constant light, but intensity of light can also be a factor. Logan (1954) placed female rats in a room illuminated constantly with either one or five 40 w fluorescent lamps and followed vaginal smears to determine the onset of persistent estrus, defined as the beginning of 10 or more consecutive days of cornified smears. She found that the average time required was 27.5 days among the rats illuminated with only one 40 w lamp, and 18.8 days among the rats illuminated with five 40 w lamps. Thus the intensity of light does appear to influence the onset of persistent estrus in constant light.

INDIRECT EVIDENCE FOR LIGHT INTENSITY EFFECTS IN HUMANS

While it is quite difficult to perform experiments with light intensity in humans, particularly for the long periods that would probably be required, some interesting reports have been published which correlate the amount of light perceived with the age of menarche. In the first report, Zacharias and Wurtman (1964) noted that menarche occurred earlier in blind girls than in sighted girls. In the second report, Magee et al. (1970) compared girls with minimal or absent vision to those with moderate vision and also found that those with less light perception had earlier menarche. These findings must be interpreted with some caution, since a multitude of factors could be involved which were not controlled in these retrospective studies. However, the compilation of data such as this may ultimately confirm or deny the existence of a correlation between amount of light perception and onset of puberty in the human.

The reverse side of the coin, that is, a change in the sensitivity to light as a function of reproductive cycle stage, has been investigated in the human by Diamond et al. (1972). This group found that the sensitivity to a visual stimulus was lowest in women at the time of menstruation and increased as the approximate time of ovulation was approached. Thereafter it remained at the higher level. Such differences were not seen among women rendered anovulatory by contraceptive pills, nor among men tested at intervals over a similar time period, but rather both these groups showed a continuous high sensitivity somewhat greater than that of the cycling women in the second half of the cycle. Thus the effect of a given light stimulus may vary according to the time in the reproductive cycle that it is applied.

CONCLUSION

The intensity of light from the sun reaching most terrestrial forms of life is far above the thresholds which have been identified in laboratory experiments as minimum for development of reproductive organs. However, when the effects of other physical characteristics of light are being studied, such as photoperiod length or spectrum, the importance of possible variations in intensity should not be overlooked, but rather should be carefully controlled by the investigator.

ACKNOWLEDGEMENT: The preparation of this manuscript was aided in part by a grant from the Ford Foundation.

REFERENCES: see pp. 270-271.

Effects of Light Spectra on Reproduction of Mammals and Birds

by

D. P. Cardinali

INTRODUCTION

The visible and ultraviolet light in the environment affect a variety of structures and physiological mechanisms in animals. Solar radiation, besides serving as the primary source of energy and food, acts directly on the living organisms modifying their chemical composition, controlling their maturation rate and driving their biological rhythms. This Section describes some of the effects of sunlight and artificial light sources on growth and function of reproductive organs in mammals and birds. It also discusses the characteristics of the different lighting environments to which experimental animals can be exposed.

CHARACTERISTICS OF NATURAL AND ARTIFICIAL LIGHTING

The light of the sun filtered through the atmosphere and its surrounding layer of ozone is present on the average for 12 hours out of every 24 at every point on earth. The actual number of daylight hours at any point on any particular day varies with an annual rhythm whose amplitude depends upon the angular distance between that point and the equator. Spectral composition of sunlight is modified after passing through the atmosphere; essentially no wavelengths shorter than 290 nm, that is essentially no short ultraviolet and relatively few mid ultraviolet photons, penetrate the atmospheric ozone layer. Radiant fluxes at larger wavelengths increase sharply in the near ultraviolet (320-380 nm) and visible (380-770 nm) portions of the spectrum reaching a peak in the blue-green range (450-500 nm). The spectral composition of sunlight approximates that of the light emitted by a theoretical black body radiator heated to 6000 kelvins minus the ultraviolet radiation unable to penetrate the ozone layer surrounding the earth. The average radiant fluxes of the blue, yellow and red components of sunlight are all approximately equal. While the intensity of the ultraviolet light varies markedly with season the spectral quality and quantity of radiation from the visible portion of the solar spectrum are less prone to seasonal variation.

Incandescent light sources also emit spectra that approximate those of heated black body radiators. However their color temperature is considerable lower than that of sunlight, and approximately 90% of the radiation emitted consists of infrared radiation which provide heat rather than light. Within the visible portions of incandescent spectra radiant fluxes are maximum in the red range and minimum in the blue.

Fluorescent bulbs generate light by a physical mechanism different from that of the sun or incandescent bulbs. Their light comes not as a consequence of heating but from the electrical excitation of a chemical phosphor, their spectra being characteristic of the atoms comprising it (line spectra). Insofar fluorescent bulbs have been designed on the premise that the only effect of light is enabling man to see. Radiant fluxes thus peak at wavelengths in the yellow-green range where human retinal photoreceptors are maximally stimulated (550 nm for photopic vision). The propensity of biologists to express the intensity of a light emission in terms of footcandles tends to perpetuate the notion that vision is the unique effect of light on vertebrates. Footcandles are unit of illumination defined subjectively as lumens, units of brightness per square foot. When the intensity of any light source is described in footcandles only the yellow-green emissions have actually been quantified, and for a given total irradiance, the larger the proportion of emissions in the yellow-green range, the greater the footcandle rating. Since photoreceptor mechanisms and spectral sensitivities may differ among vertebrates, the intensities of light emissions should be measured in absolute irradiance units, such as microwatts per square centimeter for each spectral band. This measurement is done with radiant energy detectors that respond proportionally to visible light at any wavelength range, e.g., thermopiles. That sort of detectors (thermal detectors) absorbs radiation as heat and their responses are proportional to absolute radiant energy regardless of wavelength. Other types of radiant energy detectors change their physical characteristics upon incidence of photons, either photochemically (retinal photoreceptors, photographic emulsions), or photoelectrically (photomultipliers, photoconductive and photovoltaic cells). In contrast to thermal detectors, their responses depend on the spectral composition of the light source.

In summary the characteristics of a lighting environment should be defined in terms of three aspects: the spectrum of the light source, the irradiance levels (in absolute irradiance units) and the number of hours per day that the animal is exposed.

DIRECT AND INDIRECT EFFECTS OF LIGHT

The effects of light on vertebrates are direct or indirect depending on the mechanisms involved (Wurtman, 1970). The direct effects of light can be defined as chemical changes in the constituents of a tissue resulting from the absorption of light energy in that tissue (e.g., effects on the skin and subcutaneous tissues, photochemical and photosensitization reactions). This subject has recently been reviewed (Cardinali and Wurtman, 1975a). The indirect effects of light are those ultimately mediated by endogenous chemical signals whose amplitude depends on the responses of specialized photoreceptor cells to light. The activation of a photoreceptor cell by light results in the induced activation of cells specialized for communications, i.e., neurons, endocrine cells, neuroendocrine transducers, causing the release of a neurotransmitter or the secretion of a hormone. In some particular cases (avian pineal organ), the same photoreceptor cell seems to produce the hormone involved in the light effect (photoendocrine transduction). The effects of light on growth and reproduction are example of indirect ffects of light.

EFFECTS OF LIGHT ON THE REPRODUCTIVE FUNCTION

In analyzing the effects of light on reproductive function, the light-induced changes in the activity of the pineal organ, the hypothalamus and the anterior hypophysis must be considered inasmuch as these structures contribute secretions that ultimately affect the function of the gonads.

A. LIGHT AND THE PINEAL ORGAN

Experiments performed largely during the past decade have provided evidence indicating that the biochemical activity of the pineal organ depends on the lighting under which the animal is maintained (Wurtman et al., 1968; Wolstenholme and Knight, 1971). Exposure of rats to continuous illumination depresses the weight of the pineal and obliterates the normal diurnal rhythm in pineal function, namely in serotonin and melatonin synthesis. In mammals these effects of light are not exerted directly on the pinealocytes as in lower vertebrates, but are mediated by a complex neuronal pathway which includes: (1) the retinas, (2) the optic nerves and chiasm, (3) the inferior accessory optic tracts, (4) the medial forebrain bundles, (5) descending fibers to the upper spinal cord, (6) the preganglionic fibers to the superior cervical ganglia, and (7) postganglionic sympathetic nerves which terminate on or near pineal parenchymal cells. The administration of melatonin, one of the pineal hormones, is followed by suppression of gonadal development and lengthening of the diestrous phase of the reproductive cycle in several species (Cardinali, 1974). Conversely light-induced changes in the gonadal activity of seasonal breeders, e.g., the hamster, the ferret, the duck, are impaired or even prevented if the animals are pinealectomized.

The cells of the pineal organ exhibit significant phylogenetic changes in the way of transducing photic information (Cardinali and Wurtman, 1975b). They are PHOTORECEPTORS in lower vertebrates converting photic energy into neural impulses; they may be PHOTOENDOCRINE TRANSDUCERS in the avian pineal (i.e., they may transform photic energy into a hormonal output), and they are NEUROENDOCRINE TRANSDUCERS in the mammalian pineal (that is, they respond to the norepinephrine released from pineal sympathetic nerve endings by synthesizing and secreting the hormone melatonin).

Recently the action spectrum for the effects of light on the rat pineal organ has been defined (Cardinali et al., 1972a). An action spectrum can theoretically be examined in either of two ways: Animals can be exposed to light sources emitting monochromatic beams or relatively narrow spectral bands, all at the same irradiance levels; alternately they can be subjected to light sources whose emissions include all the wavelengths in the ultraviolet-visible-infrared range, but in different proportions. Exposure of rats to green light caused significant greater inhibition of pineal melatonin synthesis than exposure to either blue or yellow light. Light sources emitting red or ultraviolet radiation failed to produce any significant inhibition of pineal function. The action spectrum for photic inhibition of pineal melatonin synthesis was thus similar to the absorption spectrum of rhodopsin; inasmuch as rhodopsin is the unique photopigment hitherto recognized in the rat retina, these data suggest that the same photopigment mediates the visual and the neuroendocrine effects of light. However, pineal inhibition as well as other neuroendocrine effects of light still persist in the absence of intact retinal photoreceptors in the rat, but not in the absence of eyes, indicating the presence of other photosensitive cells besides rod cells in the rat retina (Reiter and Klein, 1971; Cardinali et al., 1972b).

B. LIGHT AND THE GONADAL FUNCTION

Environmental lighting has been demonstrated to affect the maturation and subsequent cyclic activity of the gonads in almost all vertebrates (Wurtman, 1967). The particular responses of each species to light seem to depend on whether the species is seasonally estrous or polyestrous, i.e., ovulating during the breeding season, in the spring or fall, or at regular intervals throughout the year. The gonadal responses of each species to light also depend on whether its members are physically active during daylight hours or during the nighttime. Another phase of reproduction is also altered by environmental lighting, i.e., the process of blastocyst implantation in mammalian species exhibiting delayed embryo implantation.

The neural and neuroendocrine pathways connecting the retinas and the gonads are poorly defined. One such pathway seems to involve the pineal organ; another may utilize cells in the hypothalamus implicated in the control of the gonadotrophin secretion from the anterior hypophysis (Moore, 1969, 1973).

Very few published studies of photic influences on mammalian gonadal function have compared the effects of different spectra or intensities of light. Rodents kept in red light reached sexual maturity earlier than those kept in short-wavelength lighting environments (Barbanti-Silva, 1932). Marshall and Boyden (1936) observed that ultraviolet light was particularly effective in inducing premature estrus in the ferret, while infrared light was ineffective. In another study Bissonette (1933) found red light stimulatory in the same species. In these early studies however the spectral characteristics and irradiance levels of the light source were poorly controlled. Unless both wavelengths and irradiance levels are precisely defined it is impossible to establish which factor accounts for the differential responses observed.

In the rat, Wurtman and Weisel (1969) observed that gonadal growth depended upon spectral composition of the light source. Rats born and reared under Vita-Life, a full spectrum light source whose emissions approximate sea-level solar irradiation, showed at 50 days of age larger gonads than animals kept under standard cool-white fluorescent bulbs at the same irradiance level. Both the light sources tested in that experiment provided a broad spectrum of visible radiant energy; they differed only in i) the relative proportions of the various wavelengths of the visible spectrum, and ii) the presence of very little long-wave and no erythemal ultraviolet radiation in the light emitted by the cool white bulbs.

Vriend and Lauber (1973) have recently suggested that the rate of gonadal growth of deer mice depends on the number of photons rather than on wavelength or radiant energy of the light source. Among animals exposed to blue (450 nm), green (545 nm) or red (650 nm) light environments at the same irradiance level, body, spleen and testis weight was maximal at 650 nm and minimal at 450 nm. In a second experiment deer mice were exposed to blue or red environments in which the number of photons were adjusted to be equivalent, according to the formula $n = I\lambda/hc$,(where I is the irradiance level at peak wavelength λ, h is Planck's constant and c is the speed of light) that is, the irradiance level of the blue light source was increased accordingly. Under these circumstances the differential responses of testes and spleen to wavelengths of light were not observed. Since any indirect biological effect of light, i.e., mediated by photoreceptor cells containing a photopigment, should exhibit an action spectrum which coincides with photopigment absorption spectrum, it would be of interest to study in deer mice the effects of wavelengths longer than 650 nm at the same irradiance level. If 650 nm is actually the action spectrum peak, less biological effect would be obtained with a light source emitting longer wavelengths in spite of the higher number of photons involved.

The relative effectiveness of different wavelengths of light to induce gonadal changes has most often been tested in birds. Light from the red end of the visible spectrum has been shown to stimulate gonadal development in starlings (Bissonette, 1932), turkeys (Scott and Payne, 1937), sparrows (Ringoen, 1942), ducks (Benoit and Ott, 1944), and quails (Woodward et al., 1969); in contrast, blue or green light have little or no effect on the activity of the gonads in these avian species. White Leghorn pullets exposed to short wavelength environments attained sexual maturity earlier than pullets exposed to red or white light (Harrison et al., 1969). Since in this experiment the photoperiod length was decreased during the rearing period the observations were interpreted as the inhibitory effect that results from reducing daily hours of a gonadal activating radiation, i.e., red or white light (Harrison et al., 1970), rather than as a stimulation caused by green or blue light.

Photoperiodic testicular responses in birds involve retinal and extraretinal photoreceptors. Increases in gonadal function by light can be induced or maintained in several avian species after removal of the eyes. Though the exact neuroanatomical location of extraretinal photoreceptors is uncertain, there is evidence that they exist in the brain, in particular in the hypothalamus (Menaker, 1972). Moreover extraretinal photoreceptors are more photosensitive than retinal photoreceptors in mediating gonadal changes in birds. In the duck, whereas retinal receptors respond only to orange-red light, extraretinal receptors respond to light throughout the visible portion of the spectrum (Benoit, 1970).

CONCLUSIONS

There is abundant evidence that light acting through the retina in mammals, and through retinal and extraretinal photoreceptors in birds, influences a variety of neuroendocrine mechanisms in addition to those subserving vision: Light accelerates or delays, according to species, the timing of sexual maturation; it modifies the growth and secretion activity of various endocrine organs; and day-light cycles can synchronize, or even generate, daily rhythms in brain monoamine metabolism, pineal function, body temperature, sleep, feeding behavior, locomotor activity, and a host of other functions. Much of this research has been directed toward the description and analysis of the sequence of events triggered by photic stimuli and few data are available about action spectra for neuroendocrine effects of light in mammals and birds. There is some experimental evidence indicating that visual and neuroendocrine photoreceptors, though sharing a common action spectrum, may be dissociated in the rat retina. In addition the action spectra and threshold intensities for avian retinal and extraretinal photoreceptors involved in gonadal activation by light seem to differ significantly. The information already obtained on spectral dependency for gonadal effects of light provide an indication of the complexity of the mechanisms that transduce light information to the neuroendocrine apparatus in mammals and birds.

REFERENCES: see pp. 271-272.

Phototoxicity in Animals

by

T. W. Dukes

INTRODUCTION

Light can act on the skin in several ways. It is known that the carcinogenic wavelengths of ultraviolet light are 2600 to 3400 Å (Perez-Tamayo, 1961). Sunburn and skin destruction result from exposure to similar wavelengths (La Via and Hill, 1971). One of the more interesting actions is that called photosensitization or photosensitizational dermatitis that has been recognized for some time. The lesions of photosensitization have nothing to do with sunburn, which is less severe, but is the result of the action of light upon a fluorescent pigment accumulated in the tissues. Three general sets of circumstances result in the accumulation of these pigments including defects in porphyrin metabolism, hepatotoxic photosensitization and primary photosensitization. Less is known of a fourth miscellaneous group of conditions associated with hypersensitivity to light.

The pigments accept light of one wavelength and transform it to a longer wavelength. Pigments studied so far have been chemically related to hemoglobin or chlorophyll being basically four pyrrole radicals tied with methene groups. The agents are able to obtain energy for chemical action by the absorption of light. The agents may not themselves enter into the chemical reaction which results in dermatitis but instead pass on their energy to other molecules which then become chemically reactant. The final chemical reaction proceeds only in the presence of oxygen. The photosensitive dyes are thought to concentrate within lysosomes and then, following exposure to light will catalyze a photooxidation of the lysosomal membrane, resulting in the bursting of lysosomes and cell death (La Via and Hill, 1971). Two types of sensitization occur. One is due to increased permeability of lysosomal membranes with secondary effects from released lytic enzymes and the second is due to an increased permeability of the cell membrane itself. The essential part of the process has been suggested as an oxidation product containing free radicals of the peroxy type. Lipid peroxidase is known to have damaging effect on lipid rich structures like the cell membrane system. Intracellular formation of significant quantities results in damage to mitochondria and endoplasmic reticulum, lysosomes and the external cell wall itself (Slater, 1966). The action spectrum (i.e., the magnitude of the response at different wavelengths) parallels the absorption spectrum of the agent.

CLINICAL OBSERVATIONS

There is burning itching, erythema, and inflammatory edema often with vesicle formation, necrosis hemorrhage and sloughing of the epidermis. The changes are often sharply limited to areas of the body that are in a position to receive direct sun or that are unprotected by either pigment of dark skin or a thick hair coat, clothes or blankets etc. In cattle the teats, udder and escutcheon are frequently involved. In sheep the head and ears are primarily affected with severe edema leading to a big head (hence the name geeldikop, yellow thick head). The ears are often so swollen they bend forward and down. Sometimes the face is deformed and a bottle jaw is evident. White swine used experimentally were primarily affected on face, ears and back. The face and back are often affected in horses. Lesions tend to heal in days or weeks but occasionally shock or other systemic signs occur, and death may result from secondary infections and gangrene.

DEFECT IN PORPHYRIN METABOLISM

This metabolic defect in porphyrin metabolism was described very well by Kaneko (1970). Readers are referred to his work for details as well as the work by Schmid (1966). It can be hereditary or acquired and is a defect in the synthesis of normal heme pigment (ferraprotoporphyrin). Excess accumulations of free protoporphyrin, uroporphyrin and coproporphyrin occur in serum and tissues as a result of this error of metabolism. The degree of photosensitivity will vary with the extent of porphyrin deposition and exposure to sunlight. The condition in cattle is a recessive hereditary characteristic. Homozygous animals have characteristic discolouration of teeth and urine, photosensitivity of lighter areas of skin. They lack condition and are weak. In swine the condition was inherited as a dominant characteristic and photosensitization was not seen even in white pigs. It was also seen to be inherited in cats as a dominant characteristic and again no photosensitivity was seen (Tobias, 1964).

A dominant erythropoietic porphyria has been seen in man in which photosensitivity is seen. Hepatic porphyrias have only been seen in man thus far. It is probably inherited as a dominant characteristic and may be precipitated by drugs such as barbiturates and alcohol. Several divisions, based on clinical signs have been suggested, but only the cutaneous form is accompanied by photosensitivity.

MISCELLANEOUS

Xeroderma pigmentosa, transmitted as a recessive gene in man, is characterized by light dry skin, areas of pigmentation and hypersensitivity to sunlight. After exposure, the skin shows hyperkeratosis, atrophy and ulceration. Neoplasia is associated with this condition (Peres-Tamayo, 1961). There is a tendency in collie dogs to develop a desquamative or ulcerative dermatitis of the skin on the bridge of the nose. The cause is not known but is thought to result from solar radiation in animals somehow hypersensitive to light.

HEPATOTOXIC PHOTOSENSITIZATION

This is the most common type of phototoxicity in animals. These conditions are associated with interference with excretion of phylloerythrin and as such are associated with chlorophyll rather than hemoglobin. It is dependent on phylloerythrin (a product of plant absorption) and toxic injury to the liver. Phylloerythrin is normally eliminated through the liver with the bile. A ketone on the sixth carbon atom distinguishes phylloerythrin from phylloporphyrin. The pigment reaches the skin via the circulation when liver or ductal system is injured. There are usually signs of acute toxic hepatitis and often but not always, icterus is present. Occasionally other signs of liver malfunction are present (e.g. nervous signs). No specific morphologic pattern exists in the liver. In "facial eczema" of sheep and the alfalfa associated photosensitivities there is a pericholangitis leading to occlusion of bile ducts (Glenn et al., 1965).

Numerous plants have been incriminated. Clare (1955) in his monograph listed many disorders as being accompanied by photosensitivity and hepatic dysfunction and icterus. It is recommended for further reading. Outbreaks are observed to occur at highly variable periods but often seasonal.

New Zealand investigators have discovered that under certain conditions of growth, temperature and humidity, fields of forage became heavily infested with fungus known as *Sporodesmium bakeri* or *Pithomyces chartarum*. A mycotoxin was demonstrated in these cases to be the toxic substance responsible for the photosensitization. A similar mechanism has been suggested for the photosensitive dermatitis associated with Bermuda grass in the Southern U.S. and for hepatogenous photosensitivity caused by feeding flood damaged alfalfa. Further investigation is needed to determine the extent of analogous situations with other plants.

Toxic hepatic injury and icterus with associated photosensitivity have been seen with treatment of some trypanocidal drugs (phenantridinum) and also with some cases of Carbon tetrachloride poisoning and hepatic fascioliasis. Rift valley fever can also lead to hepatotoxic photosensitivity.

An hereditary defect in hepatic phylloerythrin excretory mechanism occurs in Southdown sheep that can lead to photosensitivity associated with hyperbilirubinemia. It is seen in N.Z. and U.S., but there are no liver lesions (Cornelius and Gronwell, 1968). A syndrome in Corriedale sheep similar to Dubin-Johnson Syndrome in man is also associated with failure to excrete phylloerythrin and to photosensitization. (Cornelius et al., 1965).
Quin (1933) could produce photosensitivity ligation of bile duct with resulting obstructive jaundice if fresh green plants were fed.

PRIMARY PHOTOSENSITIVITY

This type of light sensitivity can result when pigments are ingested, absorbed and enter the circulation with no hepatic injury or any form of porphyrinemia. Fluorescein dye has been known to be involved. Various chemicals and drugs can produce the syndrome. Chemicals from plants produce fagopyrism and hypericism. Fagopyrism is buckwheat poisoning caused by fagopyrin, a derivative of naphthodianthrone found in buckwheat (*Fagopyrum esculentum*) and hypericism is caused by hypericin, a red pigment from the plant *Hypericum perforatum* or closely related species commonly called St. John's wort, Tipton weed, goat weed or cammock. In both these a red fluorescent pigment is found in the plant and also in the blood. The responsible wavelength is in the 5400 to 6100 Å range differing from that causing the hepatogenous type. The absorption spectrum is narrow in fagopyrism (about 5900 Å).

The anthelmintic phenothiazine is probably the best known drug causing problems. It is oxidized in the gut to form phenothiazine-sulfoxide a photodynamic pigment metabolized in the liver. If this substance gets to the general circulation, a dermatitis results on exposure to sun. The effective wavelengths are about 3600 Å. This is especially true if the animal is young or has liver disease since elimination is less efficient. Keratitis can also be seen when the chemical is present in tears and aqueous humour. The condition is common in swine and cattle. Sheep are more resistent since they more readily convert sulphoxide to sulphate. The keratitis has been reported in sheep, swine, cattle and birds with pigs developing the skin lesions more often than the others. Intravenous injection of rose bengal predisposes to photosensitization.

Other drugs can also produce the condition and many have been shown to be potentially phototoxic. Among those used in man are the phenothiazine tranquillizers, chlorothiazide diuretics, hypoglycemic agents, sulphanilamide and sulphonylurea, dimethylchlortetracycline and coal tar derivatives in pitch. Certain antimalarial compounds were tested in Texas (Bay et al., 1970). Many of these produced photosensitivity in the skin of swine. Eight methoxysporalen can readily produce the classical signs in conjunction with light.

EXPERIMENTAL METHODOLOGY

Descriptions of some methods have been published (Bay et al., 1970; Hakim, 1961; Sams, 1966; and Allison, 1966). Allison's technique involved the suspect photosensitive substance being placed on or injected into the skin of animals and exposed to various wavelengths. Erythema and weal with itching and burning sensation are found. Various animal species have been used in experimental studies. White swine proved very successful in studies of drug induced phototoxicities. The drug (8 MOP) 8-methoxypsoralen, known to be phototoxic in several species, can be used for experimental studies or as a positive control. Fluorescent blacklight lamps or filtered sunlight can be employed. Readers are referred to Bay et al. (1970) and Sams (1966) for more details. The intensity of the radiation can vary with the distance from artificial light sources, with the wavelength of light and with the source used. The time of day, type of filter and location geographically when using natural light should also be considered. One must also keep in mind that animals kept in filtered sunlight must also have adequate ventilation to prevent heat prostration during exposure to filtered sunlight.

It should be possible to isolate and identify photodynamic agents in blood or tissues of photosensitive animals and to specify the specific wavelengths of light which can activate the process. This has been done for some substances. More study of the pathogenesis of the condition is needed at the cellular, subcellular and biochemical levels. Photooxidation of lysozymal membranes is thought to be involved.

Further investigation is needed with the plant associated hepatic phototoxicities especially relative to pathogenesis. Care should be taken to establish the exact wavelengths of light involved in any experimental work.

SUMMARY

Phototoxicity in animals can be the result of ingestion of photosensitive agents (primary), the accumulation of endogenous pigments (porphyrins) or the accumulations of exogenous pigments secondary to liver damage (hepatotoxic). It should be possible to isolate and identify photodynamic agents in blood or tissue of photosensitive animals and to specify the specific wavelengths of light which can activate the process. Some substances have been worked with but others remain. More work needs to be done with the pathogenesis and biochemistry of the specific agents especially at the cellular and subcellular levels.

REFERENCES: see pp. 272-27 .

Effects of Light and Temperature on Certain Physiological Functions of Shrews

by

M. Gębczyński

INTRODUCTION

Shrews (Soricidae, Insectivora) are very small carnivorous animals. The smallest mammals in the world, *Suncus etruscus* and *Sorex minutissimus*, which weigh approximately 2 g as adults, are members of the Soricidae family. The great activity and considerable food requirements of shrews have attracted the attention of physiologists. The first measurements of oxygen consumption by shrews showed that this is very high, and this led Pearson (1948) to form the hypothesis that this index is higher than in rodents. Although there is no complete agreement (Martinsen, 1969) as to whether the metabolism of shrews is in fact higher than in other small mammals, a large number of new data have been collected which confirm this hypothesis (e.g. Morrison et al., 1959; Buckner, 1964; Gebczynski, 1965; Neal and Lustick, 1973).

Apart from certain physiological differences in shrews, astonishing phenomena of morphological changes, first described in relation to *Sorex araneus* (Dehnel, 1949), were found. Young shrews leave the nest (June-October) on their 21st day of life, by which time they are fully developed and completely adapted to an independent way of life. The young usually overwinter as sexually immature individuals and it is only in the spring of the next year that they are ready for reproduction. They live until December at most, and therefore the life span of a young adult is up to 9 months and of an old adult rarely more than 6 months. During this life cycle many of the morphological indices were found to change e.g., weight of brain, heart, kidneys and other internal organs (see comprehensive review by Pucek, 1970). All these organs become considerably smaller in autumn, and particularly in winter, then increase in size again in animals which survive the winter.

Also, shrews are of great importance since they eat amounts of insects and insect larvae up to as much as 300 kg/ha a year (Judin, 1971) some of which are pest species. They are therefore of considerable importance as predators, and in consequence have attracted the attention of forest ecologists. Despite the interest aroused, the physiology of Soricidae is not sufficiently known, particularly not compared with that of small rodents. This may be partially due to the difficulties arising in work with shrews as experimental animals, since they are generally intolerant of artificial rearing. Only white-toothed shrews reproduce in captivity, and therefore continuous trapping is required in the majority of studies in order to replace stocks of test shrews.

Species of the Soricidae family inhabit temperate zones but have successfully occupied both arctic and tropical areas. It may therefore be presumed that the various species have different metabolic reactions to the effect of the ambient temperature.

EFFECT OF TEMPERATURE

METABOLIC RATE

The Masked shrew, *Sorex cinereus*, which is encountered ove the whole of the North American continent, exhibits a distinct metabolic response to cold (Fig. 1). The average slope of the curve relating metabolic rate and ambient temperature (2°-11° C) gives a value of 0.62 cc O_2/g hr per 1° C (Morrison et al., 1959). This same index in the Common shrew (*Sorex araneus*), which inhabits areas in Europe and the northern part of Asia, is similar to that in *S. cinereus*. As the Common shrew has been studied both in summer and winter it was possible in addition to observe seasonal differences. For

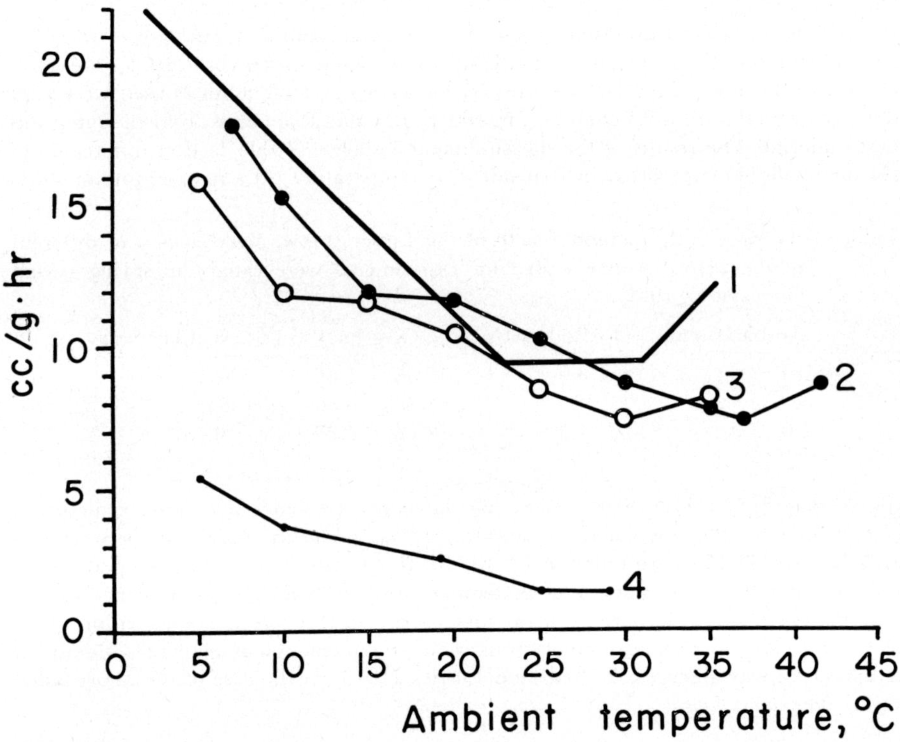

Fig. 1. The relation of oxygen consumption of some shrews to ambient temperature. 1: minimal oxygen consumption of *S. cinereus*, body weight 3.3-4.0 g (After Morrison et al., 1959); 2: resting oxygen consumption of *S. araneus*, summer, body weight 7.8-8.2 g (Gębczyński, 1965); 3: resting oxygen consumption of *S. araneus*, winter, body weight 6.9-7.2 g (Gębczyński, 1965); 4: *Suncus murinus*, minimal oxygen consumption, body weight 42.0-54.0 g (Dryden et al., 1974).

instance, summer variations in oxygen consumption are most intensive for temperatures from 7° to 15° C, the average being 0.75 cc/g hr per 1° (Gebczynski, 1965), whereas in winter, at a low temperature (5-10° C), this value is 0.45. It was found that the metabolic rate of *S. araneus* at the same temperatures is higher in summer than in winter (Fig. 1).

In the case of *Suncus murinus*, a shrew which lives in tropical zones, the metabolic reaction to the effect of cold is slightly different (Dryden et al., 1974). The metabolic rate is far lower in this species than in *Sorex cinereus* and *S. araneus* (Fig. 1), but reduction of the ambient temperature from 9° to 5° C evokes an increase of 0.39 cc/g hr per 1° C in oxygen consumption. If we take into consideration that the metabolic rate of the Asian house shrew (*S. murinus*) is lower than in the Common shrew and the Masked shrew, it will be clear that the relative increase rate due to the effect of cold is very great. In *Suncus* oxygen consumption at a low temperature increases by about 10% per 1° C, but by 5% and 6% respectively in *S. araneus* and *S. cinereus*.

In another species of Euro-asian shrew, *Sorex minutus*, ambient temperatures were observed to have a longlasting effect on oxygen consumption (Gebczynski, 1971). Measurements of this index lasted uninterruptedly for 24 hrs and the animals used were kept during this period in a 9 l chamber, in which they had free access to food, water and nest material. The results of the measurements indicate (Table 1) that differences in the metabolic rate are distinctly dependent on temperature. At a temperature of 10° C

Table 1. Average daily metabolic rate of the Lesser shrew, *Sorex minutus* in different temperatures. Animals used for experiments were caught in spring (From Gebczynski, 1971).

Ambient temp. °C	Body wt/g	cc O_2/g hr ± S.D.	kcal/kg$^{0.75}$day	n
10	3.9	16.86 ± 1.66	677.9	6
15	4.5	13.98 ± 0.80	558.1	6
20	4.0	12.52 ± 0.89	485.6	12

the average daily energy requirement calculated per metabolic body size (kcal/kg$^{0.75}$) is 25% higher than this value measured at 15° C, while the difference between 15° and 20° is only 15%. Stimulation of the metabolic rate by ambient temperature is therefore not the same, since a lower temperature exerts a stronger influence than a temperature nearer the thermoneutral zone, which in this species comes within limits of 25-30°. A similar phenomenon of stronger metabolic reaction at a low (2° C) ambient temperature was observed by Neal and Lustick (1973) in the case of the Short-tailed shrew *Blarina brevicauda*.

BODY TEMPERATURE

The effect of ambient temperature on the body temperature has been measured for *Sorex cinereus* (Morrison et al., 1959) and *Blarina brevicauda* (Neal and Lustick, 1973). The average body temperature in *S. cinereus* is 38.8° C and no significant differences are seen between 5° and 32° of ambient temperature, but a temperature of about

−10° C lowers the body temperature to as low as 27° C after 15 min. Body temperature changes even more quickly when the Masked shrew was exposed to the simultaneous effect of cold and wind.

In *B. brevicauda* body temperature was 1° lower in individuals exposed to a temperature of 0-2°, but at ambient temperatures from 5° to 30° the differences were negligible. Only a temperature of 35° influences body temperature, by raising it to 41.4° C.

EFFECT OF LIGHT

Crowcroft (1964) kept 6 Common shrews under laboratory conditions during the winter. Three of them were subjected to the effect of artificial lighting for 16 hrs per day. After four weeks these individuals proved to be sexually mature, whereas the control shrews remained sexually inactive.

It has also been shown (Borowski, 1964) that the length of day may affect moulting processes in shrews of the genus *Sorex* to some extent. It proved to be possible to evoke the autumn moult in this way, but not the spring moult. The studies made so far on the effect of light on shrews have shown that apart from activating sexual maturation and some degree of regulation of moulting, this factor has no important effect on the animals (Pucek, 1964).

SEASONAL CHANGES

LEVEL OF METABOLISM

Average daily oxygen consumption in the Common shrew (Gebczynski, 1965) and the Lesser shrew (Gebczynski, 1971) is greatest in young adults in summer. In autumn this index becomes lower in both species, while the metabolic rate is the same in the Common shrew in winter as in autumn. In the Lesser shrew, on the other hand, oxygen consumption in winter is higher than in autumn, although it is slightly lower than in summer. In overwintered animals there is considerable reduction in metabolic rate and this is more distinct in summer than in spring. Old adult shrews thus differ markedly from young adults in respect of this index, and consequently the effect of the season cannot be examined without taking the specific life cycle of shrews into consideration.

TISSUE METABOLISM

The metabolism of different tissues exhibits pronounced seasonal differences (Malzahn, 1974). The tissues of the liver and kidneys reduce the rate of oxygen consumption in successive seasons from the summer onwards (Fig. 2). In old adults this value is half that found for young adults. The seasonal relation between the metabolic rate of heart tissue and muscle tissue is different, since in both these tissues the rate of oxygen consumption is independent of the shrew's age, being the same in summer in young adults as in old adults. On the other hand metabolism in individuals caught in winter is markedly higher, in the case of heart the differences can be 60%, and even 150% in the muscles of the leg.

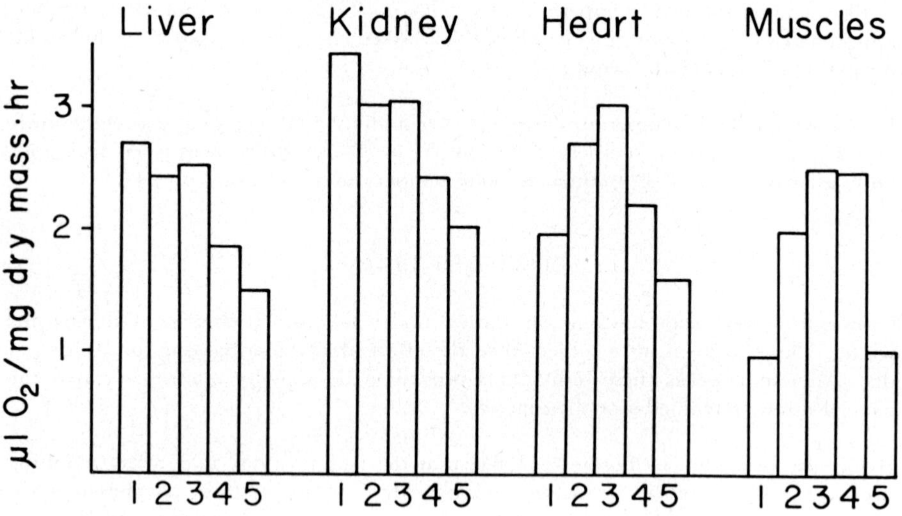

Fig. 2. Oxygen consumption by tissue of the Common shrew in different seasons (Malzahn, 1974). 1: summer, young adult; 2: autumn, young adult; 3: winter, young adult; 4: spring, old adult; 5: summer, old adult.

CONTENT OF CYTOCHROME C

The cytochrome c content in the liver, kidneys and brown fat has been measured in the Common shrew in different seasons (Hyvärinen and Pasanen, 1972). The level of cytochrome c differs in different tissues but is subject to seasonal variations in all of them (Fig. 3). Its highest level is reached in autumn, which suggests that it plays a special role in the animal's adaptation to colder weather. It must also be mentioned that the level of cytochrome c is in fact higher in the organs and tissues of *S. araneus* than in small rodents (Hyvärinen and Pasanen, 1972). This confirms the hypothesis that in comparison with other mammals, the shrews have a different metabolism.

ACTIVITY OF ALKALINE PHOSPHATASE

In summer and autumn the alkaline phosphatase activity in the kidneys of *S. araneus* is about 30 μl P liberate/mg of fresh tissue whereas in the winter months this value exceeds 50 μl P liberate/mg; in spring the level is the same as in autumn (Hyvärinen, 1968). The activity of alkaline phosphatase in brown fat also fluctuates similarly in the same species of shrew (Pasanen, 1969), as does the activity of alkaline phosphatase and glucose-6-phosphatase in the liver, depending on the season (Hyvärinen, 1969).

MORPHOLOGICAL BLOOD INDICES

Blood indices in the Common shrew, *Sorex araneus*, were measured in different seasons during the life cycle of this species (Wolk, 1974). Changes in the various indices are not identical, for instance red blood count (RBC) is subject to considerable fluctuations. RBC is one-third lower in shrews in summer than in winter, when the number of erythrocytes exceeds $23 \times 10^6/\text{mm}^3$ (Fig. 4), whereas the diameter of an erythrocytes

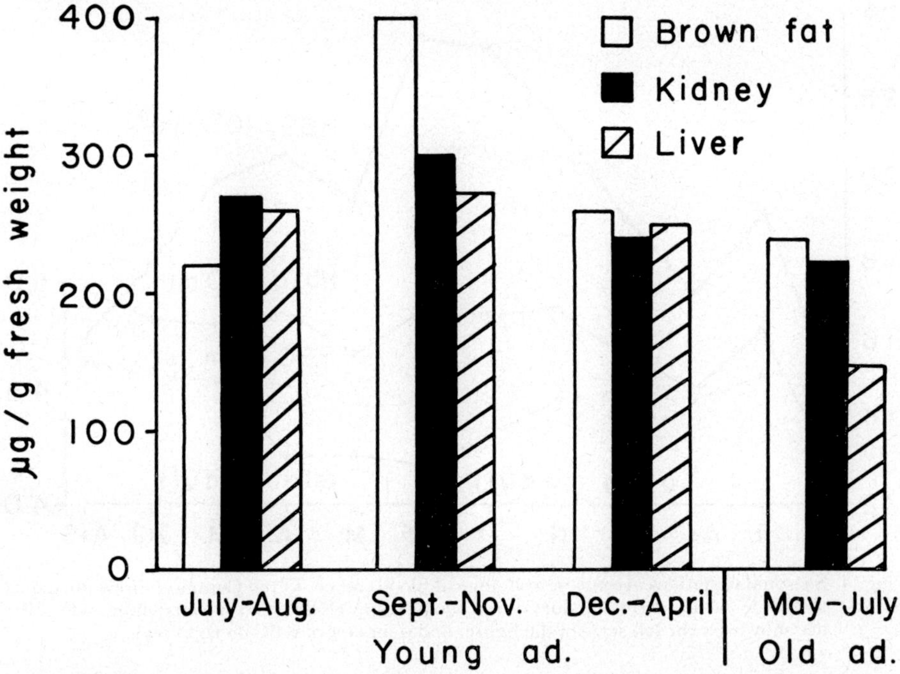

Fig. 3. Seasonal changes in cytochrome c content of the brown fat, kidney and liver of the Common shrew (Hyvärinen and Pasanen, 1973).

is smaller in winter and early spring than in the other seasons of the year. Unlike these indices mentioned before haemoglobin content is not subject to such distinct seasonal variations and is only slightly lower in young adults in summer and autumn, but in young adults in winter and an old adult in spring, summer and autumn it remains on a similar level.

GENERAL EFFECT OF BIOMETEOROLOGICAL FACTORS ON SHREWS

Examining the geographical distribution of shrews in Siberia, Mezhzerin (1964) considered that it is directly dependent on ambient temperature. For example *S. araneus*, *S. minutus*, *S. unguiculatus* and *Neomys fodiens* only inhabitat areas where the average temperature for January is above −30° C, while the smallest shrew *S. minutissimus*, lives only in areas with a more severe climate and occurs in places where the average January temperature is lower than −10° C. Regions with lowest winter temperatures are inhabited only by the smallest representatives of the genus *Sorex* (Judin, 1971).
An example of strict dependence on habitat conditions has been found in the life cycle of *Sorex* (Pucek, 1965). Seven phases of development can be distinguished during the

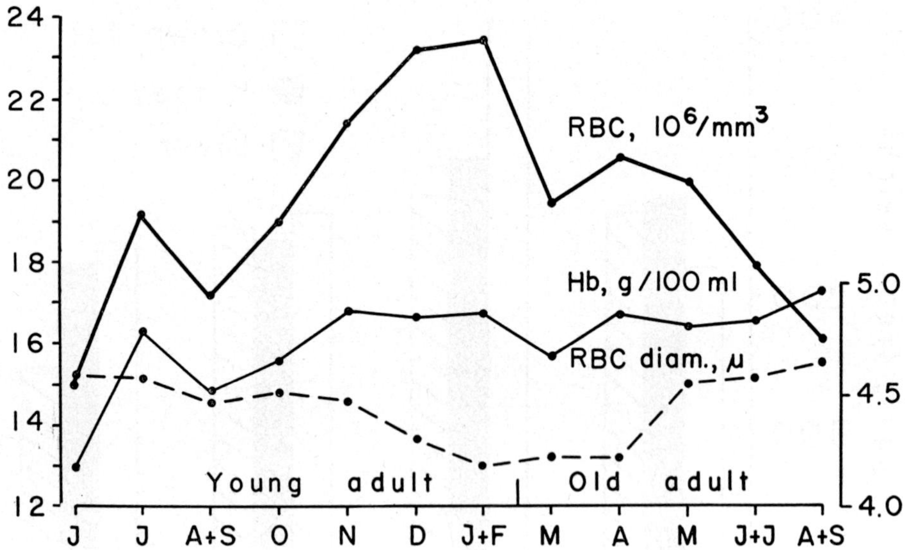

Fig. 4. Seasonal variations in some morphological blood indices in the Common shrew during its life cycle (Wolk, 1974). Values of red blood count (RBC) and haemoglobin level (Hb) are shown on the left scale of the figure, and diameter of RBC on right scale.

life of the shrew. The autumn regression and winter depression, following on a period of morphological and physiological stabilization in young adults in summer, constitute a form of adaptation to the severe conditions of these periods in temperate and arctic zones. This hypothesis is supported by the fact that the winter depression is more distinctly evident in northern and eastern populations (Pucek, 1970). Growth of the whole body, change in many of its physiological functions and attainment of sexual maturity takes place in spring when weather conditions are favourable.

CONCLUSIONS

Despite the fact that the effect of several meteorological factors on some of the physiological indices in shrews is well known, the knowledge we possess is totally insufficient to understand their interrelationship. We do not for instance know what physiological mechanism governs the geographical distribution of shrews depending on climate, neither do we know what effect altitude has on members of the Soricidae family. We do not understand either why light has so small an effect on these animals. The concentration of oxygen and carbon dioxide is different in the burrows of small mammals than in the atmospheric air, but the way these conditions affect shrews has not yet been ascertained.

REFERENCES: see pp. 273-274.

CHAPTER 3

EFFECTS OF HIGH ALTITUDE ON ANIMALS

Domestic Animals at High Altitude

by

R. F. Grover

INTRODUCTION

At the outset, it should be emphasized that satisfactory adaptation to high altitude has been achieved for virtually all species of domestic animals. Abundant proof of this fact is evident on the altiplano of the Andes where one now finds large numbers of cattle, sheep, goats, pigs, horses, as well as dogs and cats, in addition to the native populations of llamas and man, all adapted to living at altitudes above 4,000 m. This is not to say that the introduction of domestic animals to high altitude has been without problems. During the first decades of the Spanish occupation of the Andes, reproduction within the human population as well as in the animals they brought with them was very difficult. In the city of Potosi, Bolivia, at 4,000 m altitude, there stands a monument to the first child of Spanish parents to survive from among all those who were born there in the 53 years after the founding of that city (Monge-M., 1948). As recently as 1950, Cuba (1956) reported that the mortality rate was still 25% among lambs born above 3,500 m. Thus, it probably takes several generations before the majority of individuals in a population become well adapted. However, once such a population is established, then survival of that species at high altitude is assured.

High altitude environments are encountered in the various mountainous regions and high plateaux of the world. They pose the universal environmental stress of atmospheric hypoxia resulting from the reduced atmospheric pressure. However, these lofty regions often have a rigorous climate such that cold rather than hypoxia becomes the environmental stress which usually determines habitability. Cold also limits the vegetation which can survive at high altitude, and thereby the availability of food is determined. The less dense atmosphere at high altitude provides less filtering of solar radiation which can become intense. Also, humidity is often low. Thus, the animal and human populations found in the various high altitude regions of the world are determined more by climatic factors than by the elevation above sea level. The Andian altiplano situated above 4,000 m altitude has permanent populations consisting of many species of domestic and wild animals, as well as several hundred thousand humans. This is permitted by a climate lacking severe cold since the region is near the equator. However, as one moves farther from the equator, either north or south, the snow line (and the timber line) occurs at progressively lower altitudes, and consequently animal populations are restricted to lower altitudes also. In the Rocky Mountains, the highest permanent human population is located at 3,100 m (Leadville, Colorado) and most animal species spend the winter months below 3,000 meters, although they migrate up to nearly 4,000 m as the snow recedes in the summer. In this region, clearly climate rather than altitude determines habitat.

Nevertheless, the atmospheric hypoxia at high altitude constitutes a significant stress per se. Most species cannot tolerate exposure to altitudes above 5,500 m indefinitely. Furthermore, cold acts synergistically with hypoxia to lower the altitude limit tolerable in mountain environments. However, since cold stress is dealt with elsewhere in this text, the following discussion will consider only the physiological and pathological effects of atmospheric hypoxia on various animal species.

HYPOXIA

The earth is enveloped in a blanket of air which is held to the earth by the force of gravity. This layer of air, like all fluids, exerts a pressure determined by its height and density. It follows that atmospheric pressure will be greatest at the bottom of this layer of air i.e., at sea level, and will be progressively less at increasing altitudes above sea level. Several factors influence the atmospheric pressure which actually exists in various mountainous regions on earth. Although air density tends to decrease at high altitude, this is partially offset by the lower ambient temperature generally found at altitude which tends to increase air density. The net result is shown in Figure 1, which is a compilation of atmospheric pressure readings made at many high altitude stations relatively near the equator (Dill and Evans, 1970). However, an additional factor has been recognized in the mountains of the antarctic. The rotation of the earth causes the atmosphere to accumulate to a greater height around the equator than over the poles. Consequently, at any absolute altitude above sea level, the atmospheric pressure is significantly less at the poles than at the equator. At a station at 2,800 m altitude in the antarctic, the observed atmospheric pressure was 510 torr (Guenter et al., 1970), whereas at the same altitude nearer the equator, the pressure would be 545 torr (Fig. 1). In other words, in terms of ambient atmospheric pressure, 2,800 m altitude in the antarctic is equivalent to 3,400 m altitude in the Andes.

At all terrestrial altitudes, the atmosphere contains approximately 21% oxygen (actually 20.93%). Consequently, the partial pressure of oxygen, P_{O_2}, will be 21% of the total atmospheric (barometric) pressure, P_B. At sea level the normalized P_B is 760 torr giving a P_{O_2} of 160 torr (Smith, 1969).

Animals do not sense the P_{O_2} of the atmosphere, but rather the P_{O_2} of the arterial blood, Pa_{O_2}. As the ambient air is inhaled into the trachea and bronchi, it is heated to body temperature and saturated with water vapor. For many animals, normal body temperature is 39° C and the resulting water vapor pressure, P_{H_2O} is 52 torr (in man at 37° C, P_{H_2O} is 47 torr). Since P_B is unchanged, the sum of the partial pressures of the gases in the air is reduced by P_{H_2O}. Consequently, the partial pressure of oxygen in saturated inspired air, P_IO_2, will be 21% of (P_B-P_{H_2O}); at sea level, $P_IO_2 = 0.21$ (760-52) or about 150 torr (Fig. 1).

The primary function of the lung is the exchange of gas across the alveolar-capillary membrane. Once the inspired air enters the alveoli, oxygen diffuses into the blood, and carbon dioxide diffuses from the blood into the alveolar gas. Normally, this exchange of gases proceeds with such rapidity that blood and gas come into equilibrium

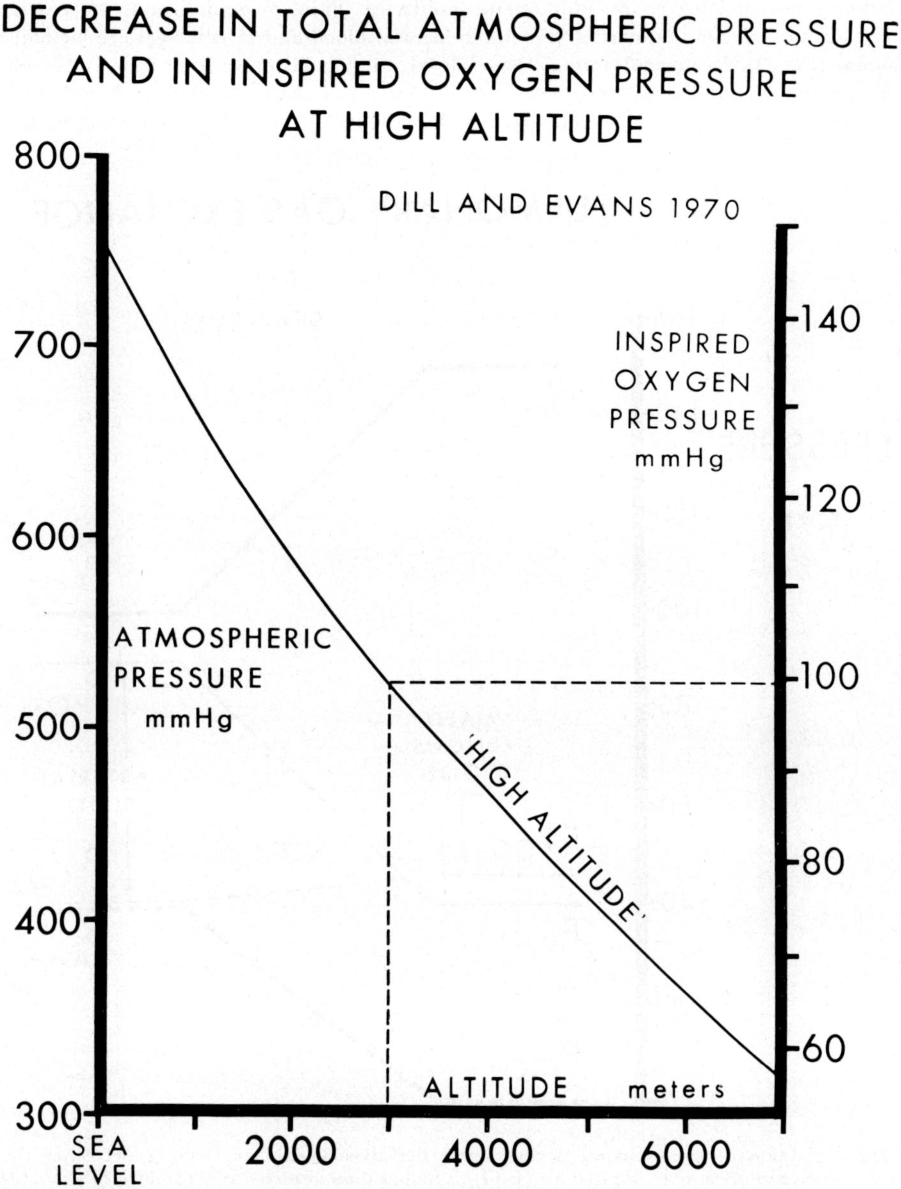

Fig. 1. With ascent from sea level to high altitude, there is a progressive decrease in total atmospheric pressure. When the associated decrease in the oxygen pressure of inspired air reaches 100 mm Hg (torr), significant physiologic effects appear. Therefore, the physiologic threshold for "high altitude" is about 3,000 m (From Dill and Evans, 1970).

before the blood leaves the pulmonary capillaries. Thus, at equilibrium, the partial pressures of oxygen and carbon dioxide in arterial blood and alveolar gas are virtually equal (Fig. 2). In normal animals at sea level, the partial pressure of carbon dioxide in arterial blood, Pa_{CO_2}, and hence in alveolar air, P_ACO_2, is 40 torr. Metabolism determines the respiratory exchange ratio, R. If R is 0.8, then CO_2 production is only

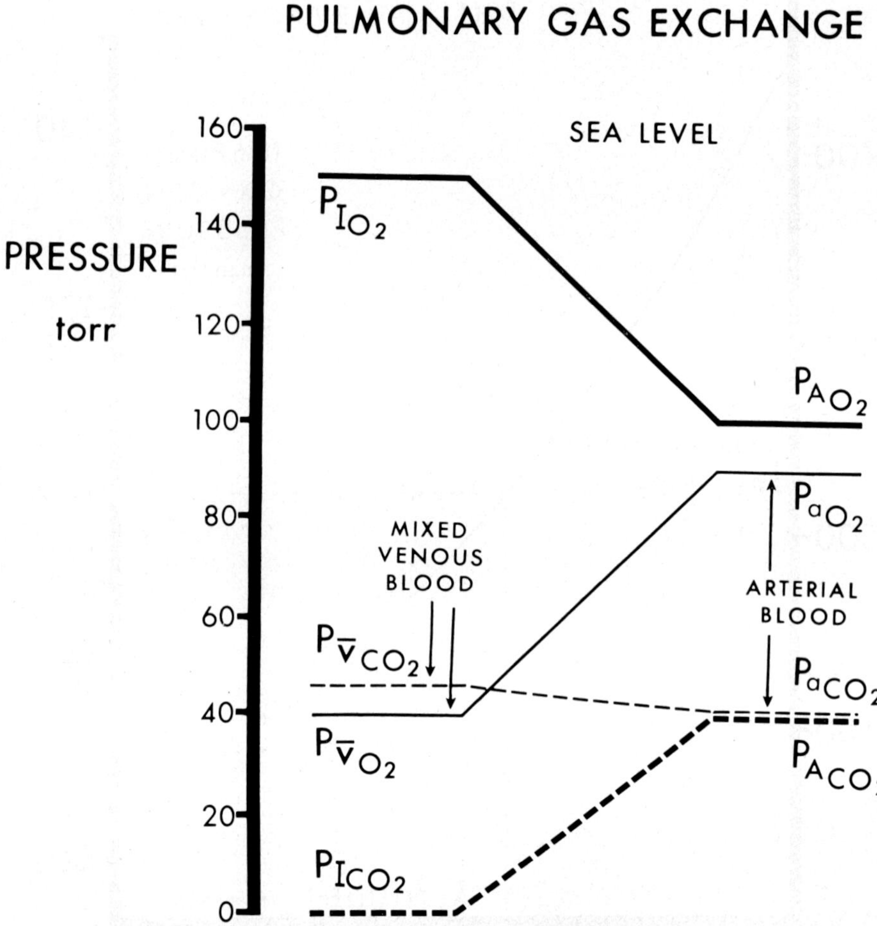

Fig. 2. Pulmonary gas exchange occurs across the alveolar-capillary membrane. Since the oxygen pressure of inspired air (P_IO_2) is higher than in mixed venous blood (P_VO_2), O_2 diffuses from air to blood. The O_2 pressure in blood rises, while the O_2 pressure in air falls. At equilibrium, the O_2 pressure in blood which will flow into the arterial system (PaO_2) approaches the O_2 pressure in alveolar air (P_AO_2). Concurrently, carbon dioxide (CO_2) diffuses out of the mixed venous blood (P_VCO_2) into the inspired air (P_ICO_2) which is free of CO_2. The CO_2 pressure rises in the air and falls in the blood. At equilibrium, the CO_2 pressure in alveolar air (P_ACO_2) equals that in arterial blood ($PaCO_2$).

0.8 of oxygen uptake. In the alveoli, then, a rise in P_ACO_2 of 40 torr will be accompanied by a greater decrease in P_IO_2 of about 50 torr. Thus, at equilibrium, the partial pressure of oxygen in the alveolus, P_AO_2, at sea level will be approximately 150-50 = 100 torr. These calculations of average P_AO_2 are expressed by the "alveolar air equation" which is:

$$P_AO_2 = 0.21\ (P_B - P_{H_2O}) - Paco_2\ (0.21 + 0.79/R)$$

Since there are regional variations in pulmonary gas exchange, the partial pressure of oxygen in arterial blood, Pao_2, is always somewhat less than P_AO_2 (A-a difference); at sea level Pao_2 is 80-90 torr.

From the foregoing it is evident that an animal's Pao_2 is determined primarily by total atmospheric pressure, P_B, and is approximately 70 torr less than P_IO_2 at sea level. When the animal is taken to high altitude, he is exposed to a decreased P_B, and the associated reduction in P_IO_2 will produce an equal reduction in Pao_2 (Fig. 3) unless modified by compensatory mechanisms. The mechanisms potentially available to lessen the fall in Pao_2 include hyperventilation to decrease $Paco_2$, a metabolic alteration to increase R, and increased uniformity of regional lung function to minimize the A-a difference in PO_2.

ADAPTATION OF OXYGEN TRANSPORT MECHANISMS TO HYPOXIA

PULMONARY VENTILATION, HYPERVENTILATION

As just described, the gas exchange function of the lung occurs at the alveolar-capillary membrane. Here, blood takes up O_2 and releases CO_2. Consequently, alveolar ventilation must be regulated to match pulmonary capillary perfusion if the resulting blood gas tensions are to remain normal. To accomplish this regulation of ventilation, the arterial blood gases are monitored by two sets of chemoreceptors. Arterial PO_2 is monitored exclusively by the carotid bodies located at the bifurcation of the carotid arteries. These peripheral chemoreceptors also provide a portion of the monitoring of arterial PCO_2, the remainder occurring at the central chemoreceptors located on the floor of the fourth ventricle of the brain.

Under normal conditions at sea level, Pao_2 is 80-90 torr and $Paco_2$ is 35-40 torr. The chemoreceptors are very sensitive to acute changes in $Paco_2$, being stimulated by small elevations which in turn produce a prompt increase in ventilation. This augments the removal of CO_2 from the alveoli, lowering P_ACO_2 and thereby returning $Paco_2$ to normal. Thus, ventilation responds directly and linearly to acute deviations in $Paco_2$ from the normal range of 35-40 torr. The sensitivity to CO_2 is defined and quantitated by the slope, S, of the straight line relating minute ventilation $\dot{V}_E$ (BTPS) to $Paco_2$ (Fig. 4). If the level of $Paco_2$ deviates from normal, this reflects a change in effective alveolar ventilation; a rise in $Pacoo_2$ indicates hypoventilation, while a fall in $Paco_2$ indicates hyperventilation.

In contrast, the carotid bodies are relatively insensitive to changes in Pa_{O_2} near the

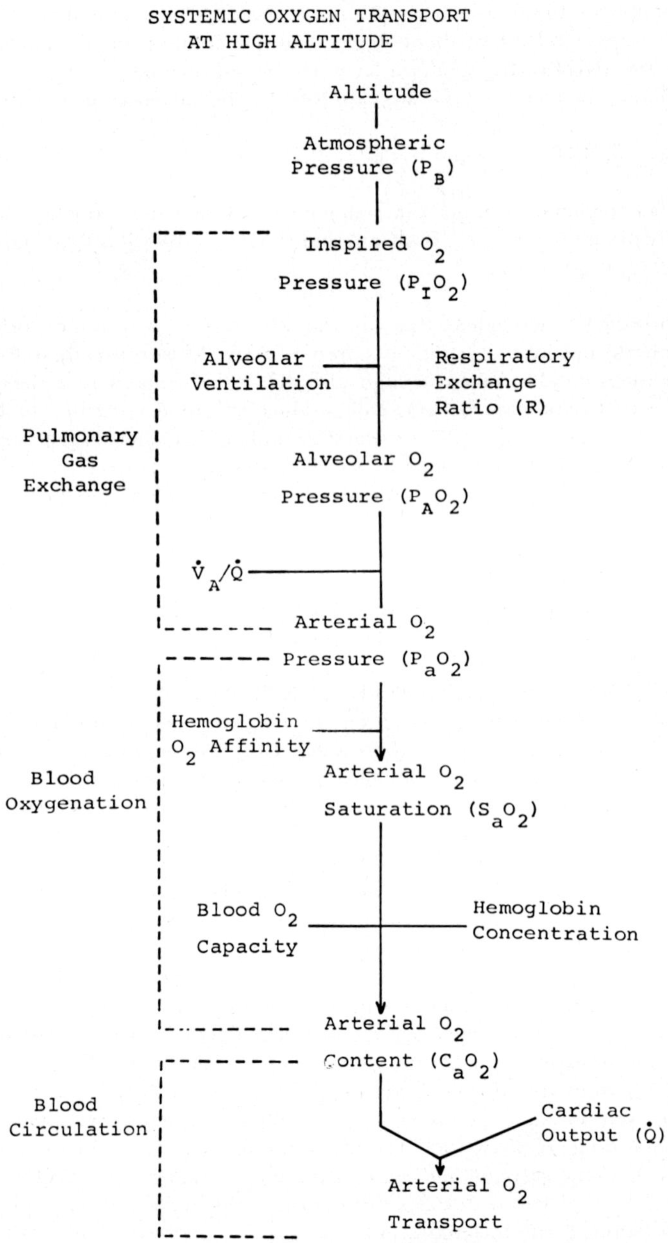

Fig. 3. The transfer of oxygen from the atmosphere to the tissues of the body. Pulmonary ventilation brings the oxygen into the lungs. Pulmonary gas exchange then occurs across the alveolar-capillary membrane. Within the capillary, the oxygen is then bound to hemoglobin. This oxygenated blood is then circulated throughout the body by the heart.

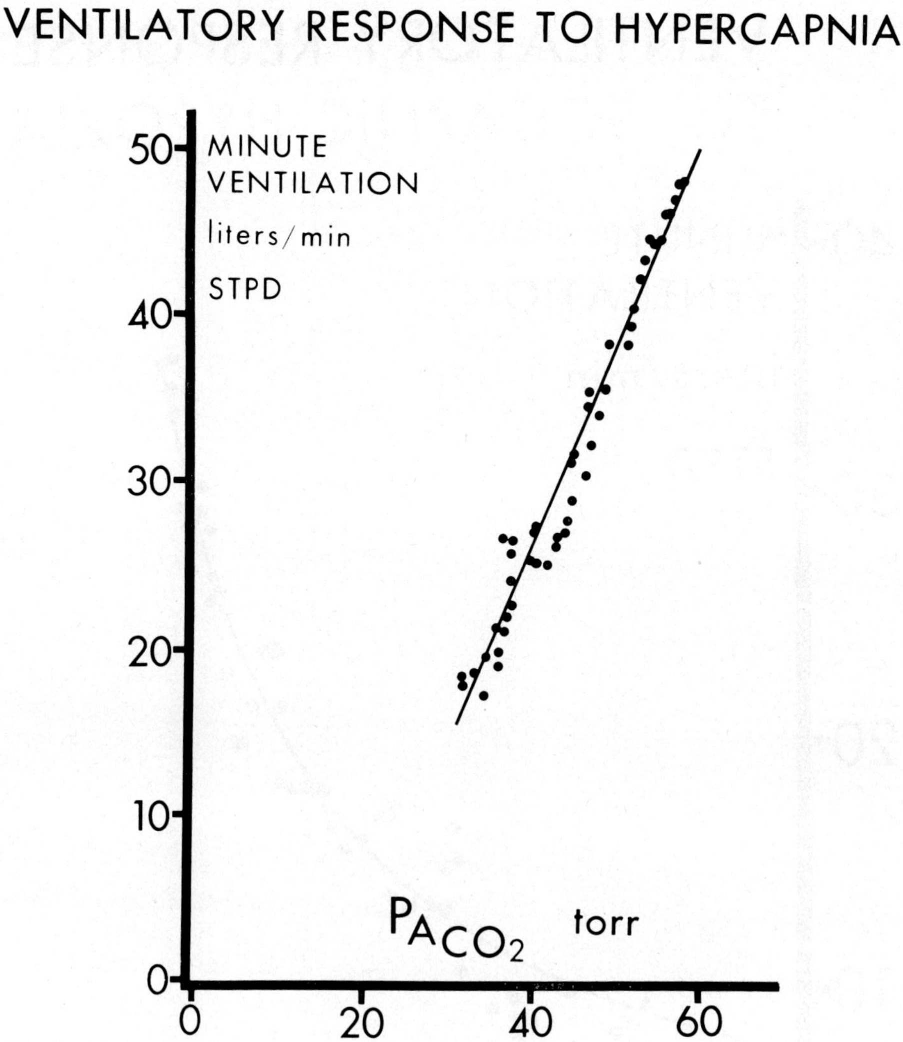

Fig. 4. In the calf, an increase in the carbon dioxide pressure of alveolar air (P_ACO_2), and hence in arterial blood ($PaCO_2$), produces a linear increase in minute ventilation.

normal range of 80-90 torr. Stimulation clearly begins only when Pa_{O_2} is lowered to 65-70 torr, and as hypoxia becomes progressively more severe, ventilation increases at an accelerating rate. The response to hypoxia is therefore described and quantitated by a hyperbolic curve (Fig. 5) whose equation is

$$\dot{V}_E = \dot{V}_{E0} + \frac{A}{P_{aO_2} - K}$$

where $\dot{V}_{E0}$ is extrapolated minimum ventilation at infinitely high Pa_{O_2}, and A is the

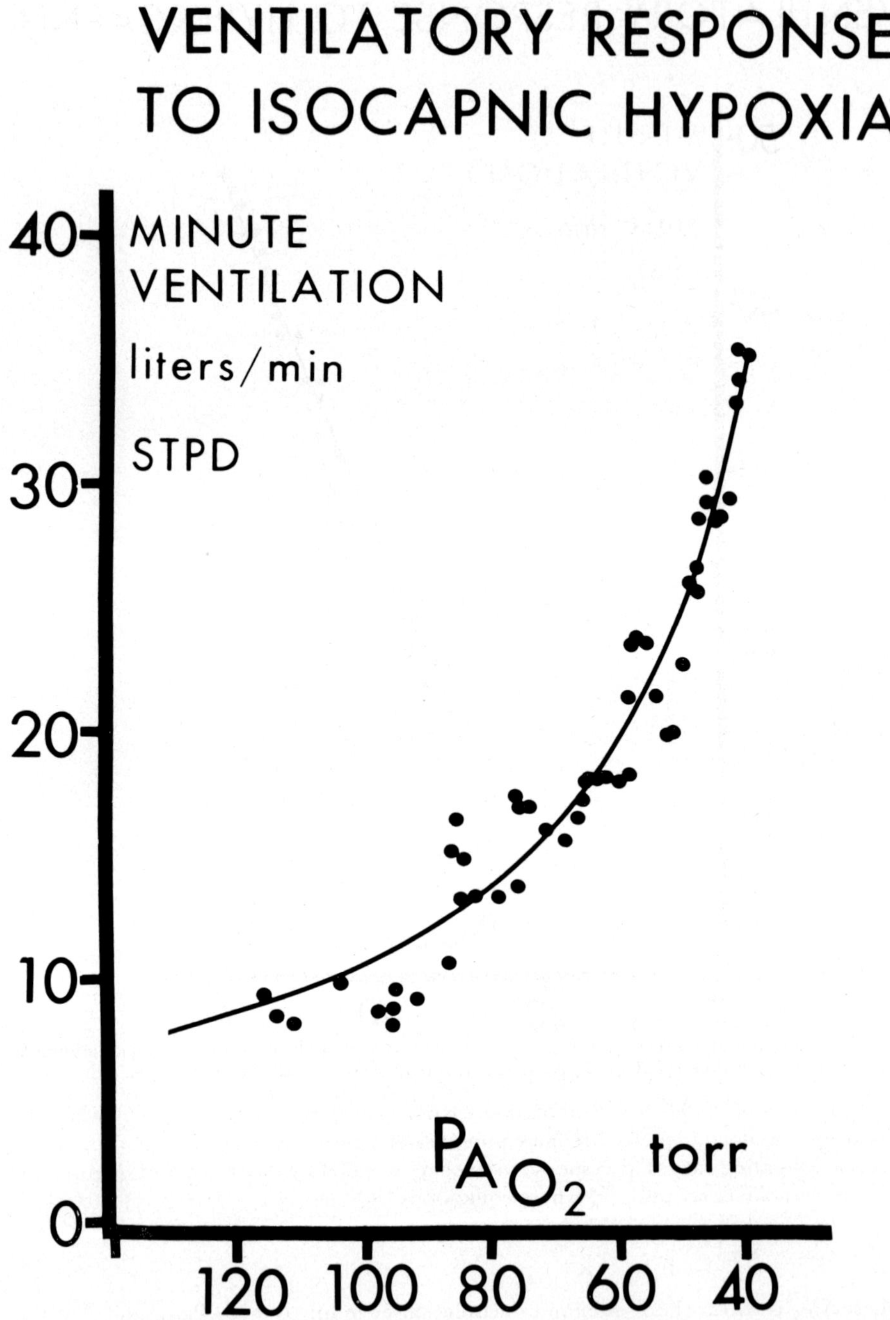

Fig. 5. In the calf, lowering the oxygen pressure in alveolar air (P_AO_2) and hence in arterial blood (PaO_2), stimulates ventilation. Minute ventilation increases progressively in a hyperbolic fashion as the hypoxic stimulus becomes more severe.

sensitivity to hypoxia such that the larger the A value, the greater the sensitivity. The constant K is characteristic of each species and represents a theoretical minimum value for Pa_{O_2}; as actual Pa_{O_2} approaches K, the rate of increase in $\dot{V}_E$ approaches infinity

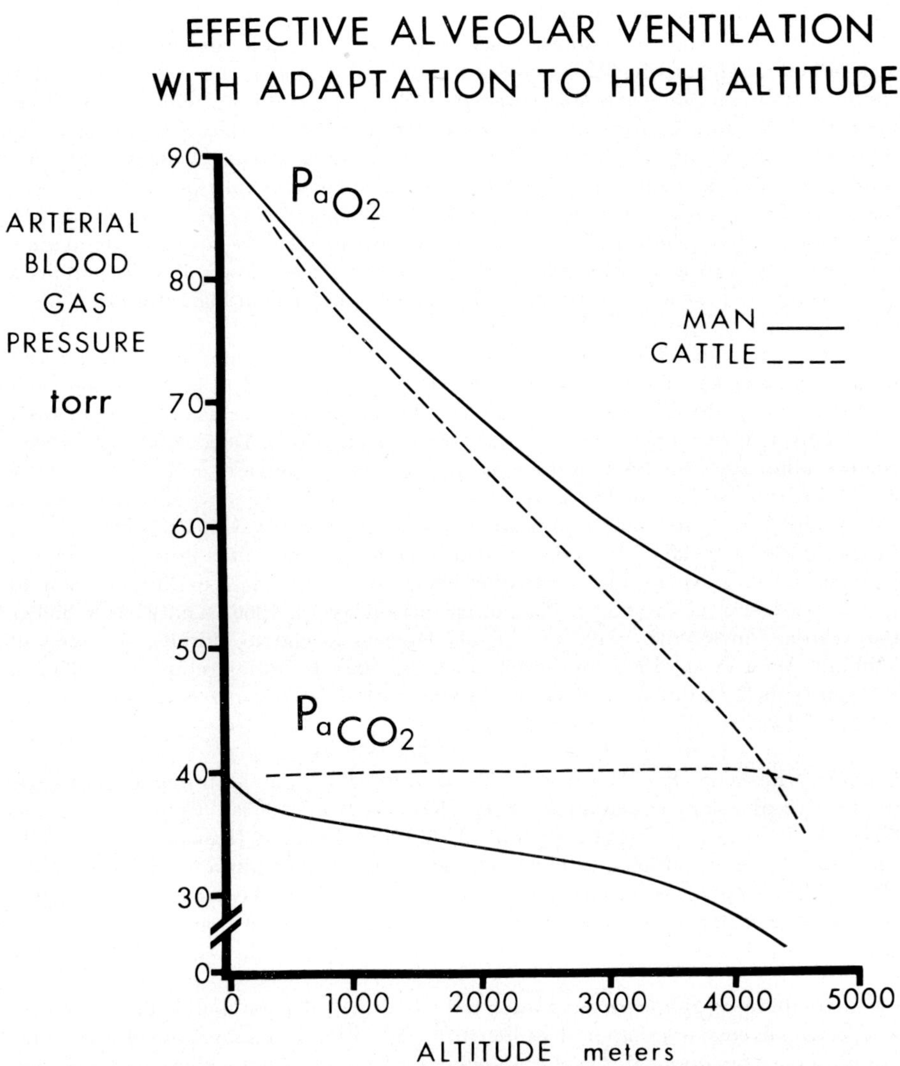

Fig. 6. With increasing altitude, the oxygen pressure of inspired air (P_IO_2) decreases as in Fig. 1. Consequently, the O_2 pressure will also fall in alveolar air (P_AO_2) and arterial blood (PaO_2). The magnitude of this fall can be lessened if effective alveolar ventilation is increased to lower arterial carbon dioxide pressure ($PaCO_2$) as in man (solid lines). Cattle (dashed lines) do not exhibit this compensatory hyperventilation, and $PaCO_2$ remains constant. Consequently, at any given altitude, cattle are more hypoxic than man.

This equation describes the ventilatory response to a selective acute lowering of Pa_{O_2} while Pa_{CO_2} is held constant (Weil et al., 1970).

When man ascends to high altitude, the decrease in total atmospheric pressure and the associated decrease in atmospheric P_{O_2} lowers P_AO_2 and hence Pa_{O_2} (Fig. 3). As Pa_{O_2} reaches 65-70 torr it stimulates the carotid bodies which in turn increase effective ventilation. Consequently, Pa_{CO_2} begins to decrease. Initially, this reduction in CO_2 stimulus attenuates the ventilatory response to hypoxia. However, with continued exposure to high altitude, adjustments occur in acid-base balance (bicarbonate excretion) to correct for the alkalosis produced by the decrease in Pa_{CO_2}. Consequently, after some days, man at high altitude exhibits sustained hyperventilation with his Pa_{CO_2} maintained at a lower level than at sea level (Kellogg, 1963). In other words, the reduction in Pa_{CO_2} indicates a sustained increase in effective alveolar ventilation. Teleologically, this is a useful adaptive mechanism, since a lower P_ACO_2 permits a higher P_AO_2 and, hence, a lessening of the degree of hypoxia at high altitude.

Knowing man's ventilatory adaptation to high altitude, it comes as a surprise that many animal species fail to show hyperventilation when taken to high altitude. This is true in spite of the fact that an hypoxic ventilatory drive can be demonstrated clearly when Pa_{O_2} is lowered acutely (Fig. 5) (Bisgard et al., 1973). Detailed studies of ventilatory adaptation to high altitude in bovine calves (Bisgard et al., 1974) have established no decrease in Pa_{CO_2} and hence no increase in effective alveolar ventilation at altitudes up to 3,850 m, in spite of marked hypoxemia with Pa_{O_2} 43 torr (Fig. 6) (Grover et al., 1963). In contrast, man adapted to this same altitude (3,800 m) increases his effective alveolar ventilation which lowers his Pa_{CO_2} to 28 torr (from 40 torr at sea level), giving him a Pa_{O_2} of 64 torr (Klausen, 1966), i.e., 21 torr higher than cattle. Put another way, man would have to ascend to altitudes in excess of 5,000 m (West et al., 1962) to become as hypoxemic as cattle are at only 3,850 m, and under such conditions man would respond with marked hyperventilation whereas cattle exhibit virtually no hyperventilation.

Chronic hypoxemia (Pa_{O_2} 42 torr) at 3,850 m altitude also fails to stimulate an increase in effective alveolar ventilation in sheep (Reeves et al., 1963a). Goats (Mines and Sorensen, 1970), swine (McMurtry et al., 1973), and the llama (Banchero et al., 1971; Brooks and Tenney, 1968) do hyperventilate somewhat at altitude (fall in Pa_{CO_2}), although the response is considerably less than in man. Only the dog exhibits a hyperventilatory adaptation to high altitude comparable to that of humans (Bouverot et al., 1973).

These comparative studies of adaptation to high altitude suggest that while an increase in effective alveolar ventilation does lessen the fall in Pa_{O_2}, this component of adaptation is apparently optional. On the Andean altiplano at altitudes above 4,000 m, one finds large herds of domestic animals including cattle, sheep, pigs and, of course, the native llama. Direct and systematic measurements of Pa_{O_2} and Pa_{CO_2} in these fully adapted herds have not been made, but there is no reason to suspect that they ventilate differently from animals taken from sea level to high altitude for study. The optional nature of hyperventilation is also indicated by the fact that humans who have lived at high altitude for 10 to 20 years lose their ventilatory response to acute hypoxia (Weil et al., 1971). Such long-term residents and natives hyperventilate less than adapted newcomers, as indicated by slightly higher levels of P_ACO_2 (Sorensen and Severinghaus, 1968a). Since 10 to 20 yrs of chronic hypoxia is required to depress hypoxic ventilatory drive (Byrne-Quiun et al., 1972), the normal life span of domestic animals is too short for such a depression to develop (Brooks and Tenney, 1968; Hornbein and Sorensen, 1969). Consequently, an animal model of this interesting human phenomenon has not been discovered.

In concluding this discussion of ventilatory adaptation to high altitude, it should be pointed out that while the peripheral chemoreceptors and the so-called hypoxic ventilatory drive are responsible for the ventilatory response to acute hypoxia, other factors seem to be of greater importance in adjusting the level of ventilation in chronic hypoxia. Thus, cattle have a clearly demonstrable increase in ventilation during acute hypoxia (Bisgard et al., 1973), and yet they fail to hyperventilate during chronic hypoxia (Bisgard et al., 1974). Conversely, among humans, high altitude natives have minimal hypoxic ventilatory drive at rest, and yet they develop chronic hyperventilation at high altitude (Pa_{CO_2} 32 torr at 4,300 m compared with Pa_{CO_2} 41 torr at sea level (Sorensen and Severinghaus, 1968a, b). Furthermore, ventilatory adaptation to high altitude occurs in goats even after surgical denervation of the peripheral chemoreceptors (Sorensen and Mines, 1970).

METABOLISM. INCREASE IN RESPIRATORY EXCHANGE RATIO (R)

The respiratory exchange ratio (R) is defined as the ratio of CO_2 production to O_2 consumption ($\dot{V}_{CO_2}/\dot{V}_{O_2}$). During normal aerobic metabolism, R is determined predominantly by the chemical structure of the materials being utilized for energy production (Hausen et al., 1972). For example, the carbohydrates have an R (RQ) of 1.00 whereas fatty acids have an R of 0.71.

As indicated earlier by the alveolar air equation, P_AO_2 is inversely related to R. This means that when CO_2 enters the alveoli until P_ACO_2 reaches a value of 40 torr (Fig. 2), the reciprocal uptake of O_2 will lower P_AO_2 by 48 torr if R is 0.8, whereas P_AO_2 will be lowered by only 40 torr if R is 1.0. In other words, with P_ACO_2 constant at 40 torr (normal for sea level), increasing R from 0.8 to 1.0 will increase P_AO_2 by 8 torr. At high altitude, this higher P_AO_2 would reduce the degree of hypoxia.

This theoretical advantage of increasing R at high altitude has actually been demonstrated in man. In three subjects exposed to a simulated altitude of 3,350 meters (P_B 502 torr), mean values for Pa_{O_2} were 41.3 torr on a low carbohydrate diet, and 48.2 torr on a high carbohydrate diet (Hansen et al., 1972). Thus, a deliberate manipulation of the basic foodstuffs in man's diet can significantly alter blood oxygenation. Little information is available on the effect of high altitude on R in animals. In 1960, we studied steers and lambs at 1,600 m and during adaptation to 3,850 m (Grover et al., 1963; Reeves et al., 1963). Even though the composition of their diet was identical at both altitudes, R increased at high altitude, from 0.87 to 0.99 in the steers, and from 0.87 to 0.98 in the lambs. Since P_ACO_2 remained unchanged, the calculated P_AO_2 at high altitude was 50 torr with R at 0.99, whereas P_AO_2 would have been only 45 torr had R remained 0.82. Therefore, it appears that in these steers and lambs, a metabolic change occurred which resulted in a higher value of R, and this in turn lessened the severity of hypoxia significantly at high altitude. The nature of this metabolic change is unknown, although the increase in R to nearly 1.00 implies a shift to almost exclusive carbohydrate metabolism. We do not know if the metabolic change (shift in R) observed in this study is typical for ruminants adapting to high altitude.

BLOOD OXYGENATION. HEMOGLOBIN BINDING OF OXYGEN, AND INCREASE IN HEMOGLOBIN CONCENTRATION

For each species, the ventilatory adaptation to high altitude determines the partial pressures of O_2 and CO_2 in the alveoli and hence in the arterial blood. However, the resulting quantity (ml) of O_2 which is taken up by the blood is dependent upon two factors; first, the O_2 carrying capacity of the blood which is determined primarily by the hemoglobin concentration (since relatively little O_2 goes into physical solution in plasma), and second, the affinity of hemoglobin for O_2. The quantitative relationship between O_2 pressure and the O_2 saturation of hemoglobin (Hb-O_2) is described by the Hb-O_2 dissociation curve. By virtue of the sigmoid shape of this curve (Fig. 7), Pa_{O_2} can decrease from 95 torr to 50 torr and the Hb-O_2 saturation (Sa_{O_2}) will remain above 85%; this relationship favors the loading of blood with O_2 in the lungs. As PO_2 decreases further, SO_2 begins to decrease more rapidly; this steeper portion of the dissociation curve favors the unloading of O_2 to the peripheral tissues. These relationships are very favorable for the transport of O_2 near sea level where Pa_{O_2} is relatively high. Under resting conditions, approximately 25% of the O_2 in arterial blood is delivered to the body without lowering venous blood P_{O_2} below 40 torr, thus insuring adequate tissue oxygenation and avoiding tissue hypoxia.

At high altitude, the decrease in atmospheric pressure and P_IO_2 taxes this system. Man compensates by hyperventilating so that at 4,000 m, he achieves an Sa_{O_2} above 85%. Cattle do not hyperventilate, and as a result Pa_{O_2} is lowered to 45 torr at 3,850 m, but the corresponding Sa_{O_2} is still above 80% (Grover et al., 1963). This illustrates that the lack of adaptive hyperventilation at high altitude does not constitute a serious handicap until the individual reaches relatively extreme altitudes.

This conclusion is valid for those species having hemoglobin with an affinity for O_2 similar to the human. However, certain species of sheep have hemoglobin Type B with a low affinity for O_2 (Dawson and Evans, 1962). One would consider this a distinct disadvantage to the loading of O_2 at high altitude, especially when combined with the sheep's known lack of adaptive hyperventilation. Just as in cattle, sheep at 3,850 m also have a Pa_{O_2} of 45 torr but the corresponding Sa_{O_2} is only 65% (Reeves et al., 1963a). In view of this, it is remarkable but true that sheep continue to thrive under these conditions.

While sheep are at the low affinity end of the hemoglobin spectrum, the llama is clearly at the high end of the spectrum. Llamas adapted to 3,400 m hyperventilate somewhat but still have a Pa_{O_2} of only 51 torr. However, the corresponding Sa_{O_2} is 91% due to the high Hb-O_2 affinity (Banchero et al., 1971). This contrast between llama and sheep in terms of Hb-O_2 affinity was illustrated dramatically when both species were

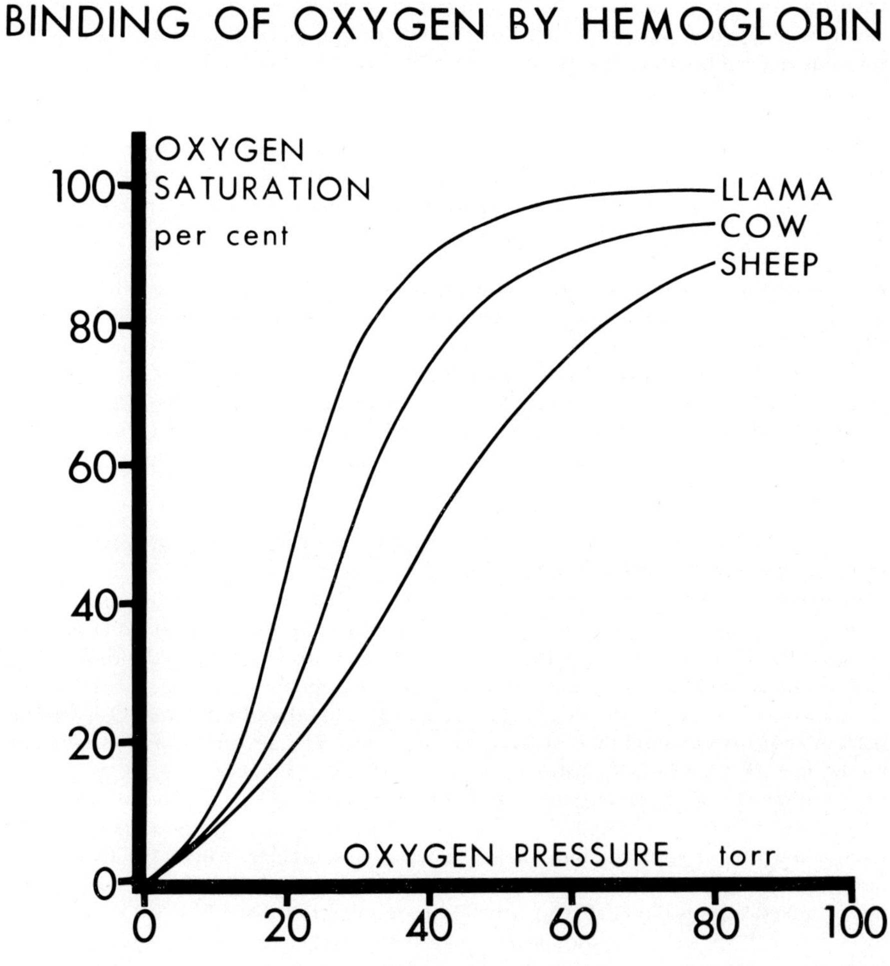

Fig. 7. Within the red blood cell, oxygen (O_2) is bound to hemoglobin. The affinity of hemoglobin for O_2 varies markedly among animal species. Consequently, at any given O_2 pressure, the degree of hemoglobin saturation is much less in sheep than in the cow, and in both it is less than in the llama. A low affinity of hemoglobin for O_2 impairs O_2 uptake in the lung but favors O_2 release to the peripheral tissues.

exposed to a simulated altitude of 6,800 m (P_B 335 torr) in a hypobaric chamber. The llamas were quite content with an Sa_{O_2} of 71%, whereas the sheep lost consciousness with an Sa_{O_2} of only 33% (Banchero et al., 1971). However, the role of Hb-O_2 affinity in adaptation to high altitude should not be overemphasized when one recalls that llamas and sheep graze together at 4,300 m on the Andean altiplano.

Once Sa_{O_2} has been determined by Pa_{O_2} and the Hb-O_2 affinity, then the actual quantity of O_2 in the arterial blood (arterial O_2 content, Ca_{O_2}, in ml O_2 per 100 ml blood) will depend upon the hemoglobin concentration (g Hb per 100 ml blood) which determines the O_2 carrying capacity of that blood; 1.0 g Hb binds 1.36 ml O_2. Thus, if an animal's blood contains 10.0 g Hb per 100 ml, then if Sa_{O_2} is 96%, Ca_{O_2} will be 13.0 ml O_2/100 ml; if Sa_{O_2} is reduced to 81%, Ca_{O_2} will be decreased to 11.0 ml O_2/100 ml. To counteract the effect of a reduction in Sa_{O_2}, many species respond with an increase in red cell production which increases the Hb concentration. If, in response to an Sa_{O_2} of 81%, Hb concentration increases to 16.0 g per 100 ml, then Ca_{O_2} will be restored to 13.0 ml O_2/100 ml.

This polycythemic response to high altitude is fully developed in man so that at 4,300 m, Ca_{O_2} is as great as at sea level even though Sa_{O_2} has decreased from 95% to 85% (Banchero et al., 1966). Polycythemia sufficient to compensate for arterial desaturation is also observed in cattle, swine, and dogs. Only sheep seem to have an inadequate polycythemic response to altitude. When adapting to 3,850 m from 1,600 m, the increase in Hb concentration from 11.5 to 13.5 g/100 ml does not compensate for the decrease in Sa_{O_2} from 85% to 60% with the result that Ca_{O_2} is reduced from 13.3 to 11.0 ml O_2/100 ml (Reeves et al., 1963a).

SYSTEMIC OXYGEN TRANSPORT. CIRCULATION OF OXYGENATED BLOOD BY THE HEART

The amount of O_2 carried to the peripheral tissues each minute is determined by the quantity of O_2 in each unit of blood, Ca_{O_2}, and the number of such units of blood pumped by the heart each minute, i.e., the cardiac output ($\dot{Q}$). In adapting to high altitude, most animals including man preserve Ca_{O_2} remarkably well. Consequently, O_2 transport will be determined primarily by $\dot{Q}$. The absolute level of $\dot{Q}$ is less important than its value relative to tissue O_2 demands. The adequacy of $\dot{Q}$ is indicated by two factors. First is the quantity of O_2 extracted from each unit of blood, as reflected by the quantity of O_2 remaining in the venous blood (C_VO_2), i.e., the arteriovenous (a-v) difference in O_2 content. Upon ascent from low to high altitude, if the a-v difference narrows (less extraction) then $\dot{Q}$ is relatively greater, whereas if the a-v difference widens (greater extraction) then $\dot{Q}$ is relatively less. The second factor is the P_{O_2} of mixed venous blood (P_VO_2) which is an indicator of tissue P_{O_2}. If $\dot{Q}$ is adequate to maintain P_VO_2 at prealtitude levels, then tissue oxygenation is probably adequate. However, if P_VO_2 decreases, this may indicate tissue hypoxia.

In man from sea level, adapted to 3,100 m altitude and studied both at rest and during exercise, a-v difference widened significantly indicating a decrease in $\dot{Q}$ under all conditions (Alexander et al., 1967). However, the reduced $\dot{Q}$ must be considered "adequate" since there was no decrease in P_VO_2. These changes in O_2 transport resulted in part from a decrease in Hb-O_2 affinity (Grover et al., 1970).

Cattle from 1,500 m taken to 3,000 m for 5 months had values of a-v difference not different from control animals remaining at low altitude, i.e., no change in $\dot{Q}$ (Will et al., 1962). Another group of cattle exposed to more severe hypoxia for two months at 3,850 m developed a significant narrowing of a-v difference indicating that $\dot{Q}$ was relatively greater at high altitude. In spite of this, P_VO_2 which was 40 $\pm$ 6 torr at

1,600 m decreased significantly to 28 ± 2 torr implying tissue hypoxia and, hence, an "inadequate" $\dot{Q}$ at high altitude (Grover et al., 1963). This interpretation is supported by the poor weight gain which these animals exhibited at altitude.

Lambs 3-5 months old taken from 1,600 m to 3,850 m had no significant change in a-v difference indicating no change in $\dot{Q}$. However, as in the cattle, P_VO_2 was lowered from 42 ± 7 torr to 29 ± torr (Reeves et al., 1963a).

Three llamas taken from sea level to 3,400 m showed no change in either a-v difference of P_VO_2, indicating no change in $\dot{Q}$ which remained "adequate" (Banchero et al., 1971). It is noteworthy that because the llama has a high Hb-O_2 affinity, the unloading of O_2 must occur at lower values of P_{O_2} than in other species. Consequently, P_VO_2 for the llama is 30 torr, compared with 40 torr in man, cattle, sheep and dogs.

Four dogs taken from 1,500 m to 3,400 m had no significant change in $\dot{Q}$ at rest or during exercise after three weeks at high altitude. Resting P_VO_2 of 40 ± 6 torr at 1,500 m did not change at high altitude (Vogel and Hannon, 1966).

From these limited data, it appears that with adaptation to moderately high altitude, systemic O_2 transport is well preserved in the dog, llama, cattle, and man, but less so in sheep. However, the decrease in P_VO_2 in cattle and sheep suggests that in these two species, tissue O_2 requirements may not be met adequately at high altitude. This interpretation seems justified in cattle, since they gain weight more slowly and cannot build fat stores at high altitude. On the other hand, growing lambs do very well at high altitude in spite of both a decreased systemic O_2 transport and a decreased P_VO_2.

TISSUE OXYGENATION. INCREASE IN CAPILLARY DENSITY

It is pressure which supplies the energy for O_2 to diffuse from one region to another. In the lung, P_{O_2} is greater in the alveolar air than in the pulmonary capillary blood, and hence O_2 diffuses from air to blood. When that oxygenated blood reaches the systemic capillaries, the blood P_{O_2} exceeds the P_{O_2} of the surrounding tissues, and with this pressure difference, O_2 diffuses out of the blood into the tissue. The magnitude of this pressure difference determines the distance O_2 molecules will diffuse, and thereby limit how far a cell can be from a capillary. At high altitude with the inescapable fall in Pa_{O_2}, the length of this diffusion pathway can become critical for tissue oxygenation. It follows that if the density of capillaries were increased, the diffusion pathway would be shortened. Unfortunately, it is technically difficult to determine tissue capillary density, and consequently there is little information on this potentially useful adaptive mechanism. Valdivia (1958) has demonstrated an increase in capillary density of skeletal muscle in guinea pigs from high altitude compared with sea level controls. Whether or not this occurs in other animals adapted to high altitude remains to be determined.

Other potentially adaptive mechanisms include a decrease in Hb-O_2 affinity augmenting the release of O_2 (Grover et al., 1970), and an increase in cellular mitochondria to facilitate O_2 utilization.

MALADAPTATION TO HIGH ALTITUDE

When a large group of individuals of a given species native to sea level moves to high altitude, most of those individuals will eventually adapt satisfactorily. However, the acute response to high altitude may be quite unpleasant. Most humans taken rapidly to above 4,000 meters suffer "acute mountain sickness" with symptoms of malaise, headache, drowsiness, anorexia, nausea, and vomiting (Roy and Singh, 1963). These symptoms are usually most severe during the first night at altitude, and then subside rapidly. Whether or not animals experience "acute mountain sickness" cannot be known, although dogs have been noted to decrease their food intake during their first days at high altitude (Vogel and Hannon, 1966).

Rarely, humans develop acute pulmonary edema within 72 hrs of ascent to high altitude (Singh and Roy, 1969), which can be fatal if not treated promptly. The pathogenesis of this condition is unknown. It has never been reported in any species other than man, and this absence of an animal model makes investigation of this important condition difficult.

Another form of human maladaptation to high altitude is known as "chronic mountain sickness" or "Monge's Disease" or "chronic soroche" (Penaloza and Sime, 1971). It develops in persons who have been living at high altitude for many years, and is generally thought to indicate a loss of adaptation. Specifically, the individual seems to have complete loss of hypoxic ventilatory drive (Fig. 5) (Sorensen and Severinghaus, 1968a). The hallmark of this condition is extreme polycythemia as a consequence of marked hypoventilation with severe hypoxemia and C_{O_2} retention. A similar condition has been created in dogs in which hypoxic ventilatory drive has been abolished by surgical removal of the carotid bodies (Gilfillan et al., 1967). Although extreme polycythemia rarely occurs spontaneously in animals at high altitude, it has been reported in lambs (Cuba, 1956), pigs (McMurtry et al., 1973), rabbits (Reeves et al., 1963b), and young cats (Weidman et al., 1965).

HYPOXIC PULMONARY HYPERTENSION

Pulmonary hypertension develops at high altitude in most animal species including man. When the elevation in pulmonary arterial pressure is only mild to moderate, the condition is benign. However, in some individuals, pulmonary hypertension may be of such severity as to cause right heart failure, and this certainly constitutes maladaptation.

It is well established that when P_AO_2 is lowered acutely, pulmonary arterial pressure rises. This is due primarily to a constriction of small muscular pulmonary arteries, although the mechanism which links airway hypoxia to contraction of pulmonary vascular smooth muscle remains elusive. Nevertheless, this vasoconstriction increases the resistance to blood flow through the lung. To overcome this resistance, the right ventricle generates a higher pressure, i.e., pulmonary hypertension.

When the airway hypoxia persists for weeks, months, and years, as at high altitude, the pulmonary vascular resistance usually remains increased producing chronic pulmonary hypertension. This is generally considered to be a perpetuation of the response observed during acute hypoxia. However, it is noteworthy that the dog shows a brisk response to acute hypoxia and yet does not develop pulmonary hypertension during chronic hypoxia (Tucker et al., 1975).

There is marked variability among species in the degree of pulmonary hypertension which develops at high altitude. As stated, the response is minimal in the dog and also in sheep. A moderate response is observed in man, goats, llamas, rats, rabbits, and guinea pigs. Cattle and swine complete the spectrum, both developing severe pulmonary hypertension. Although the basis for this variability among species is not known, recent studies (Tucker et al., 1975) have demonstrated a strong correlation between the normal thickness of pulmonary vascular smooth muscle and the severity of pulmonary hypertension which develops under the stimulus of chronic hypoxia. In other words, the response of a given species is dependent upon the structure of its pulmonary vasomotor apparatus.

In addition to inter-species variability, there are also impressive differences in response among individuals of the same species. This has been demonstrated repeatedly when groups of low altitude cattle are taken to high altitude. In response to the same hypoxic stimulus, some individuals develop only a modest elevation in pulmonary arterial pressure while concurrently other individuals develop severe pulmonary hypertension (Grover et al., 1963; Will et al., 1962). Again, the basis for this individual variability is not known. Perhaps this, too, reflects differences in the original degree of muscularization of the pulmonary arterial tree. It is true that following chronic hypoxia, steers with the highest pulmonary arterial pressures have the greatest medial thickness of their small muscular pulmonary arteries (Alexander and Jensen, 1963; Grover and Reeves, 1963).

Whatever the basis for this individual variability, it appears to be inherited. Cattle at high altitude can be separated into two groups, one susceptible to severe pulmonary hypertension and the other resistant to it. When animals within each group are bred, the offspring of susceptible stock are also susceptible, and the offspring of resistant stock are resistant. This genetic trait has now been demonstrated twice in second generation offspring of the original two groups (Weir et al., 1974).

Does pulmonary hypertension at high altitude have any adaptive value? While this is a teleological question, it is reasonable to investigate. In animals with a large thorax, such as cattle, gravitational forces acting over the vertical distance from sternum to spine create significant hydrostatic pressures within the pulmonary circulation. This is also true in erect man, although here the hydrostatic pressure increases from apex to base. As a consequence, the human lung is not perfused uniformly, regional flow being greater at the bottom of the lung than at the top. Since ventilation is distributed more uniformly, regional ventilation ($\dot{V}_A$) is not well matched to regional perfusion ($\dot{Q}$), $\dot{V}_A/\dot{Q}$ is greatest at the apex and decreases toward the base of the lung. As a consequence, the efficiency of pulmonary gas exchange is reduced, and this accounts for the fact that the P_{O_2} of arterial blood is less than the P_{O_2} of alveolar air (Fig. 2).

If the pulmonary arterial pressure were increased, this would counteract the hydrostatic pressure, and the lung would be perfused more uniformly. This, in turn, would increase the efficiency of pulmonary gas exchange, which would be desirable when P_AO_2 is reduced as at high altitude. In men living at high altitude where pulmonary hypertension is usual, lung perfusion is indeed more uniform than at sea level (Dawson and Grover, 1974). Based on these observations, it is tempting to conclude that this is the "reason" for pulmonary hypertension at high altitude. However, similar studies in cattle indicate that even at sea level, dorso-ventral perfusion is surprisingly uniform (Ruiz et al., 1974), so greater uniformity would not result from increasing pulmonary arterial perfusion pressure. In fact, severe pulmonary hypertension in cattle probably renders regional $\dot{V}_A/\dot{Q}$ less uniform and thus interferes with pulmonary gas exchange. Thus, if pulmonary hypertension has any adaptive "value" at high altitude, it is obscure.

"BRISKET DISEASE" OF CATTLE

The right ventricle must generate an increased pressure to maintain blood flow through the constricted pulmonary vascular bed at high altitude. In response to this increased work load, the right ventricle hypertrophys. Anatomical studies have demonstrated that the increased weight of the right ventricle is directly proportional to the increase in pressure which it develops in the pulmonary artery (Grover and Reeves, 1963). However, the capacity of the right ventricle for work hypertrophy is not unlimited. Consequently, when chronic hypoxia causes a relentless progressive increase in pulmonary vascular resistance, this eventually overloads the right ventricle and heart failure develops. As part of the picture of heart failure, there is fluid retention which accumulates in the thoracic and abdominal cavities (ascites), and in the loose tissues along the dependent portions of the trunk. This subcutaneous edema is particularly prominent in the intermandibular region and beneath the sternum. Since the parasternal muscles are the brisket of beef, the appearance of edematous swelling in this region is called "brisket disease". Actually, then, "brisket disease" is a sign of congestive heart failure, and when due to pulmonary hypertension, it is an example of cor pulmonale. At high altitude, the etiology is atmospheric hypoxia. However, "brisket disease" has also been reported at low altitude in cattle with chronic pneumonia (Spörri and Schlatter, 1959), but in this condition there is also airway hypoxia which again is probably the cause of the pulmonary hypertension.

As mentioned earlier, when cattle are taken to high altitude, some individuals prove to be susceptible to pulmonary hypertension while others are resistant. Obviously it is in the susceptible individuals that pulmonary hypertension may progress to such severity that "brisket disease" develops. Among herds of cattle native to sea level and introduced into high altitude areas, the incidence of "brisket disease" may be 20-30%. However, among herds established at high altitude, the incidence is only 1-2%. We suspect that this lower incidence results from a process of natural selection. "Brisket disease" is fatal unless the animal is removed to a lower altitude. Consequently, individuals susceptible to severe pulmonary hypertension do not reproduce at high altitude and so the genetic trait is not transmitted (Weir et al., 1974). Conversely, resistant animals do reproduce, and hence the resistant trait becomes predominant. Assuming that the susceptible trait is genetically recessive, this would tend to account for the fact that the herds of cattle established above 4,000 m in the Andes are mostly cross-bred, i.e., the in-breeding inherent in pure-bred stock is avoided.

A somewhat analogous process of natural selection may operate in human populations, although the determining factor is certainly not heart failure since the pulmonary hypertension which develops in man at high altitude is less severe than in cattle. Within the human population of Leadville, Colorado, at 3,100 m, there is marked individual variability in the degree of elevation of pulmonary arterial pressure. As a consequence, the mean pressure for that population is much higher than in the Andean populations at the same (or even higher) altitudes (Grover, 1968). Also, in the latter populations there is much less individual variability. Since Leadville was settled only 100 years ago, the current residents are relative newcomers who have been at high altitude less than four generations. It is tempting to postulate that this population still exhibits the full spectrum of pulmonary hypertensive response to altitude, from resistant to susceptible. In contrast, susceptible individuals apparently have disappeared from the Andean populations leaving a predominantly resistant group, i.e., persons with lower pulmonary arterial pressures. This hypothesis implies that moderate pulmonary hypertension in man contributes in some way (other than frank heart failure) to altitude intolerance, and such individuals elect to return to lower altitudes.

Is "brisket disease" seen exclusively in cattle? Theoretically, individuals of any species could develop high altitude pulmonary hypertension of such severity as to precipitate right heart failure. In the Andes, among lambs born above 4,000 m, neonatal mortality of 25% has been reported (Cuba, 1956), and the cause of death is said to be heart failure. This appears to be analogous to "brisket disease" in cattle, although if so, it is particularly remarkable in view of the fact that three month old lambs born at low altitude develop no pulmonary hypertension when taken to high altitude (Reeves et al., 1963a). Swine taken from low to high altitude develop pulmonary hypertension as severe as that seen in cattle (McMurtry et al., 1973). Since swine are abundant on the altiplano above 4,000 m, one would expect that some individuals would die of heart failure due to pulmonary hypertension, and yet this has not been reported. Only cattle show the dramatic adverse reaction to high altitude known as "brisket disease".

REFERENCES: see pp. 275-277.

Altitude and Exercise Effects on Animals

by

P. D. Altland

INTRODUCTION

Exposure of animals and man to a natural high altitude environment usually involves altitude (hypoxia), cold and exercise stresses. In addition, emotional stress may develop if the individual has not previously experienced these physiological stresses. The response of the individual is largely determined by degree of acclimatization. In expeditions to the high Andes or the Himalayas for example, the scientists are usually acclimatized to exercise and possibly cold and the mountaineer guides and the beast of burden to altitude, exercise and cold. Many of the altitude experiments included in standard reference books such as "Hypoxia" (Van Liere and Stickney, 1964) and "Handbook of Physiology, Adaptation to the Environment" (Dill, 1964), were conducted at room temperature with animals at rest. Recently Fregly (1970) and Hale and Mefford (1970) have reviewed the current status of cross adaptation between stresses and pointed out that our knowledge of this subject is fragmentary and that no generalizations are possible. In this Section emphasis is placed upon the interaction between exercise and altitude both in the laboratory and in the natural environment.

PHYSIOLOGICAL CHANGES DURING EXERCISE AT ALTITUDE IN ANIMALS AND MAN

In order to meet the metabolic demands of exercise at altitude, the oxygen uptake is increased. This is accomplished by increasing ventilation and augmenting blood transport of gases. There is an increase in cardiac output and widening of the arterio-venous difference in oxygen concentration. During an expedition to Mount Everest (Pugh, 1958) it was found that ventilation and oxygen consumption were related to intensity of the work performed and were largely independent of altitude up to 20,500 ft (6,248 m). In studies with dogs (Piiper et al., 1966) and man (Balke, 1960) maximum breathing capacity and oxygen uptake at a work load below the sea level maximum were reached at altitudes 14,000 ft (4,267 m) and 16,713 ft (5,094 m), respectively. Circulatory and biochemical factors were also below their normal maxima at these altitudes. Due to such divergent results further study of this subject seems warranted.

Acclimatization to high altitude tends to restore normal work capacity by increasing the oxygen carrying capacity of the blood by elevating the hemoglobin concentration. This makes it possible for oxygen to be transported and delivered to the tissues with smaller changes in blood concentration. After the complex physiological and biochemical changes associated with altitude acclimatization take place, man is able to live and work normally at altitudes between 14,000 and 17,000 ft (4,267-5,182 m) where even inactive newcomers often remain ill for days (Balke, 1960).

ACTIVITY AND ALTITUDE TOLERANCE OF AMPHIBIANS, REPTILES AND BIRDS

Data on the activity of vertebrates other than mammals at altitude is relatively sparce. Frogs, toads and salamanders have been observed at altitudes up to 9,843 ft (3,000 m) in the Rocky Mountains. In the Himalayan Mountains the presence of 11 species of frogs above 12,140 ft (3,700 m), 5 species of lizards and 4 species of snakes above 11,812 ft (3,600 m) and 1 species of snake at 18,045 ft (5,500 m) has been recorded. The common European viper has been reported to live as high as 9,187 ft (2,800 m) in the Alps. Hypoxia does not appear to restrict the distribution of amphibians and reptiles, but availability of food and water is a serious limiting factor, as is sunshine (Hoch, 1964). Many more data on the distribution, habits and activity of amphibians and reptilians at high altitude are needed.

Little appears to be known about the altitudes at which birds fly throughout the world. Some of the earliest records on the capacity of birds to fly at high altitudes were made by Bert (1943). He reported that condors and vultures fly at 22,967 ft (7,000 m) without any evidence of distress. Gunacos, swans, turtle doves, sparrows and even humming birds are commonly found at 13,124 ft (4,000 m) in the Andes of Chile. The Bolivian goose or hyallata is found in the Andean highlands at about 11,812 ft (3,600 m). The geese have blood with significantly greater affinity for oxygen than members of their class at sea level (Hall et al., 1936).

In laboratory experiments conducted at 23° C sparrows, hawks, pigeons, and gulls became ill at altitudes between 20 and 30,000 ft (6,096-9,144 m) and showed signs of impending death at 34-36,000 ft (10,363-10,973 m) (Bert, 1943). If average resistance to decompression is considered, birds of prey appear to be almost as sensitive as other birds with the exception of the chicken. Adult chickens survive acute exposure to 16,000 ft (4,877 m) simulated altitude, but above this level their tolerance is greatly decreased until none survive at 26,000 ft (7,925 m). Rate of ascent was maintained at 2,000 ft/min (Altland, 1961). Roosters show a greater altitude tolerance than hens, a sex difference contrary to that found in mammals. Day-old chicks housed at 90-92° F (abt 33° C) withstand exposure to 20,000 ft (6,096 m) for 16 to 20 hrs, but all die at 25,000 ft (7,620 m). The rate of climb ranged between 1,000 and 5,000 ft/min (305-1,524 m) (Fraps and Olsen, 1950).

ACTIVITY AND ALTITUDE TOLERANCE OF MAMMALS

The burro is widely used as a beast of burden in mountain climbing expeditions. In the Andes the llama, and in the Himalayas the yak, are used to carry burdens at least as high as 17,389 ft (5,300 m). In the Pariacaca range in Peru (peaks around 20,000 ft, 6,096 m) it has been reported that the air is so thin that both man and animals appear anguished and sometimes the animals get so out of breath that no spurring is sufficient to budge them (Monge, 1948).

In various mountainous areas of the world it has been the practice for hundreds of years to graze livestock during the warmer periods of the year at altitude of 8,000 to 12,000 ft (2,500-3,700 m). At these altitudes many cattle have difficulty in breeding and are often afflicted with Brisket's Disease, which, in its classic form, causes intense dyspnea and leaves the afflicted animals unable to tolerate even mild exertion (Hecht et al., 1962).

The vicuña, native to the highlands of central South America, is able to maintain a speed of at least 30 miles per hour at an altitude of 15,142 ft (4,615 m) for considerable periods of time. The viscacha, a rodent of the chinchilla family, known to live at an altitude of about 12,140 ft (3,700 m), is highly active in its natural environment. Viscacha blood has a greater affinity for oxygen than that of the domestic rabbit (a close relative). The hemoglobin of the vicuña also has a high affinity for oxygen. At 9,220 ft (2,810 m), for example, it has an arterial oxygen saturation of 97.2% (Hall et al., 1936) compared with around 80% for unacclimated rats at that level (Altland et al., 1967).

Denzer (1950) reported a marked decrease in locomotion in mice exposed to 32,816-36,344 ft (10,092-11,077 m) compared with 9,843 to 13,124 ft (3,000-4,000 m). He also found that the most intense muscular work (walking in a rotating drum) did not affect onset of convulsions up to an altitude of 19,686 ft (6,000 m). Apparently the low oxygen supply at this level is sufficient to maintain oxidation, but at higher altitudes the mice fail to survive.

Studies on effects of forced exercise (walking in a revolving drum 6.9 m/min) on male rats at simulated altitude showed that when 2 min of rest was provided following each minute of exercise all survived for 2 hours at 24,000 ft (7,315 m), but at higher altitudes survival was greatly reduced (Fig. 1) (Altland and Highman, 1963). When female rats were required to exercise at altitude they all survived exposure to 28,000 ft (8,534 m) (Fig. 2). This is the only case recorded where female animals have a greater capacity than males to exercise at altitude. Both male and female rats showed a higher percentage survival at altitude when not exercised (Figs. 1 and 2).

ACCLIMATIZATION TO MODERATE ALTITUDE PRIOR TO EXERCISE AT GREATER ALTITUDES

As noted earlier in this Section ability to work at high altitude is reduced by the debilitating effects of hypoxia. Work performance and survival can be improved by acclimatization to moderate altitude before exposure to very high altitudes. Balke et al. (1965) reported that physical training of men at an altitude of 6,563 ft (2,000 m) can restore normal capacity for maximum aerobic work. He suggested that training at elevations higher than 6,563 ft may be even more effective. In unpublished studies with rats (Altland) exposure to 20,000 ft (6,096 m) for 6 hrs daily without exercise for various intervals from 1 to 30 days increased their capacity to exercise and survive

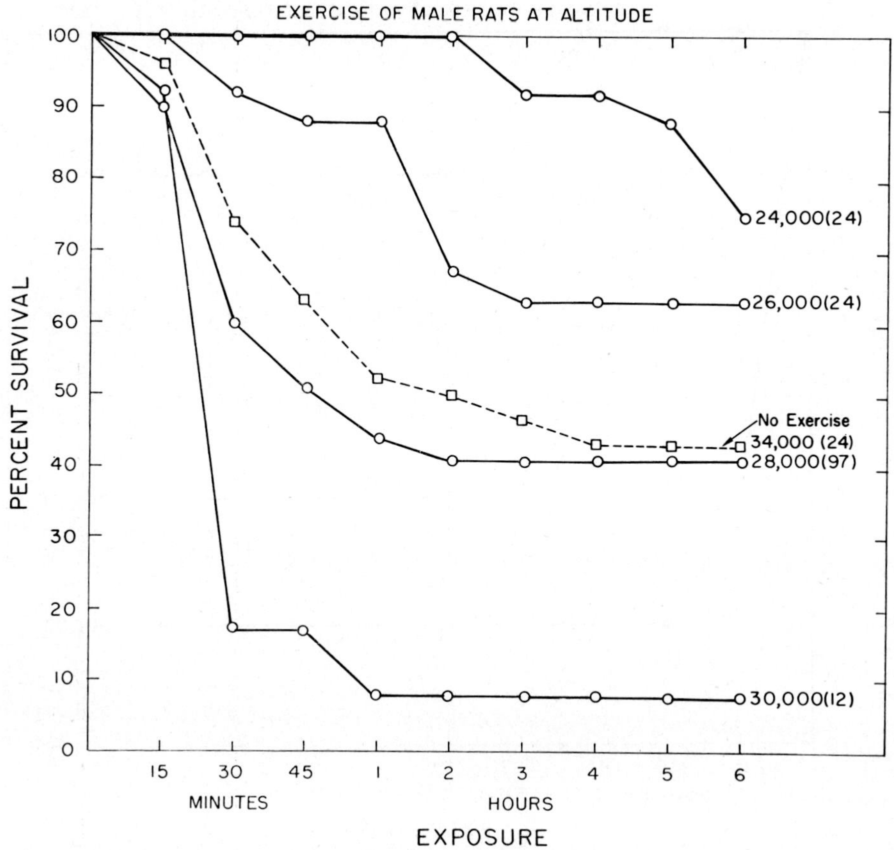

Fig. 1. Survival of young adult male rats exercised at 24,000 to 30,000 ft simulated altitude for 6 hrs. Parenthetical value is number of rats used.

during a subsequent 6-hour exposure to 28,000 ft (8,534 m) (Fig. 3). It may be noted that the greatest increase in tolerance occurred following 3 and 7 days of exposure to 20,000 ft (6,096 m). Prolongation of acclimatization up to 30 days did not prove as effective as the shorter periods.

EFFECTS OF PHYSICAL TRAINING AT SEA LEVEL ON ALTITUDE EXERCISE TOLERANCE

Denzer (1950) conducted experiments designed to determine whether or not mice accustomed to walking in a revolving drum at sea level were able to perform this work at high altitudes without experiencing any effect on their altitude tolerance. Following 4 weeks of physical training the mice were exposed to simulated altitude at a climb

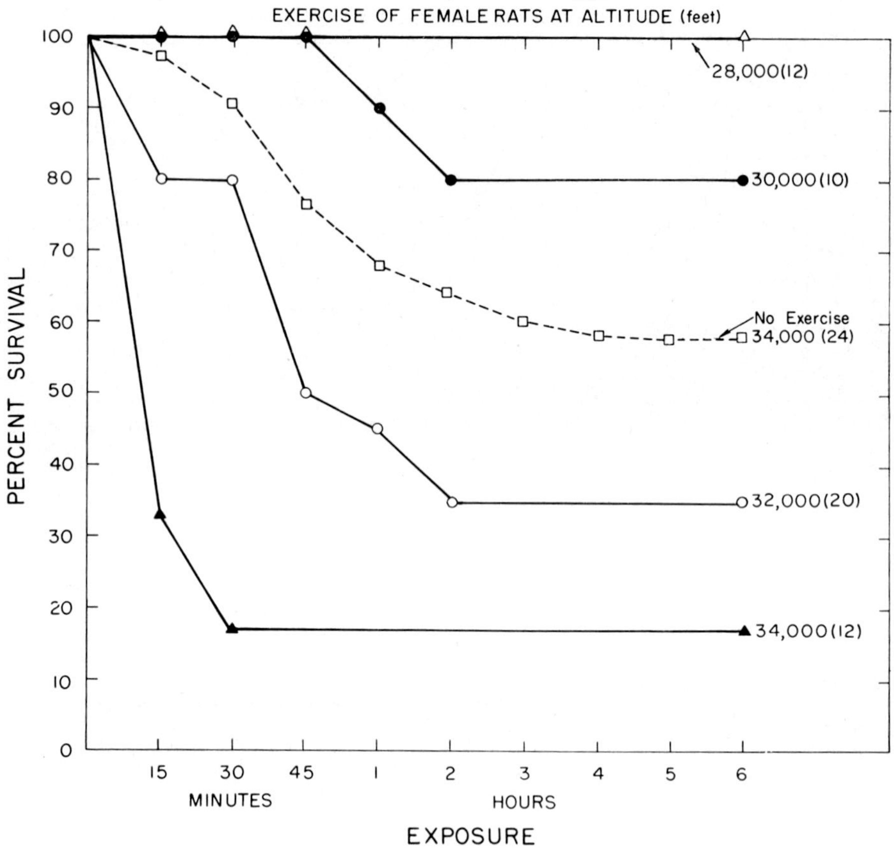

Fig. 2. Survival of young adult female rats at various simulated altitudes. Parenthetical value is number of rats used.

rate of 1,650 ft/min (503 m/min). They were then required to exercise at various simulated altitudes between 23,000 and 32,800 ft (7,010-10,000 m). Thereafter they remained at rest as the ascent continued and a convulsion threshold was reached at 36,927 ft (11,255 m). In contrast to the untrained animals, their altitude tolerance had improved by about 3,300 ft (1,083 m) under otherwise similar test conditions. This improvement was said to be due to unknown physiological-chemical processes which permit the organism to utilize the oxygen supply more economically. In studies in our laboratory designed to determine the effects of different degrees of physical training at sea level on tolerance to exercise at high altitude we have shown that 1 and 3 days of exercise training had no significant effect upon the ability of rats to exercise at 28,000 ft (8,534 m) (Fig. 4), but the gain in survival of rats trained for 7 days and from 17-19 days was significant (Altland et al., 1964).

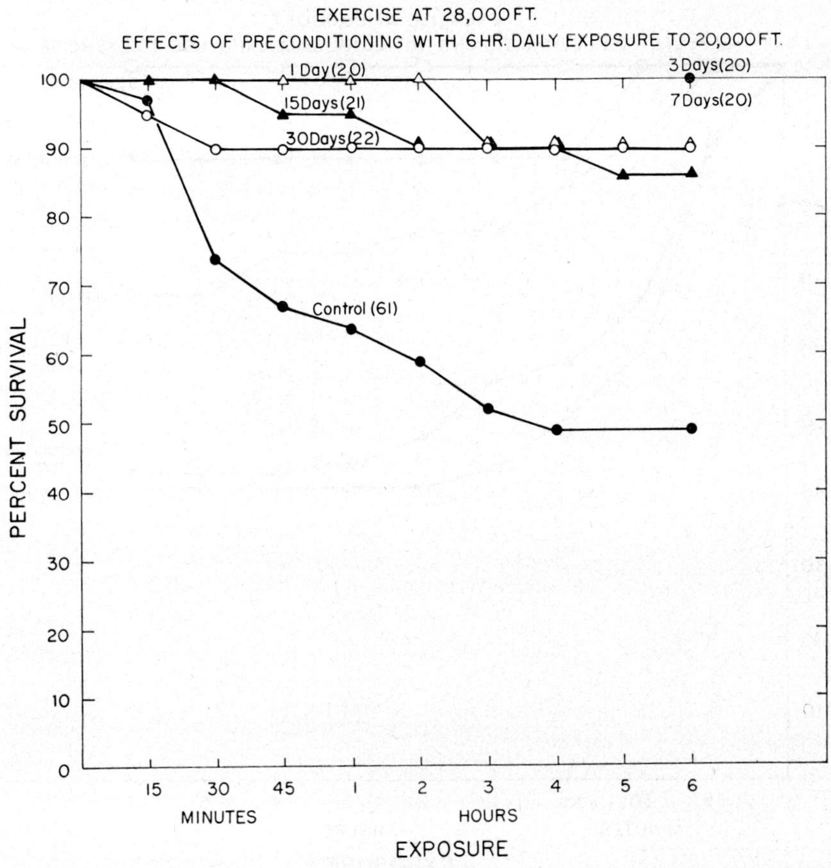

Fig. 3. Effects of altitude acclimatization at 20,000 ft on survival of young adult male rats during exercise at 28,000 ft Parenthetical value ts number of rats used.

During exercise training of rats there is a reduction in body weight, hematocrit, and serum alkaline phosphatase and elevation in serum glutamic oxalacetic transaminase, lactic dehydrogenase, and aldolase. Also there is diurnal leucocytosis with neutrophilia and lymphopenia, and necrosis of some muscle fibers. After 7 days of training these changes are less marked, but a period of from 15 to 19 days is needed to reach an advanced stage of physical training, as evidenced by the reversal of most of the above alterations. In fully trained rats there is only a small increase in serum enzymes during an exercise run at 28,000 ft (8,534 m). It has been postulated that the rise in serum enzymes after exercise is due to an increased cellular permeability permitting cellular enzymes to escape and accumulate in the blood (Highman and Altland, 1963; Garbus et al., 1964). Evidently prior exercise training lessens the augmentation of cellular permeability induced by strenuous exercise. This could be an important factor in in-

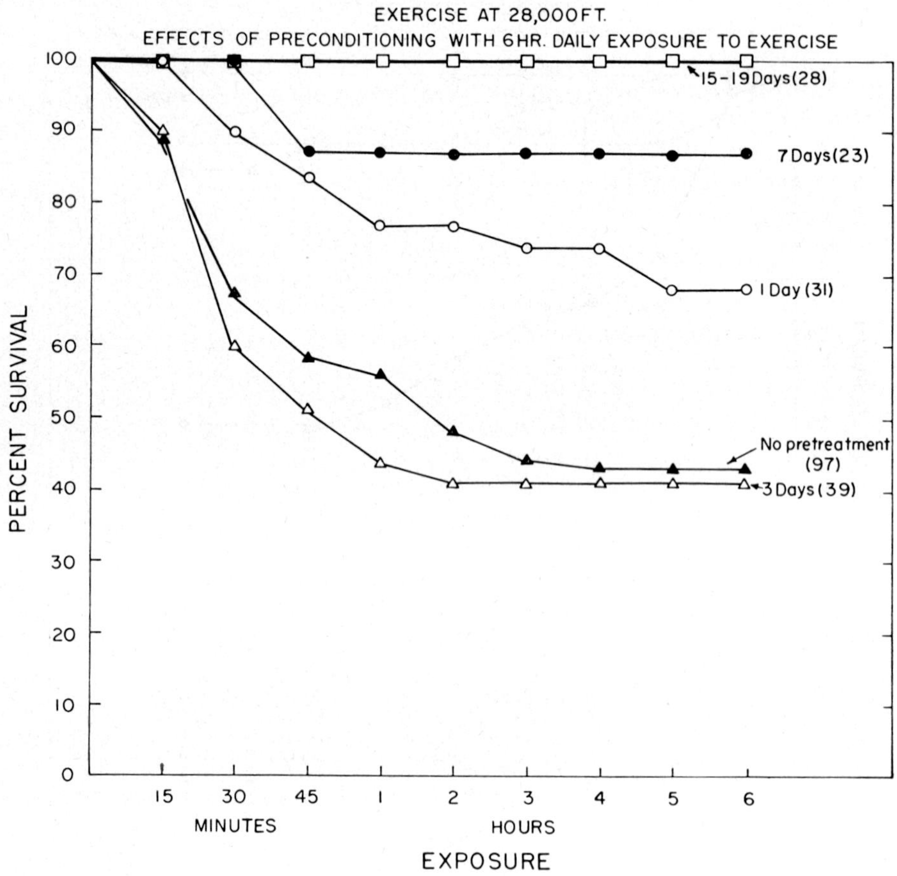

Fig. 4. Effects of physical training near sea level on survival of young adult male rats exercised at 28,000 ft simulated altitude. Parenthetical value is number of rats used.

creasing tolerance to exercise at altitude along with better known alterations such as increases in muscle myoglobin and cytochrome C and decreases in lactate and catecholamine output.

ALTITUDE ACCLIMATIZATION FOLLOWED BY EXERCISE AT SEA LEVEL

Except for one study in which natives of high altitude areas were unable to perform as well at sea level as at altitude (Velàsquez, 1966) all investigations indicate that there is a gain in performance at sea level following various periods of vigorous exercise training at altitude (Balke et al., 1965; Balke et al., 1966; Klausen et al., 1966; and

Roskamm et al., 1969). The improvement usually lasts for a few days, slowly diminishes, and is lost entirely in about 8 weeks. There are a number of reasons advanced for this improved performance such as a 40% increase in ability to elevate blood lactate after training at altitude (Klausen et al., 1966; Roskamm et al., 1969).

Experimental work with rats in our laboratory has shown that 6 weeks altitude acclimatization to 25,000 ft (7,620 m) 5 hrs/day, involving only voluntary movement, is detrimental to exercise performance on a rotating drum at sea level. The acclimatized rats developed immobility of either fore or hind limbs 3 hrs earlier than unacclimatized controls. Fatigue (total immobility of both fore and hind limbs) was exhibited by all altitude rats and in one-third of the controls after exercising 7 1/2 hrs. The poor sea level performance of the rats was considered to be due in part to the prior severe hypoxia, the polycythemia (hematocrit 80) and the high incidence of cardiac lesions produced by 6-weeks exposure to 25,000 ft (7,620 m) (Altland et al., 1969).

In studies designed to determine the relative importance of hypoxia, polycythemia and cardiac pathology in reducing exercise performance at sea level it was found that acclimatization of rats to 18,000 ft (5,486 m), a much less severe level of hypoxia than used previously, produced severe polycythemia (hematocrit 80) but no cardiac lesions and almost no fatty changes in the thigh muscle, liver and kidney arteries (Altland and Highman, 1971). Such rats had the same exercise capacity as sea level controls. Similar results were obtained when severe polycythemia (hematocrit 69) was produced by feeding rats a cobalt supplement in their diet. In this experiment no significant pathological lesions were found. These studies indicated that polycythemia was not a critical factor in reducing the capacity of the rats to exercise at sea level. Probably the most important factor was the severe hypoxia, but further study is needed to assess critically the relative effects of altitude induced myocardial lesions, per se, on exercise performance at sea level.

In summary it is evident that further study is needed to provide a better understanding of the capacity of man and animals to exercise at altitude and to determine the exact mechanisms involved in beneficial effects provided by physical training during altitude exposure. Very few data exist on the capacity of lower vertebrates to live at high altitudes. Sophisticated telemetry experiments may be useful in providing data on high altitude flight of birds and on the movement of mammals, reptiles and amphibians in the mountainous areas of the world.

REFERENCES: see pp. 277-278.

Physiological Effects of Altitude in Wild Mammals

by

Peter Morrison

INTRODUCTION

A variety of environmental influences may be operative at high altitude including cold, wind, aridity, radiation as both ultraviolet and infra red, shortages of food or water and, of course, hypoxia. This chapter considers only the latter but it is well to remember that in real life as opposed to experimental situations the effects of hypoxia may be aggravated and the physiological responses compromised by accompanying environmental stressors. However, such effects will be more notable in man (e.g., mountaineers) and domestic animals than in the adapted wild species which form our subject (Morrison, 1964, 1968). Among the various stressors hypoxia at high altitude is unique in its pervasiveness which necessitates physiological adjustment since microclimatic evasion or passive isolation is not possible.

Native mammals and birds range upward in altitude and some species occupy any region which is provided with food. The higher reaches are sterile because of infertility rather than the physiological inadequacy of the fauna. Altitudinal limits depend on sufficient moisture for plant growth but not so much as to restrict the growing season by excessive snow cover, these aspects being closely interdependent on the thermal regime. Since the permanent snow line falls continuously from low to high latitudes reaching sea level in the high arctic, the highest and most numerous examples of altitudinally adaptated species live near the equator. Thus, altitudes above 4.5 km are inhabited on Mount Kenya (Coe, 1967), above 5 km in the Andes (Pearson, 1951) and above 6 km in the Himalayas (Wollaston, 1922; Swan, 1961). Isolation from centers of science together with technical difficulties of collecting and returning live animals may account for the paucity of information on highland species. Thus, before 1960 we can cite only the studies of Hall et al. (1936) and Kalabukov (1937) on the concentration and properties of hemoglobin; and later scattered studies involve few laboratories. The following reviews can be cited: Dill (1938), Morrison (1964), Hart (1971), Bullard (1972), and Lenfant (1973).

In considering these data it is only natural to compare various aspects to the much better documentated response of man and laboratory animals. Such a comparison of wild and captive species should always be made with reservations but in this instance the problem is more serious since most observations on man and laboratory animals represent short or medium term exposure to hypoxia. This then, following Hart (1957), is *acclimatization* or *acclimation* and involves functional changes in individuals. By contrast, differences observed in native highland races or species represent *adaptation*, that is, group differences derived and presumably maintained genetically by natural selection. Such changes may be of quite different character than those following acclimatization.

EFFECTS ON PHYSIOLOGICAL FUNCTIONS

1. OXYGEN CONSUMPTION

Hypoxia impairs function by limiting the movement of oxygen from the atmosphere to its utilization in the tissues. The maintainence of metabolism at low pO_2 is probably the best general measure of resistance or adaptation to hypoxia. Studies on a series of lowland and highland Andean rodents exposed to cold (Fig. 1) defined a critical pO_2 to just maintain twice the resting metabolic rate. As shown, fourteen highland forms formed a compact grouping (p_c=54-76) as compared to the range of nine lowland forms (p_c=68-122). The respective mean values, 68 *vs* 95 Torr, define the advantage of the highland group. Segrem and Hart (1967a, b), measuring hypoxic suppression of metabolism in the white-footed mouse (*Peromyscus leucopus*), found equivalent effects in exercising and cold stimulated individuals with a critical pressure near 95 Torr or about equal to the white mouse as shown in Fig. 1.

In general, the metabolic level is maintained as pO_2 is reduced towards a critical value and then falls progressively with each further incremental reduction (Fig. 2). The response is thus defined by the critical pO_2 and the slope which, however, are both dependent on the initial metabolic level (M_0). But in the tundra vole the slope is a constant fraction of M_0 so that the response can be defined as:
$M/M_0 = 0.47—0.95\ M_0 + 0{,}0072\ P$ (Rosenmann & Morrison, 1974a).
One might suppose that an animal would absorb O_2 according to its maximum capability rather than allow a metabolic deficit (= hypothermia under a cold stimulus). Although this is the case under normoxia in small mammals – the maximum rate can be stimulated by cold (Rosenmann and Morrison, 1974b) – this is not so under hypoxia in which metabolism is clearly related to the initial metabolic level or the equivalent thermal stimulus for heat production (Fig. 2). Accordingly, performance under hypoxia depends not only on the intrinsic capacity for oxygen transport but also on the control system for mobilizing that capacity. Further, because of this dependence, comparisons between species should be made at constant metabolic stimulus. Such is shown in Fig. 3 which compares metabolism as a function of pO_2 at a metabolic stimulus (M_0) of 3.8 mets. (i.e., multiples of standard metabolism, $M_{st} = 3.8W^{.73}$) in a number of rodent species. Because of differences in slope and intercept (P_c), the relative effectiveness of different species will depend on the pO_2 or simulated altitude at which they are compared.

2. ADAPTATION FOR HYPOXIA

Having established that different species, and in particular highland species, are in fact better adapted for hypoxia the question then arises as to what physiological or anatomical differences are responsible for the observed difference in transport capacity. Oxygen transport might be facilitated by any or all of the considerable series of steps through which oxygen is carried to the tissues (Fig. 4). Thus, increased lung ventilation can compensate to a degree and help maintain aveolar pO_2. The respiration rate, the mass of the diaphram, the size of the chest and lung and the tracheal cross-section can all modify this aspect. In the movement of oxygen into the blood stream, the alveolar area and the related capillary area are directly involved in the transfer rate. Separate listing for blood, erythrocyte, hemoglobin, and heme indicate the involvement of at

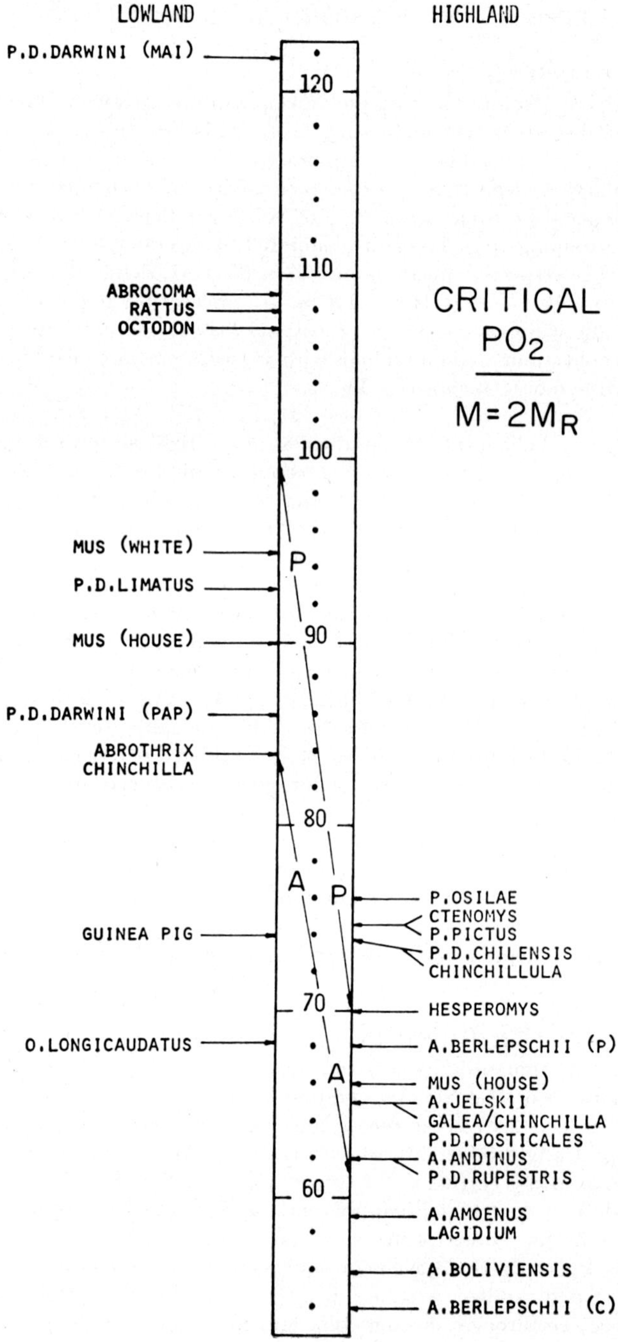

LOWLAND
HIGHLAND
P.D.DARWINI (MAI)
120
110
ABROCOMA
RATTUS
OCTODON
CRITICAL
PO2
M=2MR
100
MUS (WHITE)
P
P.D.LIMATUS
MUS (HOUSE)
90
P.D.DARWINI (PAP)
ABROTHRIX
CHINCHILLA
80
A
P
P.OSILAE
CTENOMYS
P.PICTUS
GUINEA PIG
P.D.CHILENSIS
CHINCHILLULA
70
HESPEROMYS
O.LONGICAUDATUS
A.BERLEPSCHII (P)
A
MUS (HOUSE)
A.JELSKII
GALEA/CHINCHILLA
P.D.POSTICALES
A.ANDINUS
P.D.RUPESTRIS
60
A.AMOENUS
LAGIDIUM
A.BOLIVIENSIS
A.BERLEPSCHII (C)

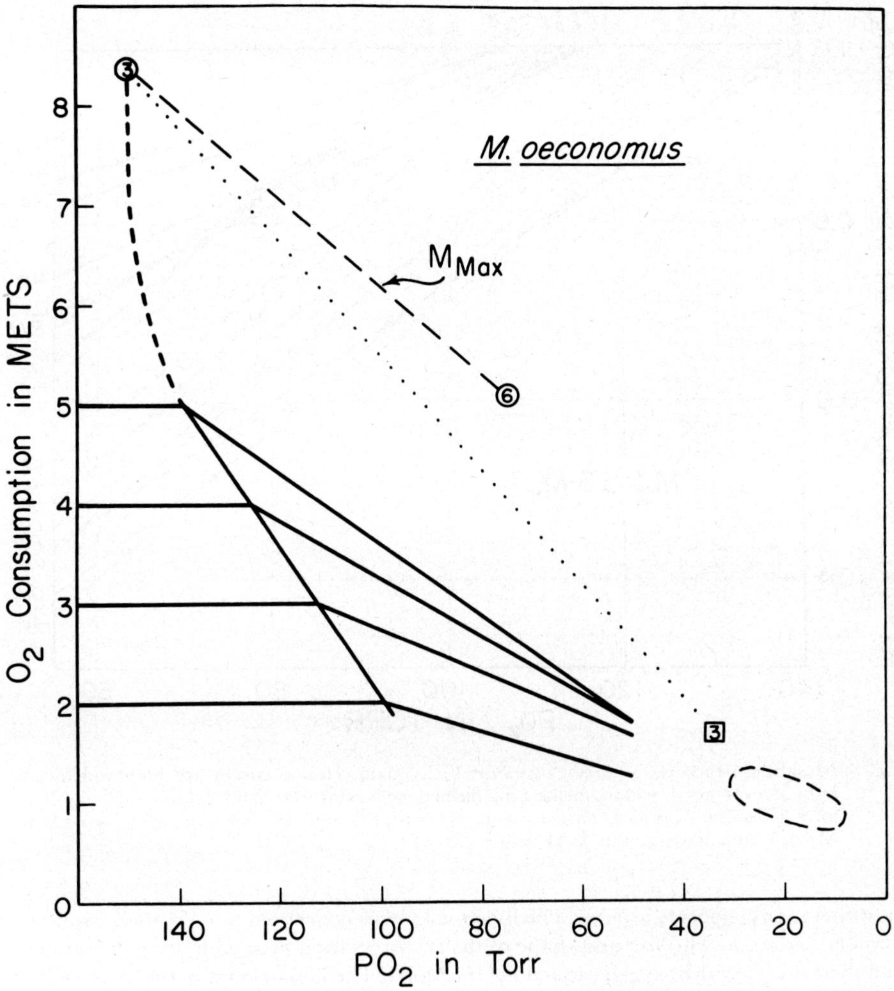

(at left)

Fig. 1. Critical pO_2 for maintaining twice resting metabolic rate in lowland and highland rodents: *Akodon* (A), *Phyllotis* (P), *Oryzomys* (O). (*Hesperomys* = *Calomys*) *Abrothrix* is taken as the reference for lowland *Akodon*. (Observations with B. K. McNab; also Morrison, 1964).

(above)

Fig. 2. Influence of pO_2 and metabolic stimulus (cold) on metabolism in the tundra vole after Rosenmann & Morrison, 1974a). Mets are multiples of the standard (mean) metabolic rate for mammals, $M_{st} = 3.8\ W^{.73}$ (cc O_2/g.hr, g). Square shows lethal values at thermal neutrality, Elipse compares lethal values in other rodent species (Hall, 1966). A higher lethal pO_2 has been correlated with the higher basal metabolic rate in microtine rodents (Morrison and Rosenmann, 1975).

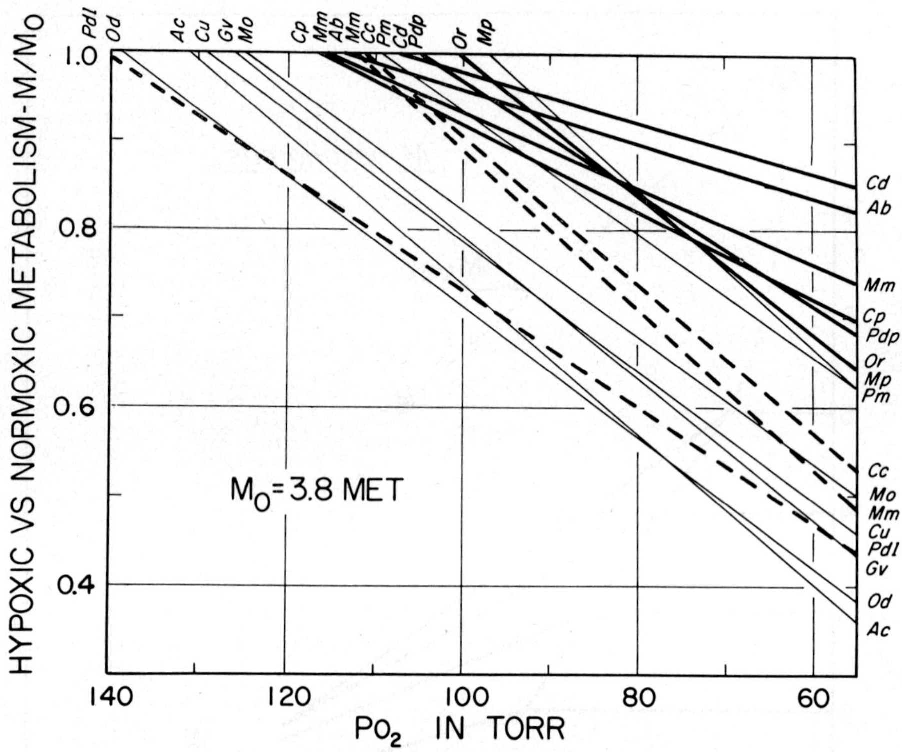

Fig. 3. Metabolic response of various rodents to hypoxia. Heavy curves are highland forms, light curves are lowland forms and dashed curves are lowland relatives of highland forms: *Calomys* (Cd/Cc), *Phyllotis darwini* (Pdp/pdl) and *Mus musculus brevirostris* (Mm/Mm). (After Rosenmann & Morrison, 1975).

least that many separate aspects. The blood volume is concerned in increasing capillary number and area. The size and shape of the erythrocytes may modify oxygen transfer; and their number, the oxygen capacity of the blood. The hemoglobin in the cells may be more or less concentrated, and its constitution and configurations will modify oxygen binding by the heme. The heart, by its rate and stroke volume, will directly pace oxygen transport to the tissues and will require a matching arterial tree to carry the flow. Diffusion again limits transfer across tissue capillaries and perhaps even more limits movement into the cells in which myoglobin must play an important though not fully defined role. Finally the oxygen is transferred and utilized through cellular enzyme systems.

3. BLOOD: CAPACITY FOR OXYGEN

In persons moving from sea-level to high altitude perhaps the most conspicuous acclimatization is the increased hematopoietic activity which results in elevated erythrocyte and blood hemoglobin levels. Similar differences have been observed in permanent residents at high altitude as compared to lowland man. Kalabukov (1937) observed higher erythrocyte counts in mountain races of wood mice (*Apodemus*) and in lowland races exposed to natural or simulated high altitude, and Hock (1964) similarly found higher hemoglobin in *Peromyscus* at 3.8-4.3 km as compared to 1 km at White Mountain in California. Since these altitudes are lower than the habitats of many Andean rodents, we might expect the latter to show an even more striking response. On the other hand the vicuña has an ordinary level of blood hemoglobin (Hall et al. 1936). Indeed, the hematocrit is well below the ordinary level of mammals but compensated for by the unique high concentration of hemoglobin in the cells (*ca* 400 g/l). A similar value (423 g/l) was found in the related guanaco (Morrison et al., 1963) but highland rodents do not correspond to these camelids in cell hemoglobin. Thus the range of 280-360 g/l in six laboratory rodents (Altman and Dittmer, 1961) is comparable to that of 280-330 g/l found in some dozen Andean species and subspecies (Morrison et al., 1963). Although a considerable diversity in hematocrits was seen in these Andean rodents, there was no

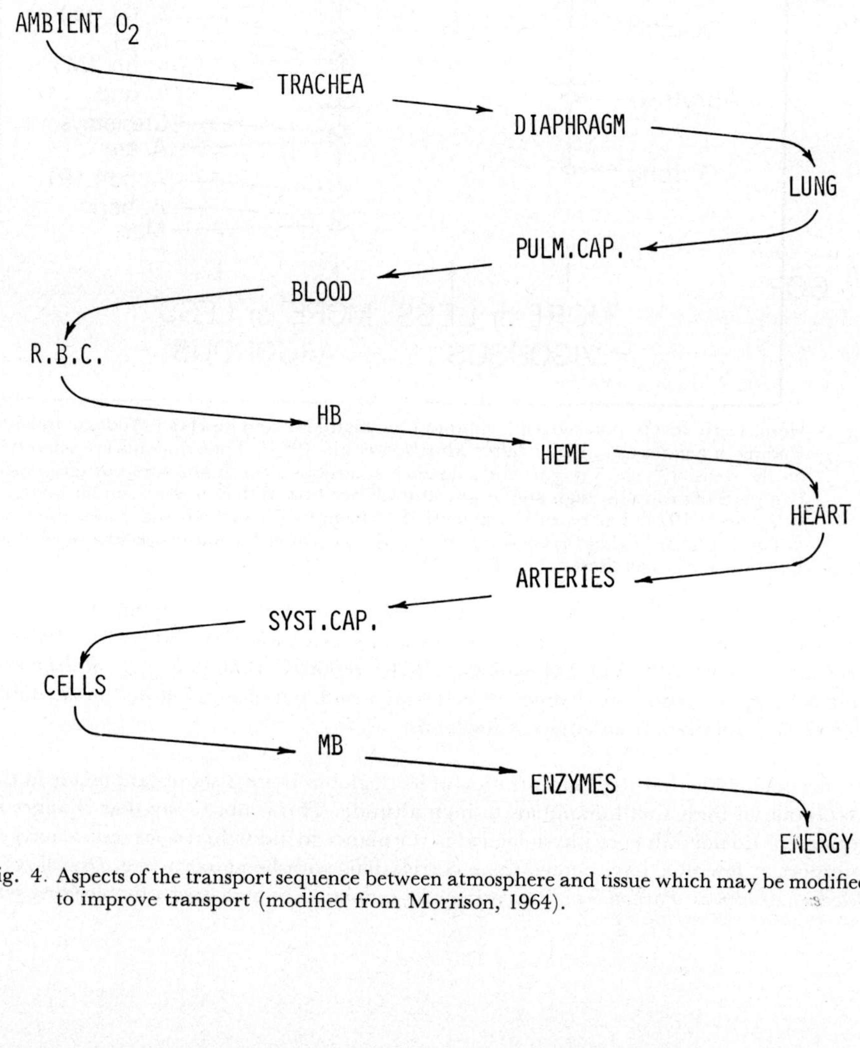

Fig. 4. Aspects of the transport sequence between atmosphere and tissue which may be modified to improve transport (modified from Morrison, 1964).

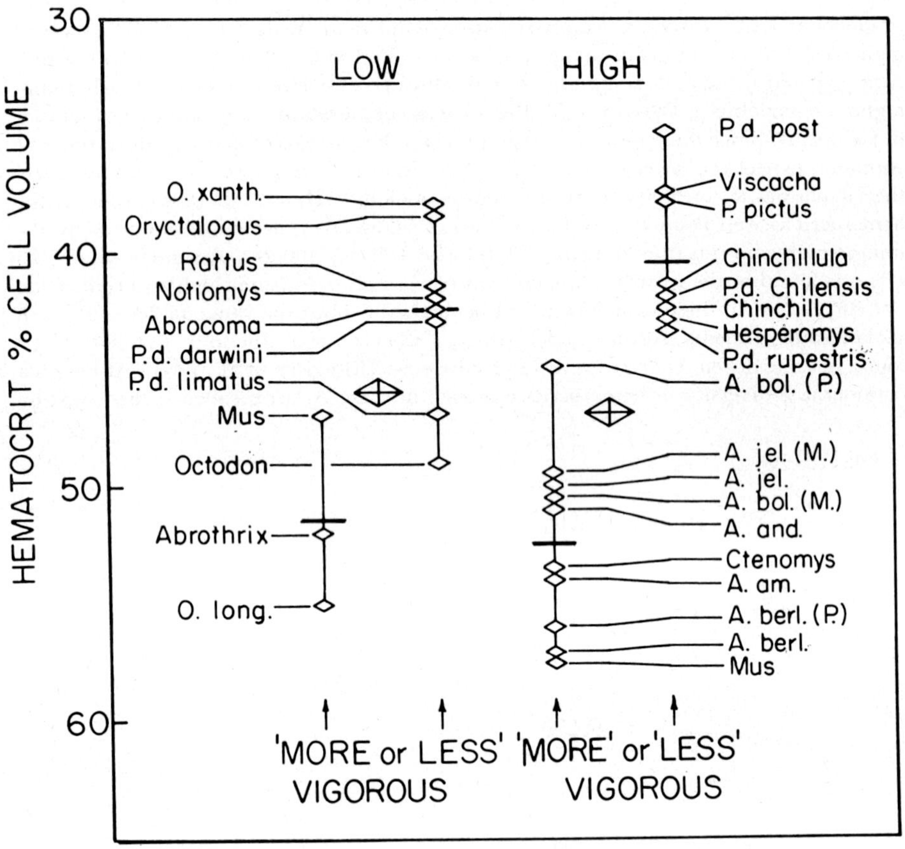

Fig. 5. Hematocrit levels (packed cell volumes) in highland and lowland Andean rodents. Coding of genera as in Fig. 1 (after Morrison et al., 1963). These rodents are separated on the basis of "vigor" to give fairly distinct subgroups at both altitudes but the respective pairs of groups at high and at low altitude are little different. In a similar manner, Carpenter (1975) has recently reported that hematocrits of highland birds show no difference from lowland species but that those of strong fliers are significantly higher than those of weak fliers.

correlation with the altitude of origin. Thus the range (by species) in highland forms was 350-570 (ave. 470) cc/l and in the lowland species was 380-550 (ave. 460) cc/l. Recently Monaskirsky and Palovinkina (1970) reported that voles (*A. alticola* and *Microtus gregalis*) from the Pamirs (3.9-4.3 km) had significantly lower hemoglobin indices than relatives from adjacent lowlands.

We may conclude that the concentration of hemoglobin is not a significant factor in the adaptation of these small mammals to high altitude. This is not to say that changes in hematocrit do not influence physiological performance in individual since tests of oxygen transport at low pO_2 have shown some correlations with hematocrit (e.g. *Oryzomys vs. Octoden; Akodon vs. Phyllotis*) and experimental reduction in hematocrit by bleeding will reduce performance. But in *adaptation* other means than the augmentation of oxygen capacity have been found. It further implies that while raising the erythrocyte level would appear to be the simplest means of improving transport, it must present some related disadvantage, perhaps associated with the increased viscosity.

4. HEMOGLOBIN: AFFINITY FOR OXYGEN

Hall et al. (1936) observed an increased oxygen affinity in the blood of several Andean species as compared to lowland forms. For example, the T_{50} of 18 Torr in the vicuña is the lowest observed in any mammal, some 9 Torr below that for man. An adaptation which facilitates loading from the oxygen-poor atmosphere may be even more desirable in rodents in which oxygen binding is characteristically weaker than in larger mammals. (Schmidt-Nielsen & Larimer, 1958). Hall (1966) measured lethal pO_2 at thermal neutrality in a series of rodents and showed a general correlation with the T_{50} of the bloods but the respective metabolic levels were not considered.

Measurements on Hb solutions from a spectrum of small highland and lowland mammals (Fig. 6) show a broad overlap in affinity for oyxgen (T_{50}) quite unlike the discrete grouping for overall transport (Fig. 1). However, it must be kept in mind that the response of Hb inside the cell is greatly modified by levels of cellular components notably hydrogen ion and organic phosphates such as 2,3-DPG. When *related* highland and lowland species and subspecies of *Phyllotis* are compared, the former have a clear loading advantage in the higher oxygen affinity of their hemoglobins (Fig. 6). Aste-Salazar & Morrison (1962) identified differences in electrophoretic mobility and alkali denaturation in these hemoglobins and recently we have been able to differentiate between highland and lowland pairs of *Phyllotis* and *Calomys* in terms of specific amino acid substituents (observations with P. Ramakrishnan). Such a qualitative change may be thought of as characteristic of adaptation as compared to quantitative changes in acclimatization such as in hematocrit although the distinction is not rigorous.

In the acclimatization of man to altitude it has been suggested that the opposite shift, a reduction in oxygen affinity following substantial increase in organic phosphate (2,3-DPG) facilitates unloading at the tissues (Lenfant et al., 1968). However, Rand et al. (1973) have shown in the rabbit that although hyperoxic/hypercarbic or hypoxic atmospheres quickly reduce or increase erythrocyte 2,3-DPG with a corresponding effect on T_{50} this has no effect on tissue oxygen levels during hypoxia. In four species of wild mice no significant increases in 2,3-DPG were observed after hypoxic exposures of 2-45 days at pO_2 of 68-90 Torr (observation with M. Rosenmann).

5. ORGANS AND TISSUES

Turning now to morphological features one can compare the size (relative mass) of various organs and structures directly concerned with oxygen transport. Here again it is prudent to compare similar forms as highland *vs* lowland *Phyllotis* in which we found a considerable increase in heart ratios (23%) and some increase in lung, trachea, and diaphragm ratios (12-14%) (observation with K. C. Morrison). These changes, while appreciable, are modest compared to the increase of 90% observed in the acclimation of white rats to altitude (Timiras et al., 1957), the 46% in mountain *Peromyscus*(Hock, 1964), or even to the 53% in heart ratio we found in house mice at Morococha (4.7 km).

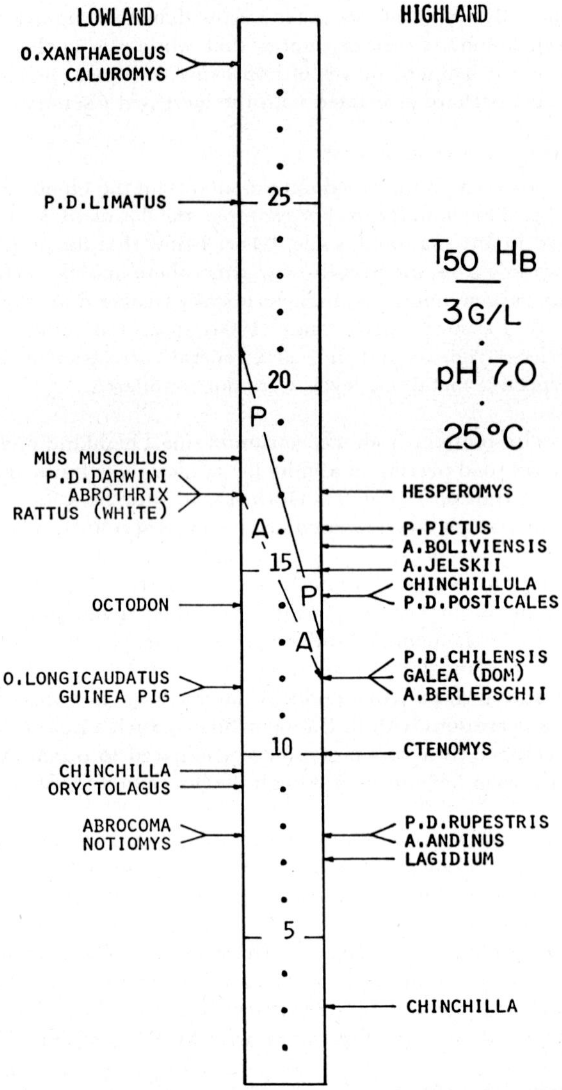

Fig. 6. pO_2 for half saturation of hemoglobins from lowland and highland rodents (3 g/l in M/15 pO_4 + NaCl at 25°C). Values at pH 7.0 interpolated from measurements at pH 6.60 and 7.40. (Observations with K. V. Kerst; also Morrison, 1964).

Adrenal weights have special significance as measures of stress as shown by increases in laboratory species during hypoxic acclimation. In like manner, the adrenals of house mice from Morococha were three times the size of their counterparts from Lima. In striking contrast, the adrenals from native highland and lowland Andean species were of equal size, and Hock (1964) also found no difference in adrenal size in mountain *Peromyscus*.

As well as differing in relative mass, tissue components may change their concentrations. Thus, highland species of *Phyllotis* showed about twice the concentration of myoglobin in diaphragm and pooled leg muscle as did a subspecies from sea-level; but heart myoglobin had the same concentration in both groups (Reynafarje and Morrison, 1962). House mice showed the reverse pattern with animals from 4.5 km having more heart myoglobin but no more diaphragm or leg myoglobin than did sea-level house mice. Other wild species conformed to the pattern in *Phyllotis*, including the active Akodon which had shown substantially higher blood hemoglobin levels. Roberts et al. (1966) reported different concentrations of several liver enzymes (serine/threonine dehydrase, DPNH-cytochrome C reductase and transhydrogenase) in deer mice (*Peromyscus maniculatus sonoriensis*) from 3.8 km and from near sea level. They also found the same changes following cold acclimation to suggest a similarity between the adjustments for hypoxia and cold.

7. HEART AND RESPIRATION

As well as their chamber sizes, the frequencies of the heart and lungs determine their effectiveness for O_2 transport. Bullard and Kollias (1966) compared ground squirrels (*Citellus lateralis*) native to 3.8 km with white rats acclimated to 3.8 km when both were exposed to a simulated altitude of 10 km (49 Torr O_2) 10°C. The results seemed to clearly confirm the advantages of altitudinal adaptation since the ground squirrels showed an initial increase followed by a modest decrease in heart rate while the white rats showed an immediate drastic reduction in heart rate. However, this example illustrates several hazards in such comparisons: (1) the conditions were extreme being, in fact, lethal for most of the rats; (2) sedentary caged animals were compared to more vigorous wild individuals; (3) ground squirrels and marmots as a group may already show some properties appropriate to hypoxia even at sea level by their burrowing mode or by their specialization as hibernators. A further uncertainty in such comparisons concerns the nature of the response which may reflect (1) the basic rate of the organs, (2) the response of the neuroendocrinal control system to moderate hypoxia, or (3) the failure of either the organs or the control system from extreme hypoxia.

Observations by Morrison and Elsner (1962) on species and subspecies of *Phyllotis* from sea level and from 4.5 km, and made at both altitudes meet some of these problems. The lowland mice increased their heart and respiration rates when transferred to 4.5 km, but not to the level of the highland mice. Highland mice, by contrast, did not reduce their rates at sea level, these therefore being inherent and not dependent on hypoxic stimulus (although altitudes above 4.5 km might have been effective). The same response was seen in highland offspring born and raised at sea level to confirm the genetic basis for this activity.

SUMMARY AND CONCLUSIONS

A number of the characteristics of highland rodents are contrary to the changes produced by experimental or short-term hypoxia. This may be accounted for in terms of two contrasting processes, acclimatization and adaptation. Both act to meet the re-

quirements of life at high altitude but the former represents a quantitative modification or distortion of an existing pattern while the latter acts to fix more favorable functional and structural patterns by natural selection. The former is a phenomenon of the individual while the latter involves the recent history of the species or population.
Thus, in highland rodents hemoglobin levels are no different than in similar lowland species, in contrast to the elevated levels found in migrants to high altitude. However, we must account for a seeming discrepancy in the elevated hemoglobin levels of mountain *Apodemus* and *Peromyscus* (Kalabukov, 1937; Hock, 1964). These species were not distinguished taxonomically from adjacent lowland forms with which they formed an unbroken distribution and since the higher areas are restricted there must be continuous admixture from broader and lower surrounding areas. By contrast, the Andes provide a broad highlands largely isolated from lowland habitats for the development of specialized adapted populations. Although acclimatization appears as effective as adaptation since house mice from Morococha exceeded several native highland mice in overall performance, they must do this at some great cost, as witnessed by their adrenal hypertrophy.

ACKNOWLEDGMENTS. The author acknowledges assistance by P.H.S. grants GM-10402 from the National Institute of General Medical Sciences and H-4458 from the National Heart Institute, and N.S.F. grant # 67073 and # 65005.

REFERENCES: see pp. 278-279.

CHAPTER 4

EFFECTS OF NOISE STRESS ON LABORATORY ANIMALS

SECTION 1: Endocrine and Neuro-Histochemical Aspects of Noise Stress in Laboratory Animals (A. ANTHONY)

Chapter 4

Endocrine and Neuro-Histochemical Aspects of Noise Stress in Laboratory Animals

by

A. Anthony

INTRODUCTION

In animals as in man there are two aspects of noise as an environmental pollutant. One relates to noise induced impairment of hearing; the second relates to harmful neuroendocrine effects of noise as a physiological stress. Effects of noise on hearing loss are largely based on extensive histopathological studies of damage to cochlear hair cells and spiral ganglion cells of laboratory animals such as the guinea pig. There is sparse information on how noise affects hearing acuity thresholds in laboratory rodents which may exhibit a peak frequency response close to 10 kHz and whose hearing range may extend to much higher frequencies than mans' (Fuller and Wimer, 1966). Thus, it is difficult to evaluate the impact of aircraft, highway or other man-made sources of noise on wildlife or domesticated animals. The dependence of wild animals on hearing acuity for avoiding prey, locating food, and mating suggests that noise induced hearing impairment could affect the survival of some species.

With respect to nonauditory effects of noise stress, there have been no studies on wildlife and few controlled studies on animals such as swine, poultry and cattle (Welch and Welch, 1970). Based on behavioral observations, swine and sheep appear to adapt readily to noise. However, recurring reports that noise elicits alarm reactions and affects milk production in cattle, causes a reduction in fertility and egg laying in poultry and increases mortality in mink kits indicate the problem is far from being resolved.

Current knowledge on nonauditory effects of noise as a stress stimulus is based largely on studies using rats and mice. These provide clear evidence that as well as damage to hearing, auditory stress can trigger a host of disturbances in body function the persistence of which could prove harmful to health. An excellent source to the considerable literature in this area is "Physiological Effects of Noise" edited by Welch and Welch (1970). Audiologists frequently stress the need for caution in extrapolating animal data to noise effects in man (Kryter, 1970). Equal recognition should be given to the well established tenet that almost all our knowledge of human physiology in health and disease is based on animal experimentation. It is significant that noise induced stress reactions are remarkably similar in animals and man, with respect to both neurological and endocrine changes. There remains considerable conjecture, however, regarding the significance of observed reactions. Some view the pattern of noise induced physiological changes as essentially adaptive in nature. Others express concern that whereas animals may adapt to short durations of noise exposure, chronic noise stimulation could lead to a breakdown of homeostatic defense mechanisms. Repeated observations that noise stress can exert adverse effects on reproductive function (Singh, 1972), endocrine and metabolic functions (Sackler et al., 1959), resistance to viral infections (Jensen and Rasmussen, 1963) cannot be discounted simply because the physiological events precipitating these deleterious changes have yet to be identified.

Noise induced changes in bodily processes of animals are currently believed to stem from an excitation or inhibition of endocrines via autonomic, cerebral and reticular nervous system pathways (Anticaglia and Cohen, 1970). These include changes in secretion of pituitary corticotropin and adrenal corticosteroids, vasoconstriction in peripheral vessels, imbalance of blood electrolytes, decreases in respiration rate, gastrointestinal motility and electrical resistance of the skin, and alterations in brain excitability. The wealth of evidence supporting the contention that noise has adverse effects on bodily functions other than hearing is beyond the scope of this report. Instead, an attempt is made to outline certain relationships between noise exposure, hypothalamic pituitary regulation of adrenocorticoid secretion and brain excitability. It is felt that this approach would aid in delineating critical areas where more research is needed and also suggest possible explanations for some of the conflicting reports regarding endocrine aspects of noise stress.

HYPOTHALAMO-PITUITARY REGULATION OF HOMEOSTASIS DURING NOISE STRESS

Although our knowledge of the precise mechanisms which control pituitary adrenocorticotropic (ACTH) activity is far from complete, there is no question that the hypothalamus plays a vital role in initiating release of ACTH in response to all stresses including noise. Recent evidence indicates that the regulatory aspects of ACTH secretion are extremely complex in that several types of feedback mechanisms involving multiple brain sites appear to be operative (Yates et al., 1971). In addition, accumulation and retention of corticoids by different brain areas may influence the extent or threshold of hypothalamic excitation.

It should be noted that in the 1950's, when interest in noise stress research was at its peak, very little was known about ACTH regulation. The working hypothesis was that harmful effects of noise stem from an overactivation of the ACTH-adrenocortical system, excessive or protracted output of corticoids and a subsequent impaired functioning of target organs influenced by adrenocortical hormones (Anthony and Ackerman, 1957). Considerable evidence has been amassed over the past 30 years which corroborates the concept that a major immediate consequence of noise is adrenocortical activation. However, some noise experiments suggest that synthesis and/or release of corticoids might be curtailed with certain types of noise stimuli (Biró et al., 1959; Anthony and Ackerman, 1959). Insufficient consideration has been given in the literature to the possibility that certain harmful effects of noise could stem from suboptimal functioning of the adrenal cortex.

To understand how noise might effect a transient or protracted suppression of ACTH and corticoid secretion, it is desirable to review some of the major feedback loops or humoral servomechanisms presumed to have an inhibitory influence on adrenocortical activity. These are schematically shown in Figure 1. The main driving mechanism of pituitary adrenocorticotropin (ACTH) is the hypothalamus which activates the anterior pituitary VIA the corticotropic releasing factor (CRF). It has been shown that a number of independent and opposing feedback loops determine and regulate the rate of ACTH secretion (Ganong and Martini, 1969; James, 1968; DeWied and Weijnen, 1970). Undoubtedly, the major immediate regulatory agent of ACTH output is the corticoid counterbalancing inhibition. Although the neurochemical basis of corticoid action remains unknown, recent work has shown that corticoid suppression of ACTH synthesis and release can occur at several points in this regulatory circuit, i.e., the cerebral cortex, midbrain, hypothalamus or adrenal cortex (Yates et al., 1971). Superimposed on this corticoid regulatory mechanism are several additional neural

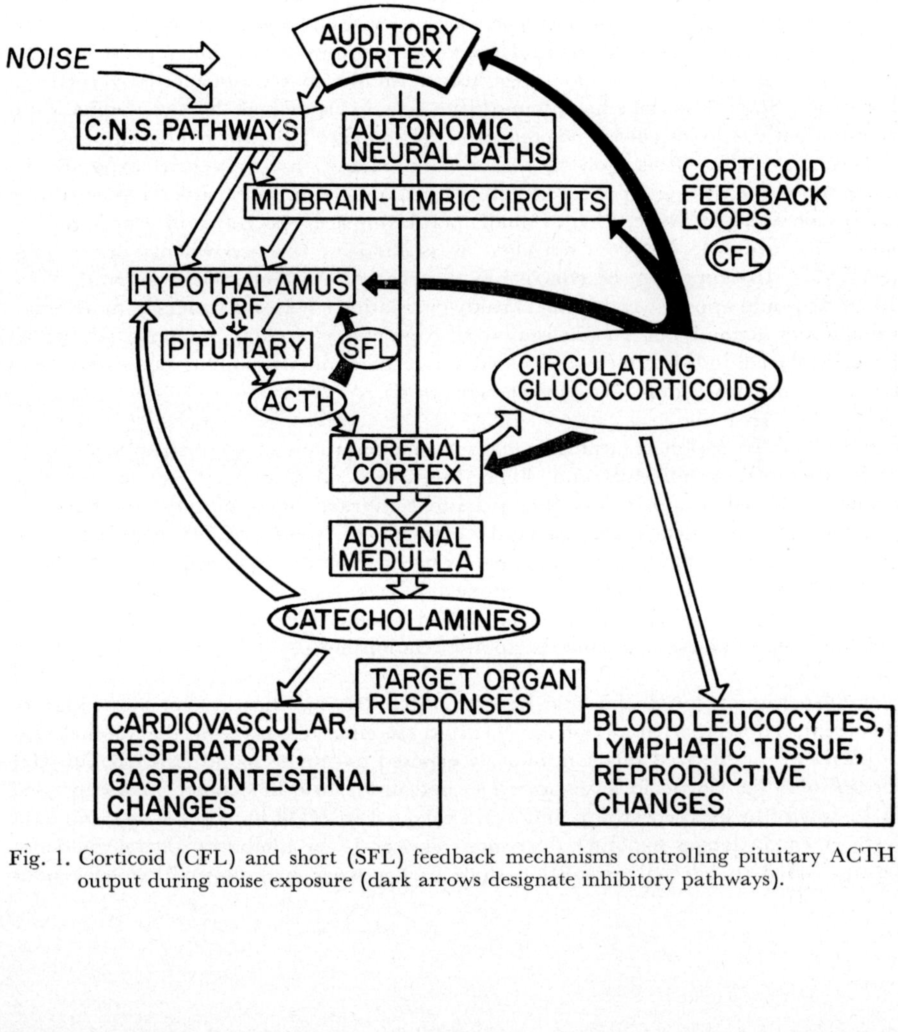

Fig. 1. Corticoid (CFL) and short (SFL) feedback mechanisms controlling pituitary ACTH output during noise exposure (dark arrows designate inhibitory pathways).

and hormonal circuits which can either complement or counteract corticoid influences. For example, the inhibitory actions of ACTH levels (in portal vessels or the general circulation) on the hypothalamus are referred to as "short" feedback loops (DeWied and Weijnen, 1970). Also, catecholamines produced by the adrenal medulla can alter the physiological state and hence the excitability of hypothalamic and auditory cortical pathways. Autonomic nervous system responses represent an integral part of the neuro-endocrine defense mechanism especially in the initial stress phase (preceding ACTH release) which involves posterior hypothalamic activation. This part of the system is held responsible for such diverse effects of noise as peripheral vasoconstriction and alterations in respiratory and heart rates.

ADRENOCORTICAL ACTIVATION DURING NOISE STRESS

The current belief that intense noise acts as a physiologic stress comes from microscopic analysis of histological and histochemical changes in the adrenal cortex and evidence of increased corticoid levels in the adrenal vein (Anthony and Ackerman, 1955, 1957, 1959; Henkin and Knigge, 1963). However, numerous conflicting reports exist regarding secondary responses in target tissues known to be influenced by corticoid hormones. Several workers have found little or no evidence of thymicolymphatic involution with acute or chronic exposures to very intense (135 dB) noise (Anthony and Ackerman, 1959). Others, on the other hand, found that very brief exposures to moderately intense sound pressure levels (85 dB) sometimes cause adrenal hypertrophy (Anderson and Van Dyne, 1966). Equally perplexing are reports that extremely intense noise exposure (135 dB) does not elicit an eosinopenic reaction whereas lesser noise levels (110 dB) appear to be effective in this regard (Anthony and Ackerman, 1955, 1959). It would appear worthwhile to study circulating corticoid changes in more detail using newer methods of corticoid assay (e.g., competitive binding, immunoassay) which have displaced indirect measures based on changes in eosinophils or adrenal and lymphatic organ weights (Vernikos-Danellis, 1965).

Some of the confusion regarding endocrine manifestations of noise stress stems from early attempts by ourselves and others to explain all secondary manifestations of endocrine imbalance in terms of a persistent overactivation of the hypothalamic-pituitary-adrenocortical system, since the existence of "short" ACTH feedback mechanisms or corticoid binding in brain centers had not been established. It now appears that some of the unexplained or paradoxical effects of noise make sense if we accept the possibility that noise activation of ACTH release does not always follow a predictable dose response or temporal response relationship.

Some experiments by Henkin and Knigge (1963) serve as an excellent illustration of this point. They reported a triphasic corticoid secretion of pattern in the adrenal vein of partially anesthetized rats continuously exposed to intense sound (132 dB, 220 Hz) for up to 48 hr. Corticoid levels were elevated in 0.5 to 8 hr exposed rats, depressed in 12 and 18 hr exposed groups and again elevated in 24-48 hr exposed rats. No data were given on longer exposure durations. The 12-18 hr inhibitory phase could not be attributed to adrenal exhaustion since earlier work had shown that exogenous

ACTH during this interval caused a marked outpouring of corticosterone. Butte et al. (1972) recently suggested that corticosterone accumulation of steroids in the brain during the first few hours of noise exposure would block control elements regulating ACTH secretion. This is supported by reports that a sudden rise in circulating corticoids can block hormonal stress responses for several hours after plasma corticoid levels return to normal, presumably because corticoids remain bound to brain receptors (De Wied and Weijnen, 1970). Most workers agree that acute exposure to intense noise usually causes a marked but transient increase in ACTH and corticoid release; however, the case for a persistent elevation of circulating corticoids with prolonged noise exposure is far from proven and merits further study.

Two indirect indices of increased ACTH activity and corticoid production which still remain in use are adrenal hypertrophy and thymicolymphatic involution. Investigations of noise effects on adrenal and thymus weight changes, however, have produced variable results, and support the contention that species, strain and sex differences exist with respect to the sensitivity of ACTH activation by noise stress. Anthony and Ackerman (1959) found female but not male C57BL/6 mice exhibited adrenal hypertrophy and thymic involution with prolonged exposure to intense noise (139 dB, 150-4800 Hz, 40 hr/wk × 5 wks). Jensen and Rasmussen (1963) found that Swiss albino mice subjected to intense noise (123 dB, 800 Hz, 3 hr/d for 1 to 6 d) exhibited adrenal hypertrophy without any weight changes in the thymus, liver or spleen. Equally important was their observation that only a very moderate and transient reduction in circulating leucocytes occurred during noise stimulation (Jensen, 1969); noise did not elicit the very marked 3-4 hr leucopenic response usually associated with exposure to stresses such as handling, confinement or trauma.

Undoubtedly one reason for observed sex differences in adrenal response to noise is that endogenous ovarian estrogens can stimulate pituitary ACTH release. This accounts for the sexual dimorphism in adrenal weights of mice and other mammals (Chai and Dickie, 1966). Thus, when females are exposed to a noise, a temporary noise induced initial outpouring of ACTH may be supplemented by an estrogen mediated ACTH release which varies during different stages of the estrous cycle.

Studies of noise effects on reproductive function provide some relevant information regarding possible relationships between pituitary secretion of ACTH and gonadotropins. Most proponents of Selye's stress theory adhere to the hypothesis that an inverse relationship exists between ACTH production and gonadotropin output. If one accepts the premise that, depending upon the severity and duration of a given sound stimulus, ACTH levels can be increased or decreased, one would anticipate reports on gonadal responses to noise to be as varied and confusing as findings dealing with adrenocortical mediated target organ responses. This appears to be the case. Studies with rats and rabbits indicate prolonged sound stimulation appears to stimulate gonadotropin output (as reflected in increased ovarian and uterine weights, greater numbers of Graafian follicles or corpora lutea and sometimes persistent estrus); (Zondek and Tamari, 1964; Singh and Rao, 1970). Sackler et al. (1959), on the other hand, reported a diminution in the number of ovarian follicles in noise exposed rats. It is noteworthy that Sackler and his associates employed single daily exposures (5 min/d × 15 d) at a relatively low frequency (375-500 Hz, 110 dB). Singh's group used continuous noise exposure (100 dB, 2 kHz for 24 hr/d for 50-150 d) and Zondek and Tamari used intermittent noise (1 min every 10 min $\times$ 24 hr (70-110 dB, 3-12 kHz) for up to two months. After reviewing the widely scattered literature in this area, Singh (1972) simply states that "much remains to be learned regarding the possible pathological effects of noise on reproductive processes."

ADRENOCORTICOIDS AND BRAIN EXCITABILITY

The concept that noise induced changes in blood corticoids can influence metabolic alterations of brain cells has gained credence from two separate lines of evidence (1) studies showing a close interrelationship of corticoids and central nervous system excitability; and (2) studies suggesting an inverse relationship between the intensity of sensory stimulation and RNA levels in specific brain areas.

Numerous studies, notably by Woodbury and Vernadakis (1967) have shown that adrenocorticoids produce marked alterations in brain excitability as measured by the electroshock seizure threshold method. Steroids triggered by ACTH secretion (i.e., cortisol in humans and corticosterone in rats), and which predominantly affect carbohydrate and protein metabolism, generally cause an increased brain excitability. Additional evidence that adrenal steroids of the cortisone type increase brain excitability is provided by reports that seizures can be induced in laboratory animals and in man by both cortisone and ACTH treatment (Woodbury and Vernadakis, 1957). One hypothesis states that glucocorticoids probably exert a direct effect on subcortical brain elements and could account for the manifold effects of corticoids on brain excitability, seizure induction and abnormal behavior. Direct confirmation of these interpretations requires further research to localize the anatomical sites of corticoid action and the neurochemical events which characterize excitation or inhibition. At present, there is essentially no information which directly bears on the relationships between noise stimulation, corticoids and brain excitability. Although some workers have postulated a relationship between audiogenic seizure susceptibility and adrenal cortical function, available data are insufficient to justify any conclusions regarding the neurochemical basis of audiogenic seizures. It has been documented, however, that prestimulation by subconvulsive doses of sound (ca. 100/dB) can permanently influence the threshold of seizure susceptibility and hence brain excitability (Fuller and Wimer, 1966). This suggests that auditory stimulation can cause permanent alterations in functioning of specific neural compartments.

NEUROCHEMICAL CORRELATES OF NOISE STIMULATION

Very little information is available on cytochemical changes in neural tissue of animals exposed to noise or other sensory stimuli. On the other hand, use of non-specific protein stains for detecting injury of nerve cells has a long history dating to the demonstration in 1892 by Nissl that chromidial substance degenerates after axon section. These stains are still in use for detection of morphological signs of damage in nerve cells. Tarmas and Grzesik (1963) reported that "tigrolysis" (dissolution of Nissl granules) and cytoplasmic vacuolization of guinea pig brain neurons following exposure to very intense noise (160 dB, 100-5,000 Hz). It is now known that what was formerly referred to as Nissl or chromidial substance largely consists of RNA. It should be noted that any exposure to a sound pressure level much above 120 dB probably causes some mechanical injury to the inner ear. If the animals were partially deafened, this could result in an extinction of noise mediated ACTH-corticoid activation via auditory neural pathways. Thus, the absence of adrenal and target organ responses observed in rodents exposed to noise intensities ranging from 120-135 dB as well as certain metabolic alterations in brain elements could be manifestations of non-excitation. This aspect of a possible abolishment of an auditory excitation requires further study since the responsiveness of neurohumoral regulatory mechanisms to other types of environmental stressors might in some cases be heightened.

Evidence indicating that cellular changes in brain RNA and protein can be used as indices of metabolic dysfunction has recently been reviewed by Talwar (1969). Investigations by Swedish workers in Hyden's and Hamberger's laboratories are especially relevant. Using microsurgically isolated nerve cells it has been shown that acoustical trauma causes profound changes in RNA and protein metabolism of spiral ganglion cells in the guinea pig ear (Hammer, 1956). One hypothesis formulated by Hamberger and Hyden (1945) is that moderate sensory stimulation is associated with increased RNA levels in specified brain areas whereas very intense stimuli can sometimes cause a progressive reduction in neuronal RNA and protein. Recent studies by Crawshaw et al. (1973) and Anthony (1973) indicate that rats exposed to moderately intense noise (100 dB, 1-5 kHz for 60 min) exhibit increased RNA levels in brain cortical neurons and in adrenal fasciculate cells. At present, however, the data are insufficient to speculate on cause-effect relationships between adrenal changes and neurohistochemical responses.

CONCLUSIONS

There is no question that intense noise exposure can elicit hormonal and neural changes which without recovery could prove debilitative. Because nonauditory physiological stress reactions to noise undoubtedly involve interactions between neurosensory processes and hormonal influences, more definitive studies are needed on possible pathological changes in brain elements. One should also recognize that the endocrine stress response of animals subjected to deafening intensities of noise (e.g., 130 dB) is probably very different from that elicited by noise levels causing partial or no hearing impairment (e.g., 80-100 dB), especially over extended exposure durations.

Viewed as a whole, there is no experimental evidence that noise per se causes "excessive" adrenocortical activation to the extent that the adrenals' capacity to function is totally abolished. On the other hand, the well documented observations that even short exposure to moderately intense noise levels can cause an imbalance in adrenocortical activation and gonadotropin output should not be dismissed as spurious findings, since a prolonged suboptimal functioning of neuroendocrine homeostatic mechanisms would certainly lead to adverse health effects in the functioning of most bodily systems.

REFERENCES: see pp. 280-281.

CHAPTER 5

EFFECTS OF ELECTRIC, MAGNETIC AND ELECTRO-MAGNETIC FIELDS AND OF IONIZATION ON ANIMALS

SECTION 1: Biological Effects of Ionization of the Air on Animals (A. P. KRUEGER)

SECTION 2: Biological Effects of Magnetic Fields on Animals
(a) Effects of Magnetic Fields on Neural Behaviour (F. RUSSO and W. E. CALDWELL)
(b) Effects of Magnetic Fields on Animal Behaviour (M. A. PERSINGER)

SECTION 3: Biological Effects of Various Physical Parameters of the Environment on Animals (R. S. POGRUND)

Biological Effects of Ionization of the Air on Animals

by

A. P. Krueger

INTRODUCTION

Soon after d'Alibard in France and Franklin, in the United States, independently obtained electricity from thunder clouds (1752) and Lemonnier (19752) proved that electricity is present in the atmosphere even in fine weather, L'Abbé Bertholon conceived the idea that atmospheric electricity might affect the course of disease in humans and animals. Bertholon's book, summarizing his interesting studies in this field, was published in 1780 and ever since that time laymen, physicians, meteorologists and others have sought links between electricity in the air and mood, affect, and physical health. The status of these inquiries was vastly improved by three sets of observations:
1. Air conducts electricity by virtue of charged gaseous molecules, often attached to water molecules—the air ions of Elster and Geitel (1899) and Wilson (1901);
2. Certain weather complexes such as the Foehn, the Sharav, the Santa Ana, etc., induce a variety of signs and symptoms in weather-sensitive people (Danov et al., 1969).
3. During these periods, the prevailing electrical balance of the atmosphere is upset and an excess of positive ions develops (Czermak, 1901).

Serious study of the effects of air ions on disease states has been made possible by the development of suitable air ion sources and instruments for measuring their concentration. Such investigations have provided acceptable evidence that air ions are capable of (1) modifying the rate of death in models of respiratory infections produced in experimental animals by challenge with measured doses of a bacterium, a fungus or a virus and (2) acting as disease incitants in weather-sensitive humans.

MODELS OF RESPIRATORY DISEASE

THE EFFECT OF UNIPOLAR POSITIVELY IONIZED AIR

on the course of Coccidioidomycosis in mice (Krueger and Levine, 1967).
This experimental model was selected because the intranasal instillation of measured numbers of arthrospores produces a very characteristic syndrome with a uniform course and mortality rate. The infection remains restricted to the lungs for at least one week (Kong et al., 1964) and despite marked fungal proliferation in the lung parenchyma, few mice succumb before the second week of the disease. The extent of extrathoracic metastasis correlates inversely with host-resistance.

The effects of three environments were tested in this series (a) ordinary laboratory air (controls), (b) air containing 3-4 $\times$ 10^5 positive ions/cm^3 and (c) laboratory air in the presence of an electrical field equivalent to condition b). By the end of 30 days approximately 30% of the control animals had died, compared with 55% of the ion-treated animals. The difference in mortality rates was significant at the 97.5% confidence interval by chi-square analysis. The electrical field had no influence on the cumulative mortality rate.

THE EFFECT OF POSITIVE AIR IONS ON BACTERIAL PNEUMONIA IN MICE

Randomly selected groups of mice were challenged by intranasal instillation of ca 160 cells of *Klebsiella pneumoniae* in a volume of 0.05 ml. One group of animals was exposed to inhalation of 1-2 $\times$ 10^5 small positive ions/cm^3, another group served as an electrical field control and the third group constituted an untreated control. Early in the disease the rate of death was significantly increased by exposure to positive air ions. The electrical field had no detectable influence on the mortality rate (Krueger et al., 1970).

THE EFFECT OF THE AIR ION ENVIRONMENT ON INFLUENZA IN THE MOUSE

This model of respiratory disease has been the subject of prolonged investigation (Krueger et al., 1970), (Krueger et al., 1971), (Krueger and Reed, 1972) and (Krueger et al., 1973).

With the cumulative mortality rate among controls ranging from 25% to 85% twelve days after challenge:

1. Unipolar low densities of positive or negative ions averaging 3 $\times$ 10^3 ions/cm^3 substantially increased the death rate.
2. Mid-range densities of either charge (17-20 $\times$ 10^3 ions/cm^3) were totally without effect.
3. High concentrations of negative ions (2.2-3.0 $\times$ 10^5 ions/cm^3) significantly decreased the cumulative mortality rate.
4. Ion-depleted air averaging 70 positive ions and 80 negative ions/cm^3 of air produced a marked rise in the mortality rate.
5. Bipolar ionized air containing 3.2 $\times$ 10^3 negative ions and 1.3 $\times$ 10^3 positive ions/cm^3 lowered the death rate.

All of the control animals breathed air representative of the ionic environment one finds in clean air over land; i.e., a total of 2.0-3.5 $\times$ 10^3 ions/cm^3 with a ratio of positive to negative ions of ca. 1.2:1.0.

More recently, we have repeated our experiments with influenza virus using instead of intranasal challenge, administration of the virus as an aerosol consisting of droplets 2.0 microns in diameter. The cumulative mortality rate by day 11 among controls was 40-55%. Low concentrations of negative or positive ions (2.7-5.0 $\times$ 10^3 ions/cm^3) high concentrations of negative or positive ions (2.3-5 $\times$ 10^5 ions/cm^3), ion-depleted air and comparable electrical fields, all were without measurable influence on the rate of death.

SUMMARY OF SPECIAL INSTRUMENTATION AND UNIQUE METHODOLOGY

The satisfactory performance of experiments designed to determinne the effects of air ions on physiological processes in animals requires attention to a number of important details; as applied to our work, they are enumerated below:

ANIMALS AND AIR ION TREATMENT CHAMBERS

We routinely employed specific pathogen-free ♀ mice of the NAMRU strain 4-5 weeks of age. They were fed Purina Lab Chow and water ad libitum and were kept in groups of 10 in plexiglass exposure chambers described by Krueger et al. (1968) (Fig. 1). The chambers, in turn, were located in three cubicles (Fig. 2). Mice were exposed

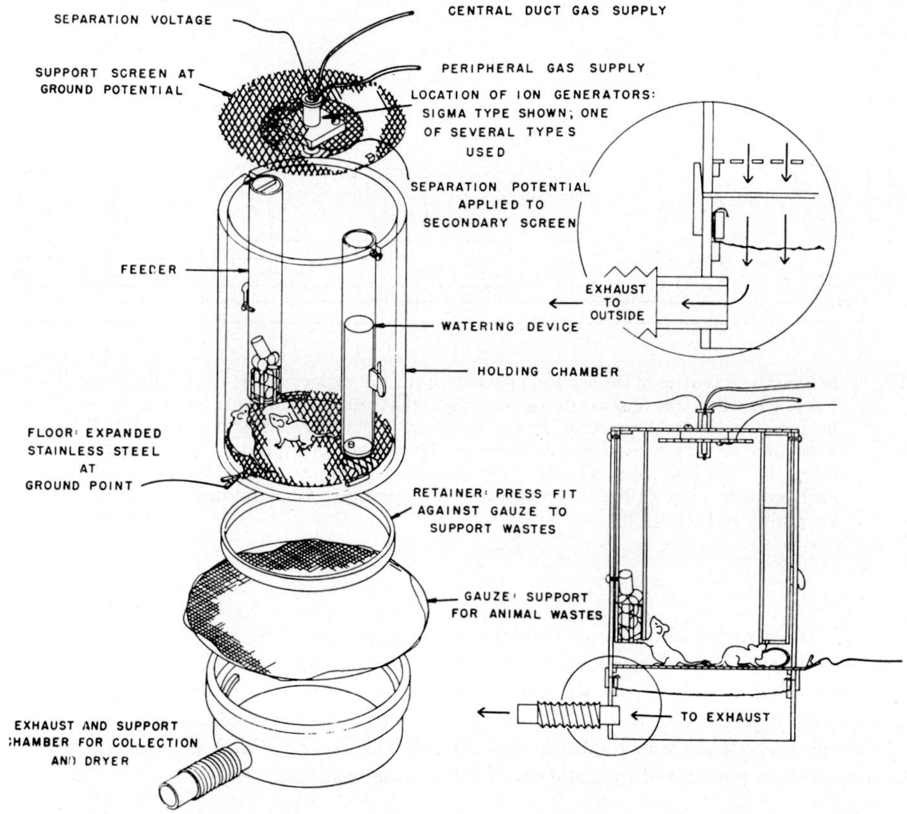

Fig. 1. Air ion exposure chamber made of plexiglass and housing ten mice. Pure air (0.178 m^3/min) flows from top to bottom. Ion generators are attached to top support screen. This screen and deck on which mice live are kept at ground potential. Feces and urine are collected on gauze below deck level and are dried in the air stream. preventing back diffusion of ammonia and water vapor into the chamber. The outside surface of the chamber is treated with static eliminator.

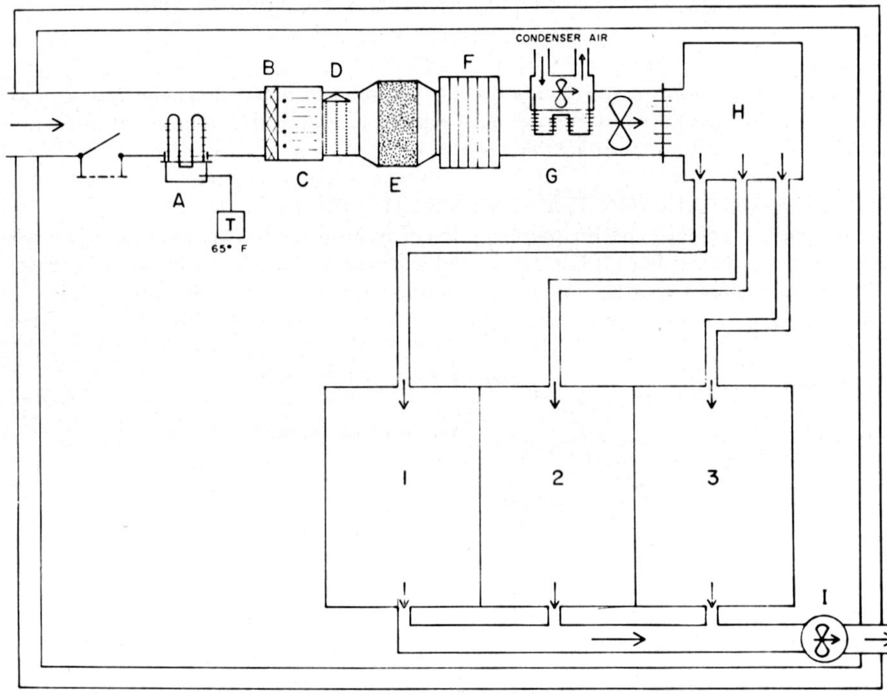

Fig. 2. Schematic drawing of three cubicles used for air ion experiments. Each cubicle measures 100 × 150 × 210 cm and contains nine plexiglass animal exposure chambers. A: Duct feeder with a room thermostat; B: Coarse filter; C: Trion electrostatic filter; D: Rectifier to remove excess positive ions produced by electrostatic precipitator; E: Activated carbon filter; F: Absolute filter; G: Air conditioner; H: Plenum; I: Exhaust fan. Air enters each cubicle at a rate of 1.84 m^3/min. Temperature is maintained at 26°C ± 1°C. Humidity is 45% ± 7%.

to a pre-determined air ion concentration for a period of 72 hrs, challenged, and returned to the initial air ion environment for periods ranging from 11-13 days. Controls consisted of mice exposed to air containing 2.0-3.5 × 10^3 ions/cm^3 with the ratio of positive to negative ions of ca. 1.2:1. Artificial light was supplied to each cubicle for 12 hrs per day. As animals died, the day of death was recorded and as soon as possible, autopsies were performed to make sure of the cause of death.

In performing intranasal challenge, the mice were lightly anesthetized with alcohol-chloroform-either mixture and 0.05 ml of the infecting agent was introduced in the nares from a micropipette. Challenge with aerosolized influenza virus was performed in a sealed chamber into which the aerosol was introduced from a Well's Reflux Atomizer.

AIR ION GENERATION AND MEASUREMENT; ELECTRICAL FIELD MEASUREMENT

Small air ions were generated from tritium-powered heads, each containing 50 mc of tritium adsorbed on zirconium or titanium and positioned on the top of the treatment chamber 27 cm above the mice. A single generator operating at 67.5 volts sufficed to produce 1×10^3-3.5×10^3 small ions/cm^3 at deck level. Five generators with an applied voltage of 900-980 volts ionized the air to a level of 2-4 $\times$ 10^5 ions/cm^3.

Ion collectors measure the spatial distribution of ions in terms of numbers/cm^3. We use an instrument designed by John C. Beckett, which consists of a series of parallel aluminum plates of known surface area. The plates are mounted on Teflon insulators in a Chamber through which air flows at a uniform rate. The plates can be charged at 22 volts, 67-1/2 volts, or 200 volts. It is assumed that each ion has one unit charge—an assumption that has been shown by Loeb (1967) to be highly probable on particles of the size now under consideration. In a uniform electrical field, a large ion with one unit charge (1.6×10^9 coulombs) will move more slowly than a small ion. Accordingly, ion mobility is inversely proportional to size. When the plates of the collector are charged at 22 volts, only the small ions impinge on them, and the very small current thus developed is measured with a femtoammeter.

In establishing exposure conditions within the cages, air ion concentrations were determined directly with a miniaturized ion collector (Krueger et al., 1966). Immediately after taking these readings the flow of current from deck to ground was measured. The metal deck served as a target probe and the current to ground reading could be equated with the numbers of ions found to be present in a unit volume of air. During the actual experiment the flow of current to ground was monitored and if any deviation from the initial value occurred, necessary remedial steps were taken.

For experiments involving mixed ion atmospheres, having ratios of positive to negative ions of $>$ or <1.0 and for those requiring unipolar mid-range concentrations of ions of either charge, we found it advisable to ionize the ambient air in the cubicles rather than in individual cages. For this purpose, multiple positively and negatively charged tritium generators were installed near the tops of the cubicles. To provide atmospheres containing abnormally low numbers of air ions, all the air entering one cubicle was passed through an ion trap. After passing through the trap, the cubicle air averaged <80 positive or negative ions/cm^3.

Electrical field strength measurements within the cages at the level of the metal deck on which the mice lived were made with a modern type of field mill.

CURRENT CONCEPTS AND HYPOTHESES OF MODE OF ACTION OF AIR IONS

A great deal of work has been done on penetration of air ions into the respiratory tract and their retention in these tissues. There is no general agreement as to the dissemination of inhaled small air ions, although the preponderence of experimental evidence suggests that penetration is shallow. There is no doubt that once air ions enter the body most of them are retained, i.e., they cannot be detected in expired air.

Regardless of the degree of ion penetration and retention, it has been demonstrated abundantly that negative and positive ions evoke a wide range of physiological and biochemical changes ascribable either to local contact or to indirect systemic effects. In well-controlled double-blind tests, Winsor and Beckett (1958) demonstrated that nasal inhalation of positive ions by humans produced dry throat, husky voice, headache and obstructed nose; maximal breathing capacity was reduced by ca. 30%. Negative ions were without effect. Palti et al. (1966) performed 41 experiments with positive and negative ions on 19 infants, 13 of whom were suffering from spastic bronchitis and/or bronchopneumonia. Treatment with negative ions induced relaxation of the smooth muscle of the bronchi and relieved tachypnea. Exposure of normal infants to positive ions induced tachypnea and spasm of the bronchial tree; the symptoms were relieved by treatment with negative ions.

In experiments with mice, guinea pigs, rats and rabbits, Krueger and Smith (1960a, 1960b) found that positive ions decreased the efficiency of the mucociliary escalator of the trachea, contracted the posterior tracheal wall, induced a state of exaggerated response of the mucosa to trauma and produced vasoconstriction, while negative ions accelerated the functional activity of the tracheal clearing mechanism and released contraction of the tracheal wall resulting from exposure to positive ions. Other investigators have failed to detect air ion effects on the mucociliary escalator (Andersen, 1969; Guillerm et al., 1966).

Quite apart from the clearing mechanism, Krueger and Smith (1960b) found that negative ions decreased the concentration of serotonin in excised strips of rabbit trachea and in the respiratory tract of mice exposed to high concentrations of negative ions. This appears to be part of a general systemic phenomenon, for blood and brain levels of serotonin are influenced by the air ion content of the atmosphere (Krueger et al., 1968, 1969; Gilbert, 1971; Olivereau, 1971). Serotonin is a powerful and versatile neurohormone known, among other actions, to cause bronchospasm and to exert multifaceted effects on the cardiovascular system, including vasoconstriction of the pulmonary vessels producing pulmonary hypertension and bronchospasm with resultant obstruction of the airway (Garattini and Valzelli, 1965). Apparently negative ions stimulate oxidation of serotonin by monoamine oxidase, for exposure of guinea pigs to high concentrations of negative ions produces an increase in urinary excretion of 5-hydroxyindoleacetic acid, a specific metabolite of serotonin.

These experimental and clinical data are helpful in analyzing air ion effects on the model of respiratory disease that has been most intensively studied, i.e., influenza in the mouse. As has been recounted above, the air ion content of the atmosphere exerts considerable effect upon the mortality rate of mice challenged intranasally with influenza virus, while no such effect occurs when the challenge dose is administered as an aerosol. Influenza virus introduced into the respiratory tract by the intranasal route is subject to all the protective mechanisms of the mucosal surfaces it traverses, and the amount of active virus persisting in the tracheobronchial tree or surviving to penetrate into the lung parenchyma will be inversely proportional to the efficiency of these functions.

It is generally agreed that small air ions impinge upon the mucosa of the upper portions of the respiratory tract. Even if one discards the controversial areas of specific actions, substantial evidence yet remains that they alter physiological processes in this region e.g., congestion and bronchoconstriction. Therefore, it is reasonable to point to air ion imbalance as the cause of shifts in mortality rate that occur following intranasal challenge with influenza virus. Such a relationship does not prevail in the case of aerosol challenge, for here, a given quantity of virus can be expected to reach the ultimate air sacs without touching the lining of the tracheobronchial tree; in consequence, the alveolar infecting dose is independent of these protective functions.

This mechanism, proposed to account for the role of air ions in altering the course of respiratory disease, is advanced as a tentative hypothesis. It is compatible with the known actions of air ions on tissue levels of serotonin and changes in the latter could lead, in turn, to the observed alterations in the mortality rate of mouse influenza. On the other hand, it neglects the possibility of air ion effects on classical elements of cellular and humoral immunity. The only reason for this omission is that Krueger and Levine (1967) could not detect any immunological impairment in the single case studied to date; i.e., the syndrome produced by instillation of measured numbers of *Coccidioides immitis* arthrospores into the nares of mice.

CONCLUSIONS

There is substantial evidence leading to the conclusion that the small air ion content of the atmosphere can influence the course of respiratory disease in experimental animals, and that states of air ion imbalance accompanying the winds of ill repute produce illness in weather-sensitive individuals.

Unipolar low concentrations of positive or negative ions, high concentrations of positive ions and ion-depletion increase the cumulative mortality rate in mouse influenza produced by intra-nasal challenge. High concentrations of negative ions or a mixed air ion atmosphere with negative ions predominating reduce the mortality rate. Mid-range concentrations of either positive or negative ions have no effect on the mortality rate. The hypothesis has been developed that air ion-induced alterations in blood levels of the powerful neurohormone, serotonin, and localized air ion effects on the upper portion of the respiratory tract account for these phenomena. This is consistent with the observation that air ions have no influence on mouse influenza resulting from inhalation of aerosolized virus—a procedure involving introduction of considerable amounts of virus directly into the lower respiratory tract. It should be emphasized, however, that the serotonin mechanism may not be the whole story, for no studies have been made of the possible involvement of other biogenic amines; e.g., acetylcholine, or of the classical immunologic elements.

The role of air ions, apparently causing disease in humans, has been suggested by two groups of Israeli investigators. Robinson and Dirnfeld (1963) discovered that air ion imbalance occurs 10-12 hrs before the drop in humidity and rise in temperature that characterize the Sharav of the Near East. It is during this early period, when the total numbers of air ions increase (with positive ions predominating) that the signs and symptoms described by Danon et al. (1969) develop in some members of the exposed population. These individuals suffer from insomnia, irritability, tension, migraine, hot-flashes with chills or sweating, a variety of allergic complaints, dyspnoea, palpitation or precordial pain, hyperperistalsis, polyuira, tremor and vertigo. Relief occurs following treatment with negative ions or administration of drugs which block formation of serotonin in the body (Sulman, 1971).

In the case of mouse influenza, air ion-induced changes in tissue levels of serotonin apparently account for the alterations in the mortality rate described above. So far as human illness, such as that developing in connection with the Sharav is concerned, excessive production of serotonin resulting from air ion imbalance* seems to be the specific physiological mediator accounting for the observed symptomatology.

Obviously, more measurements of abnormal air ion concentrations associated with established meterological patterns, such as the Sharav, Santa Ana, Foehn, etc., are needed. Also, it is both feasible and desirable to attempt to reproduce the syndromes of weather-linked illness in volunteers under controlled laboratory conditions with concommitant evaluation of physiological and biochemical parameters. In pursuing future studies of air ion effects on respiratory disease in animals, one logically should consider the possibility that these effects might involve broad shifts in endocrine function such as Olivereau (1971) and Gualtierotti (1965) independently have demonstrated in non-infected small animals. At the same time, studies are needed of air ion-induced alterations of all the known cellular and humoral factors that contribute to host-resistance.

REFERENCES: see pp. 282-283.

* NOTE BY THE EDITOR-IN-CHIEF: It should be kept in mind that the specific thermal stresses during Sharav, Foehn etc. produce serious thermoregulatory disturbances which are usually accompanied by changes in serotonin production and excretion. In view of the many contradictory reports which have appeared on the biological effects of air ions, mainly due to the very difficult methodology and the difficulty to exclude the suggestion factor during clinical studies of subjects suffering from Sharav and Foehn, it is highly recommended that readers study also the following reports: I. ANDERSEN: Effects of Natural and Artificially Generated Air Ions on Mammals. In: Biometeorology 5. Proc. 6th Int. Biometeor. Congr. Swets and Zeitlinger Publ. Co., Amsterdam: 229-238 (1972); and the comments of A. P. KRUEGER: Are Air Ions Biologically Significant? A Review of a Controversial Subject. Int. J. Biometeor., 16, 4: 313-332 (1972).

Chapter 5, Section 2a

Effects of Magnetic Fields on Neural Behaviour*

by

F. Russo and W. E. Caldwell

1. INTRODUCTION

The earth's geomagnetic field is an unobserved phenomenon (like gravitation) varying daily with the cosmic undulations (Fig. 1).
Originating from the center of the earth and influenced by the solar winds, the geomagnetosphere has taken the configuration shown in Figure 2.

The MAGNETOPAUSE extending from 8 to 14 earth radii (one earth radius equals approximately 4,000 miles) on the solar side of the earth, surrounds the earth with its protective magnetic lines of force. Just as the fragile shell of an egg protects the developing embryo within, the geomagnetosphere has for more than 450 million years (paleontological time) protected the evolving life forms on the earth's surface. Lethal cosmic rays and particles are deflected or trapped within the magnetopause, preventing their bombardment of the earth's surface.

Recent satellite excursions have brought back much illuminating information. Within the magnetopause the earth's field is rather consistent measuring about 20 gammas (one gamma equals 10^{-5} gauss). Crossing the boundary, the field suddenly increases to as much as 50 gammas and then diminishes rapidly with distance. This field was surmised no longer to be that of the earth's but rather that of the interplanetary field (Cahill, 1965). Within the TAIL, the outer boundary of the magnetopause, i.e., the 20 gamma field, was found to extend to 21 earth radii. Also, within the tail was found a fluctuating region of very weak magnetic intensity replete with charged particles. This region, called the PLASMA SHEET, is suspected as the source of many earth-bound phenomena such as the aurora borialis or northern lights, the aurora australis, etc. (Schardt and Opp, 1966).

In the midst of the upheaval of magnetic fluctuations and charged fields stands man and his companion animal life, living on from day to day, oblivious to the fluctuating cosmic events. And justly so, since for over 1.8 million years of evolution these events had determined the very essence of his being. Man and animal alike were so affected in their evolution and are yet affected. The eleven year sun spot cycle, solar storms, moon phases, and other cosmic events combine and integrate to effect varying magnetic conditions about a receptive and responsive nervous system. How and what mechanisms bring about this interaction between man, animal, and the universe are yet to be ex-

---

* See also Progress in Human Biometeorology, Part IA, Chapter 6, Section 14, pp. 392-399, paper by Madeleine F. Barnothy.

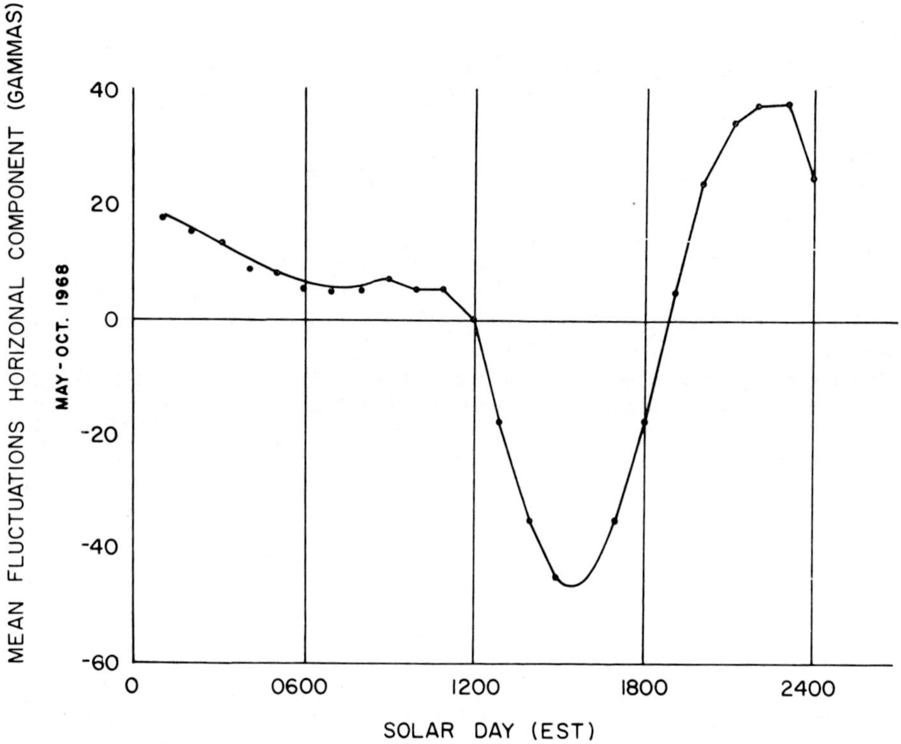

Fig. 1. Mean daily-hourly variations in the horizontal component of the geomagnetic field as recorded at the Weston observatory in Weston, Massachusetts during May through October 1968 (Adapted from Fig. 1 in Stutz, 1971).

plained. However, we shall attempt a plausible explanation in this Section, offering a possible mechanism for field interaction between the biomagnetic and geomagnetic fields.

2. GEOMAGNETIC SENSIBILITY IN ANIMALS

Since the discovery of the polar orientation of the geomagnetic field as manifested by the compass (which in Western Civilization has been attributed to Pliny the Elder, 23-79 AD, the Roman naturalist), scientists and naturalists have wondered if the same qualities were not inherent within some animals. Inquiries into the basic mechanisms directing pigeons home or gulls to their migratory grounds, or fish to their spawning grounds, have expanded into a new field of investigation called Biomagnetics. Here, the biologist, psychologist, physiologist, biophysicist and bio-engineer work side by side to unfold the manifold vagaries underlying this phenomenon in animals.

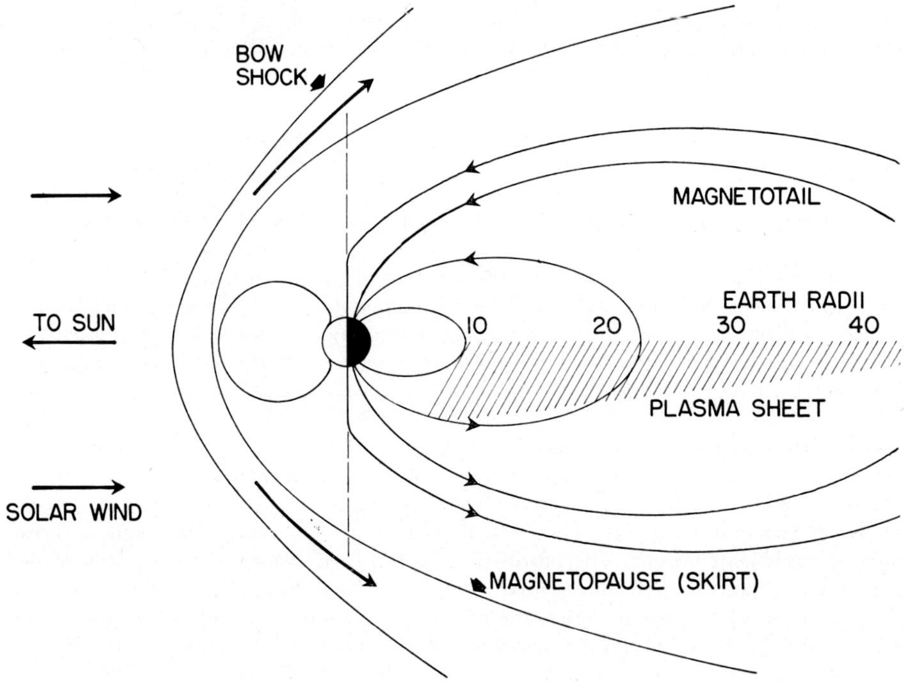

Fig. 2. General configuration of the geomagnetosphere (After Schardt and Opp, 1967, p. 257).

A. ABBREVIATED REVIEW OF THE LITERATURE

The studies into geomagnetic effects upon animals generally have taken the tack of orientation phenomenon. The results of these studies have contributed much information in the homing and migratory habits of pigeons and other animals. These studies, however, were not exclusive to the avian species; but the invertebrates were scrutinized as well.

1. *Orientation Studies*

Of these species, the work on the mud snail and planarian are well known. Brown (1971), Barnwell and Brown (1964) discovered a counter-clockwise turning response of these animals when presented within an artificial magnetic field (0.15-10 gauss). The degree of responsiveness was found to vary according to the field configuration and the "moon-phase" or time of the month. These results were questioned by Gottlieb and Caldwell (1967), who repeated the experiment and found no turning effect, but rather reported an augmented activity level in their animals when exposed to the magnetic field. The degree of augmentation was also found to vary according to the time of the month. Picton (1966), on the other hand, corroborated Brown's work using flies as subjects.

Honey-bees are known for their communicative abilities within the hive (Von Frisch, 1955). In this dark and active environment the returning forager bee relates direction and distance information of her find to her hive mates through a dance known as the WAGGLE DANCE. Lindauer (1968) found that he could vary the number of errors transmitted in the dance by varying the magnetic field influence around the hive. He found that strengthening the field increased the number of errors, whereas, weakening and even "bucking out" the geomagnetic field decreased error transmission.

In the same vein, Russo (unpubl.) in an informal experiment, effected a disorientation in the photo-taxis response of nocturnal moths introduced within a strong a.c. magnetic field (approx. 25-50 gauss). The moth's ability to respond to a pen light in a dark surround was tested before hand and found without fault. The moth's course to the light was direct and prompt. However, the photo-taxic response in the presence of the a.c. magnetic field initially, was prompt, directed toward the light, but terminated "off center", with the moth unable to "seek out" the light source.

Of the vertebrate studies, the orientation responses of the avian species are most familiar. Keeton (1971) effected a disorientation of the homing mechanism in "first flight" pigeons released over unfamiliar territory by affixing a small bar magnet to the necks of the experimental pigeons. The control pigeons carried small non-magnetic brass bars of equivalent weight. All control pigeons returned home, whereas many of the experimental pigeons were disoriented, returning home after a long delay. Some did not return at all. It appeared that the magnets did confuse the pigeons initially, but they were able to reorient using other cues, e.g., the sun, wind currents, land marks, etc.

Ring-billed gulls released in an orientation cage eight feet in diameter were reported by Southern (1971) to rest preferentially in the southeastern segment of the cage (the direction of their migratory flight). Of greater interest, Southern found that during magnetic storms, the gulls became disoriented, resting in random directions.

2. *Detection Studies*

Other than orientation behavior, the conditioned response and choice behavior are also methods for detecting magnetic sensibility in animals. McCleave et al. (1971) conditioned eels and salmon to respond with bradycardia (slowing of heart rate) when confronted with a magnetic stimulus. These animals showed no response to a weak magnetic field (1.90 gauss). Whereas, Kholodov (1964) and Yakovleva and Medvedeva (1972) reported positive responses with stronger fields (100 and 520 gauss respectively).

Free or choice response experiments demonstrated the animal's sensibility to magnetic stimuli through free approach-avoidance behavior. In an "open field" exploratory study with honey-bees for example, Caldwell and Russo (1968) found that the bees had a tendency to avoid the applied field. A subsequent study (Russo and Caldwell, 1971), where bees "chose" to occupy or vacate a compartment containing either a d.c., a.c., or pulsed magnetic field (7-450 gauss), bees chose in one instance to vacate and in all instances to spend less time within the compartment compared to controls. The a.c. and p.d.c. fields, because of their fluctuating nature, were found to have a significantly stronger effect than d.c. magnetic fields.

3. *Physiological Studies*
The physiological manifestations of magnetic sensibility have been very convincing. In ablation studies on electric fish, Kholodov (1964) reported the diencephalon of the brain stem to be most sensitive to magnetic stimuli.
Another parameter of physiological investigation closely allied to behavioral observation is the animal's activity level.

The majority of these cases report the augmentation of activity levels to magnetic stimuli. The snail work of Gottlieb and Caldwell (1967) mentioned earlier reported this phenomenon in their animals. Higher activity levels in chitons were found by Ratner (1973). The activity level varied inversely with the strength of the applied field. Bees were also observed by Russo and Caldwell (1971) to display high activity within the field compartment. Gerbils were found by Stutz (1971) highly active between 15.00 and 18.00 hr, i.e., the hours of greatest geomagnetic fluctuation (Fig. 1).

4. *Other Parameters*
Other physiological parameters of interest were those performed on human beings. Within these studies, the opposite effect, i.e., that of hypoactivity instead of hyperactivity, was noted.
Cassiamo and Audisio (1966) noted a significant depression of pulse rate in all 12 male subjects exposed to a 338 gauss magnetic field. Friedman et al. (1967) reported significant increases in reaction time (i.e., slower response time) in 30 subjects. The subject's disposition was such that his head was positioned between the poles of a 5-11 gauss a.c. magnet during the reaction time task.

The flicker fusion rate (FFR), the phenomenon which allows us to view a motion picture as a continuous event rather than a series of flickering, jerking frames, is an important neurological test for detecting subtle dysfunctions in the central nervous system (CNS). Beischer (1965) found FFR to decrease significantly from a normal 12 cps to 8 cps in a NULL geomagnetic field condition. Restoration of the field restored the FFR to normal.

B. DISCUSSION AND CONCLUSIONS
Pulling together the threads of the preceding studies would be a difficult task since there appears to be much conflicting evidence. However, there appears a continuous thread running through each. The theme of activation appears most often in one form or other. Activation is very much a function of the CNS acting in more general rather than specific manner. The orientation response, for example, a very specific response, directing the animal in a specified direction is questionable. Is orientation really specific or could it be a general response, which, on the neural level, is an attempt to reduce the "noise" impinging upon it by the magnetic field?

Undoubtedly, the aroused activity levels of the snail (Gottlieb and Caldwell, 1967; Ratner, 1973) and gerbil (Stutz, 1971) experiments in essence are general responses since the magnetic stimulus affects the total animal rather than a specific part. The disorienting effects of magnetic fields and their anomalies on homing pigeons (Keeton, 1971), gulls (Southern, 1971), and bees (Lindauer, 1968) corroborate CNS dysfunction due to interference-interference, that is, due to disruption of inhibitory control mechanisms. Interference injected within the feedback loops, basic to homeostatic control systems, resulting in runaway neural firing (hyperactivity).

Therefore, rather than information, the presence of a foreign magnetic stimulus, or anomalies of the ambient field, may act as a source of noise. Noise that blocks feedback control systems leading to pronounced activation of neural mechanisms.

It will be seen later that the inhibitory mechanism may very well be the function assigned to the glial structures surrounding every neuron.

3. GEOMAGNETIC SENSIBILITY AND NEURAL MECHANICS

A. THE SEMICONDUCTOR HYPOTHESIS OF NEURAL BEHAVIOR

Living organic tissue electrically behaves neither as a pure conductor nor as a true non-conductor. They lie somewhere between, that is, more like a semiconductor. The nervous system, specialized cells for the conduction and propagation of electrical impulses display many of the physical properties of semiconductors as contrasted to pure conductors. Some of these are shown in Table 1.

Table 1. Some properties of semiconductors common to neurons as contrasted to those of pure conductors

SEMICONDUCTORS	PURE CONDUCTORS
1. One way conduction	1. Two way conduction
2. All or none propagation (Discriminatory process)	2. Total conduction (Nondiscriminatory process)
3. Electron and ionic activity	3. Electron activity only

Therefore, a reevaluation of neural mechanisms in the light of semiconductor physics may lend further insights into neural behavior.

1. *Summary of Semiconductor Mechanics*

Semiconductors are electronic devices, e.g., transistors, made of either germanium or silicon crystals. Pure germanium or silicon in nature is a nonconductor of electricity. The introduction of an impurity, such as arsenic or indium, within the crystalline lattice structure causes an electrical imbalance or potential for current flow (Fig. 3A).

Arsenic, a donor atom which releases electrons from its outer shell into the crystalline structure as free electrons, becomes a positively charged ion. The crystal containing an excess of free electrons becomes negatively charged and is called N-TYPE material. Indium, on the other hand, an acceptor atom, removes electrons from the crystalline structure and becomes a negatively charged ion. Consequently, the crystal is imparted with a positive charge and is called P-TYPE material. The removal of electrons from the crystalline structure of P-type material is regarded as creating HOLES within the structure. These holes are viewed as freely moving particles having positive charges

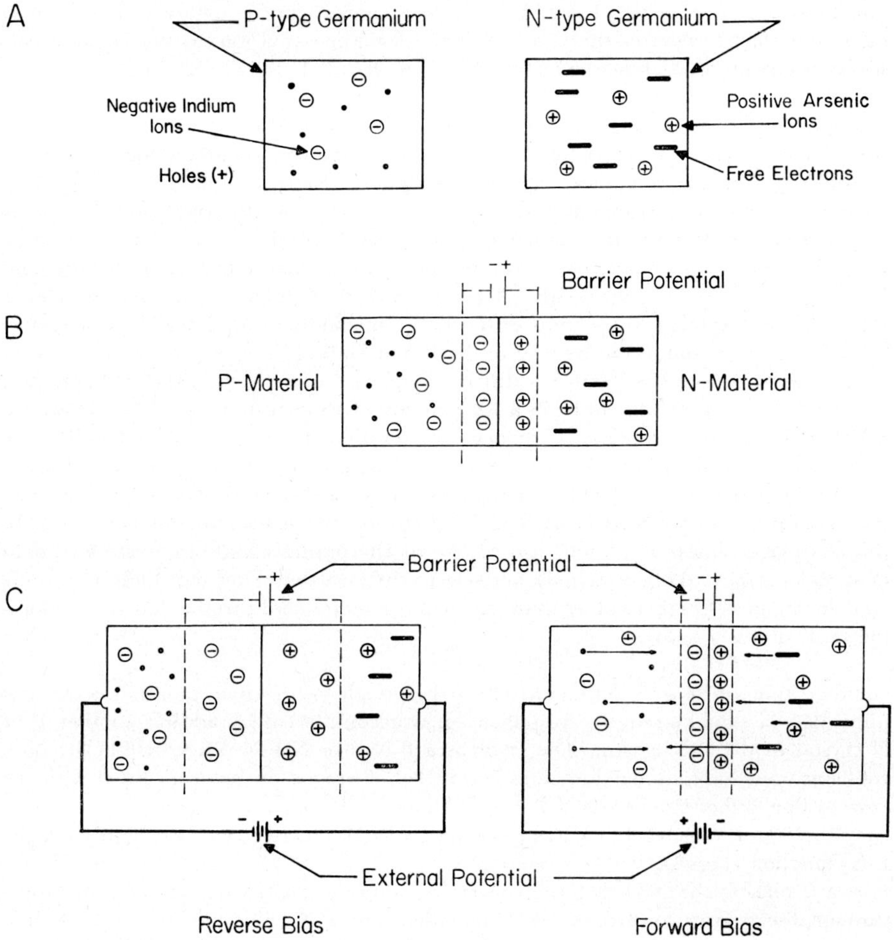

Fig. 3. Basis of Semiconductor mechanics: A: two germanium crystals injected with respective impurities; (left) indium, and (right) arsenic; B: formation of barrier potential upon joining the above crystals forming a potentially active DIODE, and C: manipulation of the barrier width by an external source, (left) widening of barrier or reverse bias condition thereby increasing resistance to current flow, and (right) narrowing of barrier or forward bias condition, decreasing resistance to current flow (Adapted from U.S. Dept of the Army, 1959).

and flowing opposite to electrons. When the two germanium crystals, for example, one of P-type and one of N-type material are joined together, there is an instantaneous rush of electrons and holes toward the P-N junction (Fig. 3B). Some of the electrons and holes cross the junction and become electrically compensated. However, the arsenic and indium ions also congregate at the junction, but because of their size, are unable to cross and therefore are left uncompensated. Consequently, the congregation of the ions form a barrier, which by repulsion of like charges, prevents further migration of holes and electrons across the junction. The uncompensated ions at the junction constitute a potential difference, which is called the BARRIER POTENTIAL.

The barrier potential is an important concept in semiconductor physics because it is the principle controlling mechanism by which the semiconductor performs functions such as rectification, switching, and amplification of electrical signals.
An increase of the barrier potential widens the barrier width which in effect raises the internal resistance of the semiconductor. A low resistance, i.e., a narrow barrier, will allow electrons or holes to cross the junction, thereby effecting a current flow. Manipulation of the barrier width is accomplished through an external voltage source, e.g., a battery applied to the ends of the P-N semiconductor or DIODE. Connecting the negative terminal of the battery to the P-material and the postive terminal to the N-material increases the barrier width by "packing" more ions at the junction. The semiconductor is then said to be in a REVERSE BIAS condition (Fig. 3C). However, if the terminals were reversed, i.e., negative terminal to N-material and positive to P-material, ions are attracted away from the junction thereby depleting the barrier and allowing electrons to flow across the junction toward the positive battery terminal. This condition is known as FORWARD BIASING the semiconductor (Fig. 3C). In this manner, a simple P-N diode can rectify an alternating signal (a.c.) into a straight or direct current (d.c.) or it may act selectively as a switching mechanism, turning "on" for signals of higher values than its barrier potential and turning "off", i.e., blocking those of lower value.

Another function of semiconductors effected through barrier manipulation is the amplification of signal strength. Amplification is accomplished by adding another P or N crystal to the P-N crystal. The result is a P-N-P or N-P-N TRANSISTOR. Both configurations have similar functions except P-N-P transistors are concerned with the current flow of holes, whereas N-P-N transistors are electron flow devices.
Amplification is effected by forward biasing the N_1-P junction and reverse biasing the P-N_2 junction (Fig. 4).
Forward biasing the N_1-P junction lowers its resistance, allowing electrons to propel through the barrier toward the P-N_2 junction. The P-N_2 junction, in a reverse bias condition, presents a wall of resistance against the rushing electrons. However, the impetus of the oncoming electrons are too great, and are able to penetrate the thickened barrier. The electrons, once through the barrier, are drawn by the attractive force of the positive battery terminal.

Ninety percent of the electrons traversing the first junction successfully penetrate the second barrier to the other side of the transistor (U.S. Dept. of the Army, 1959). Therefore, the result is amplification of power, since:

$$\text{Power (Watts)} = I^2(R_2 - R_1) \qquad 1)$$

Where I is the current in amperes, R_1 is the resistance in ohms at the N_1P junction and R_2 is the resistance in ohms at the PN_2 junction.

From formula 1 we can see that power has increased because the internal resistance

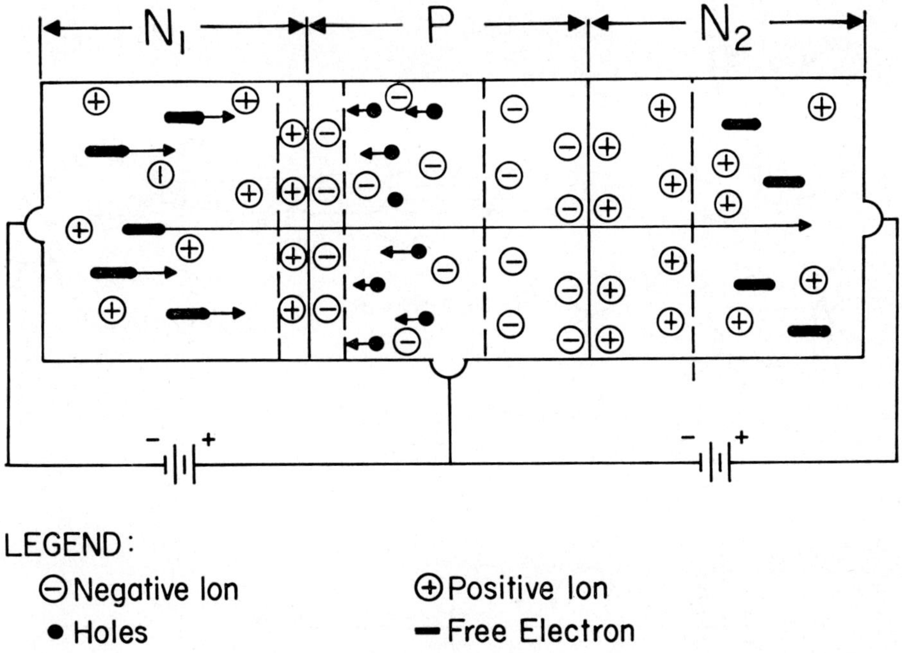

Fig. 4. The process of amplification in an NPN transistor (After U.S. Dept of the Army, 1959).

of the transistor was increased from the first junction to the second. Therefore, current being constant from input to output, an increase in resistance would affect an increase in power output which would manifest itself as an increase in voltage. In this manner, a transistor can amplify signals of the order of microwatts (1/1,000,000 watts) into milliwatts (1/1,000 watts).

2. *Morphology and Function of the Neuron*

The neuron, a living cell, specialized to conduct information from without and within the body to the (CNS) and return, comes in many sizes and shapes. Some run the length of the body or are as short as a few millimeters, e.g., interneurons within the brain and spinal cord. Generally, most neurons have the structure shown in Figure 5 (top). The neuron is composed of five parts: the DENDRITES, which receive impulses from other neurons, the CELL BODY, which carries on the physiological life processes of the neuron, the AXON or neural fiber, which carries the electrical impulse along its length to the FIBER ENDINGS where the impulse is transmitted to other surrounding neurons.

The fifth element which has been long neglected and only recently attained proper attention in neurophysiological research is the MYELIN SHEATH. The sheath, a lipoprotein structure comprised mainly of GLIAL CELLS, covers the length of the axon, broken at regular intervals, called the NODES OF RANVIER. For many years, the sheath was believed a passive structure acting only as an insulator and support

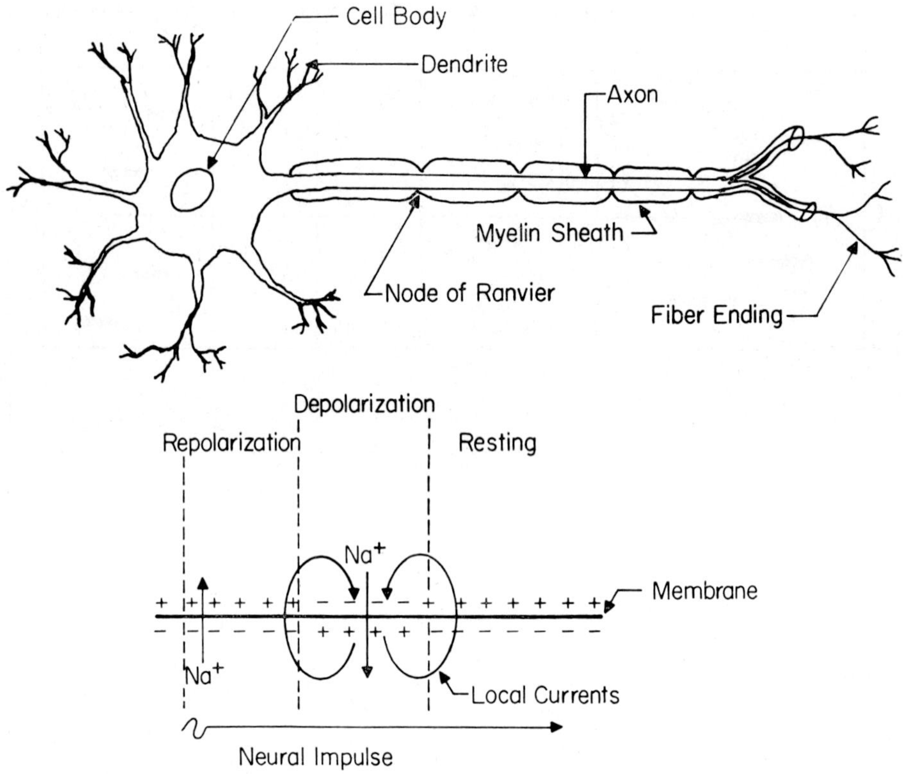

Fig. 5. Neural mechanics: general structure of the neuron (top) and "sodium pump" theory of neural impulse propagation (bottom).

for the active neuron. The importance of the myelin sheath has become evident in research of the debilitating diseases of demyelinization, such as multiple sclerosis and Devic's disease. The function of the myelin sheath is yet controversial, however, current glial research has demonstrated the active participation of glia in neural impulse propagation and metabolism. Hydén and Pigon (1960) suggested commerce between glia and neuron of RNA, succinoxidase, cytochrome, and respiratory enzyme (important constitutents for cell metabolism) during physiological stress conditions. Hydén and McEwen (1966) demonstrated that protein S-100, specific to nervous tissue, is concentrated in the glia and Deuter's nuclei of the neuron, suggesting a site for information exchange. Svaetichin et al. (1961) demonstrated the controlling effect of glia upon the neuron. They reported that only glia exhibited inhibitory post synaptic potentials (IPSP), whereas neurons exhibited only excitatory postsynaptic potentials (EPSP). Tasaki and Chang (1958) found that electrical stimulation of glial cells, in vitro and in vivo, evoked potentials similar to neurons but having a much longer time course, e.g., 1,000 times longer. Walker and Hild (1969) reported like phenomenon but also recorded stimulation of the neuron through glia stimulation. In addition, Walker and Hild reported stimulation of glial cells by neighboring glial cells.

The evidence thus far supports direct interaction between the glial cells and neurons. The scope of importance of this cell is magnified a million fold when we consider that the neuron comprises about 10 percent (or about 30 billion cells) of the nervous system and glia comprises the remainder (Vander et al., 1970; Ganong, 1969).

Generation and propagation of neural impulses according to classical neuronology is yet viewed as a "sodium pump" action (Fig. 5, bottom). The resting neuron, divested of its glial sheath, is found to have a surplus of sodium ions (Na^+) on the surface of its membrane, and potassium ions (K^+) on its inner surface. The Na^+ are impermeable to the membrane but K^+ are free to traverse the membrane in both directions, therefore a positive potential difference exists on the upper surface relative to the inner surface of the membrane. This potential difference generally rests at -70 millivolts and is called the RESTING POTENTIAL. Stimulation of the dendrites exceeding this resting potential causes the neural membrane to suddenly become permeable to the sodium ions. The sudden inward rush of positive ions causes the inner surface to become more positive with respect to the outer surface. This process is known as DEPOLARIZATION of the membrane. The interchange of positive and negative potentials continues along the length of the axon as the neural impulse. As the impulse passes a point on the membrane, REPOLARIZATION of the membrane occurs. The sodium ions are "pumped" out and the membrane returns to its resting potential. This action generally occurs in one direction only and is an all or none process. That is, the stimulus must be above threshold i.e., resting potential, to initiate depolarization of the membrane or else it does not fire. But once fired, the impulse continues the length of the neuron without any loss of energy.

The mechanism which selectively alters the permeability of the membrane during impulse propagation is unknown. The supposition of ionic transfer as the mechanism was questioned by Becker who showed that the neural impulse was actually electronic in nature (Becker, 1962). By stimulating a frog's excised sciatic nerve, frozen in liquid nitrogen ($-320°$ F.), Becker recorded a neural impulse. He concluded that the response could only be caused by electron flow since ions are immobile at $-320°$ F.

3. THE NEURO-GLIAL SEMICONDUCTOR

As stated earlier, the neuron electrically behaves very much as a semiconductor. It conducts impulses in one direction, it discriminates between strong and weak signals (all or none function), and could very well have amplifying properties as evidenced by subliminal perception experiments. Therefore, with these ingredients it seems feasible that neural impulse propagation could be explained in terms of semiconductor mechanics.

Let us assume, as proposed by Galambos (1961), that the glial sheath and neural membrane constitute the simplest unit in nervous activity and further designate the glial membrane as N-material containing an over abundance of positive sodium ions and the neural membrane as P-material with potassium ions. Consequently, a semiconductor condition would ensue. With both membranes intimately contacting the length of the neuron (Hess, 1958), a barrier potential equivalent to the neuron's resting potential would result, (Fig. 6, top). Stimulation of the dendritic zone would create a forward bias condition, i.e., a constriction of the barrier potential within the first segment, allowing current flow across the barrier (Fig. 6, centre). This current represents the local current flow generally found accompanying neural impulses. The forward bias condition would be transmitted to the next segment and so on down the length of the axon.

An inhibitory response could be likened to generating a reverse bias condition, i.e., widening the ionic barrier thereby increasing the threshold within each neural segment (Fig. 6, bottom). In this manner a neuron may selectively facilitate (forward bias) or inhibit (reverse bias) an impulse. Further, the barrier potential would be influenced by neighboring neurons through activation of their glial sheath.

The intercourse with neighboring neurons would effect a P-N-P transistor configuration, whereby signals could be amplified. Such may be the mechanism for subliminal perception.

B. MAGNETIC EFFECTS UPON SEMICONDUCTORS AND NEURONS

As discussed in an earlier Section, any current carrying device, whether pure conductor or semiconductor, generates a magnetic field. Conversely, any conductor moving through a magnetic field generates an electric current within it.

1. *The Hall Effect*

The phenomenon of magneto current induction also occurs in semiconductors and is known as the HALL EFFECT. A Hall-effect semiconductor influenced by a magnetic field decreases its resistance (i.e., reduces its barrier potential) thereby allowing current to flow through it. The amount of current flow is directly proportional to the strength of the magnetic field. When calibrated, such a device is called a gaussmeter or magnetometer and may be used to detect and measure magnetic field strengths.

2. *Magnetic Properties of Neurons*

The neuron as a current carrying device is no exception to the aforementioned principles. Seipel and Morrow (1960) recorded the magnetic field generated about a stimulated frog's sciatic nerve. Gengerelli et al. (1961, 1964) substantiated Seipel and Morrow's findings and further identified time and phase relationships of the generated field with that of the actual electric potentials. Cohen (1967) and Cohen and McGaughan (1972) detected and recorded the magnetic field generated by the heart muscle about the human torso recording an identical *QRS* complex as that displayed by the electrocardiograph (*EKG*).

Cohen (1968) successfully recorded the magnetic field generated about the head. These recordings represented the integrated firing of over 10 billion neurons. The accompanying field strength was found to measure about 7×10^{-10} gauss (i.e., 7×10^{-5} gammas). This value represents about 10^{-5} that of the earth's field and 10^{-3} that generated by the heart.

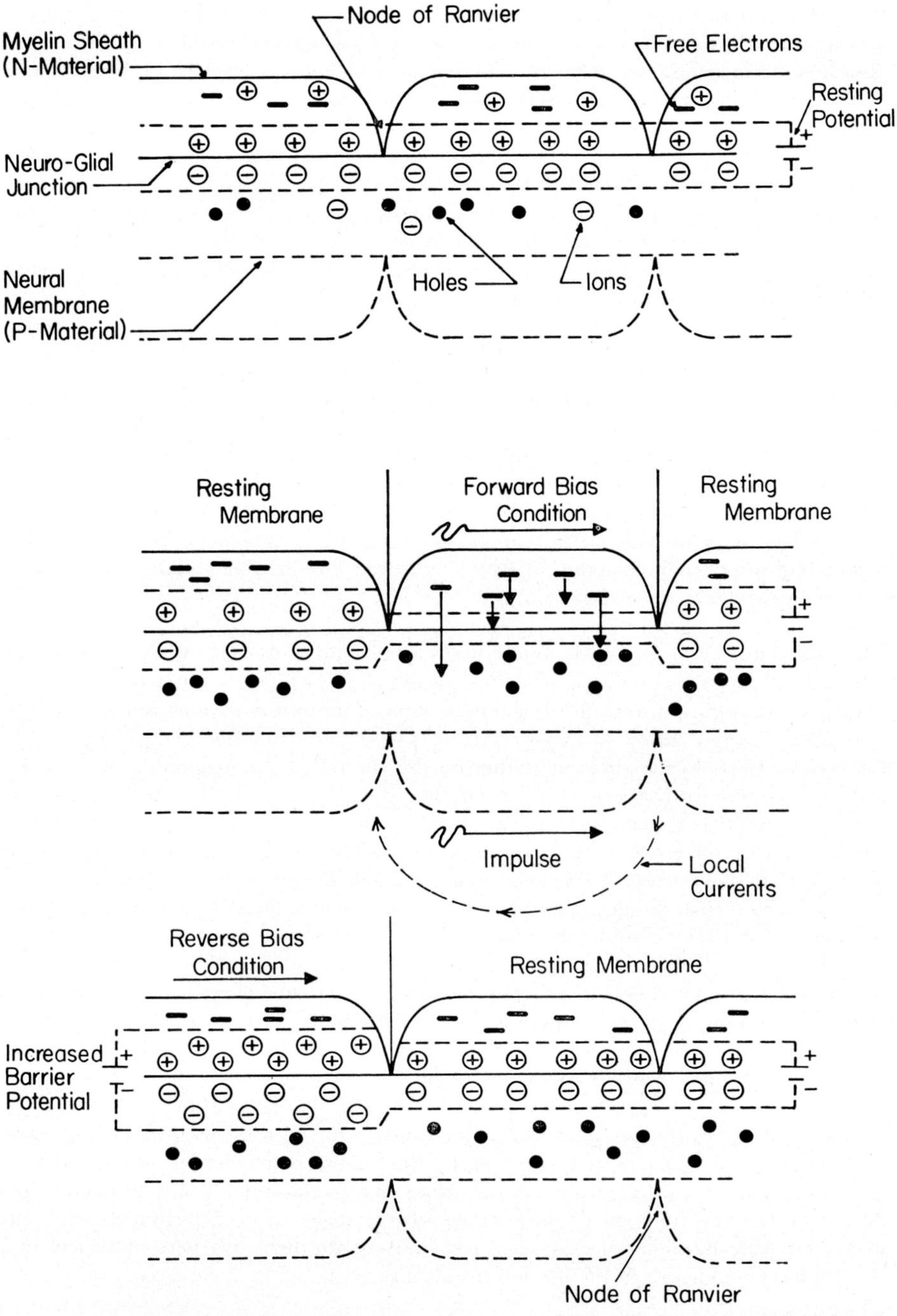

Fig. 6. Neuroglial semiconductor hypothesis of impulse propagation: neuroglial semiconductor at rest, note, barrier potential at junction (top); condition for impulse propagation, i.e., neuroglial junction in forward bias condition (centre); condition for inhibition of impulse propagation, i.e., neuroglial junction in reverse bias condition (bottom).

With the evidence of magnetic induction by neural activity could not the converse also be true, that is, induction of neural activity by a magnetic field. It is known that movement of a conductor through a magnetic field induces electrical current within it. By the same token, a fluctuating magnetic field may impose the same effect upon a stationary conductor or semiconductor. It is also known that the geomagnetic field has a daily fluctuation (Fig. 1), the magnitude of which far exceeds those generated by the nervous system. Could not this interaction be the mechanism behind "orientation" or "disorientation" phenomenon found in some animals? Could not the heightened activity level found in other animals be explained by the Hall effect phenomenon (i.e., the dropping of barrier potentials in the presence of a magnetic flux giving rise to increased current flow or neural activity)?

4. CONCLUSION—PERSPECTIVE OF GEOMAGNETIC ACTION ON NEURAL FUNCTION

The neuro-glial semiconductor hypothesis as presented in this chapter may very well hold the answer to the aforementioned questions.
The orientation studies of Brown (1971), Keeton (1971), Southern (1971), and others may well be explained as an interaction between the geomagnetic and biomagnetic fields. The resultant interaction of which "directs" the animal to take the course of least or "programmed" stimulation.

The augmented activity levels found by Gottlieb and Caldwell, Stutz, Ratner, and other investigators further support a stimulation concept possibly initiated by the Hall effect, i.e., dropping of resistance or nullification of inhibitory mechanisms.

Kholodov (1964) in his ablation studies on fish, found the diencephalon of the brain as most sensitive to magnetic stimulation. The diencephalon is that part of the brain stem containing the thalamus, hypothalamus, and parts of the reticular activating system (RAS) concerned with facilitation and inhibition of neural impulses passing through it to the cortex. All sensory neurons have processes within this part of the brain prior to their termination at the cortex. It seems quite feasible to see how magnetic stimulation of this area could result in increased activity.

Nevertheless, the activation mechanism could very well be explained in semiconductor terms. Reduction of barrier potentials as facilitation or increased barrier potentials as inhibition. In either case, we need not concern ourselves with the problems of membrane permeability or of ion flow across them.

In the course of evolution, man and animal have evolved nervous systems the design and operation of which were influenced by the environment. The geomagnetosphere was also a part of that environment. To what degree the impact of its influence had on the design and function of the nervous system is yet to be determined. With the ever increasing numbers of new scientists entering the field, we personally feel that we will have an answer in the not too distant future.

REFERENCES: see pp. 283-284.

Chapter 5, Section 2b

Effects of Magnetic Fields on Animal Behaviour

by

M. A. Persinger

There are at least two methods by which a locked door can be opened: by using the appropriate key or by breaking the lock. Both techniques achieve similar results—the door opens. The former method requires very little energy, the latter requires a great deal more. However a door with a broken lock may behave differently from a door with a functional mechanism. This analogy can be used to illustrate some of the most central and pressing issues of biomagnetic research. What is the magnetic field configuration that will open the organismic lock? Are there optimal NATURALLY occurring magnetic field intensities, frequencies and geometries that can unlock the organismic door? Do the configurations of these locks vary with the nature of the organism and system exposed? These questions are complicated by the interacting and intervening variables of: (1) short versus long term exposure differences, (2) respondent versus operant behaviour measurements, (3) stage of development during exposure, (4) stimulus operations performed on the organism before exposure, (5) interfering extraneous magnetic ambients and (6) isolation of a physical mechanism.

The methodology by which bio-effective magnetic field configurations can be determined is influenced by two approaches. One approach emphasizes that natural magnetic field frequencies, intensities and geometries should be simulated as closely as possible to produce optimal effects. A second approach assumes that a biomagnetic effect, if it exists, should be amplified by increasing the intensities of applied fields to extremely large values. This last position presumes that sufficient field intensities must be maintained to compensate for such processes as thermal agitation. An approach of this type reflects not only a theoretical framework which influences much of biophysical research, but also a mentality. There seems to be an implicit assumption among a variety of experimenters that the larger the intensity, the larger the effect. The possible existence of a biochemical reaction with narrow band specificities for magnetic field intensity and frequency is apparently not considered by such individuals.

Occurrence of specific vectorial configurations that provide maximum stimulation to bioelectrical systems would make sound evolutionary sense. It would be unfeasible as well as unadaptable, to respond to just any electrical or magnetic stimulus. Moreover, the susceptibility to magnetic field stimulation would optimally be related to a state-dependency or organismic-predisposition such that responses could be limited in their occurrence to specific instances related to adaptability. Possible state-dependencies would be the physiological changes displayed during flight in birds or hormonal changes in animals before migration or homing. The monumental work of Brown (e.g., 1966) has demonstrated that maximum modification of snail behaviour occurs with intensities that closely approximate the geomagnetic field. Becker (1967) reported similar effects with insect orientation. Several researchers have failed to experimentally support the frequently postulated thesis that magnetic fields may be used in migratory/homing behaviour in birds. Most of these studies have used field parameters which were either higher in intensity or substantially different in spatial-temporal configuration than natural geomagnetic patterns. Wiltschko (1968) has emphasized that migratory-related behaviours in static artificial magnetic fields were not significant unless the strength approximated that of the earth.

However, neither life nor environmental forces are characterized by static processes—both show temporal properties. Much of the time-varying processes in both life forms and geomagnetic-related occurrences occupy a relatively narrow band with periods of a few minutes to a hundreth of a second (100 Hz). Extremely low frequency (ELF) electromagnetic phenomena have been defined as time-varying field-waves produced in organisms or in the environment between 0.01 Hz to 100 Hz (Persinger et al., 1973). Other definitions of the ELF wave band overlap these frequencies. The ELF range includes electromagnetic signals from autonomic neurons to more global brain sources such as those measured by the electroencephalogram. Power output (EEG) from the brain shows peaks around 10 Hz. In nature, similar electromagnetic frequency bands are associated with solar disturbances, frontal (weather) movements and even a terrestrial resonance system (the Schumann resonance). Biogenic electromagnetic variations of lower frequencies are measured in minutes or hours and include d.c. potentials. Ambient magnetic perturbations associated with geomagnetic storms overlap these organismic fluctuations. Time-variation in electromagnetic processes is an intrinsic component of life forms. Ignorance of these frequency ranges in experimental research has led and will lead to premature dismissal of possible biomagnetic effects.

Very often time-varying magnetic fields and static fields of comparable intensities have opposite or different effects. Lazar and Barca (1969) noted that small intensity pulsating magnetic fields decreased sodium ions but increased potassium ions in tissue while a static field had opposite effects. Intermittent fields can delay hatching in ducks and produce unfavorable changes in albumin biosynthesis while a static field of comparable intensities has no such effect (e.g., Ossenkopp, and Shapiro, 1973; Barndsch and Jitariu, 1971). The time factor in magnetobiological effects is indeed a critical feature.

Existence of different reaction patterns to different types of magnetic fields would suggest magnetic field-organismic equivalents of a lock and key mechanism so often postulated for agonist-receptor site dynamics in drug interactions. Large or exaggerated "magnetic keys", typified by high intensity static magnetic fields, merely break or force the "lock". However, other more complex "magnetic keys" with appropriate frequency, intensity and spatial configurations that simulate biogenic magnetic field analogs of the organism would show the greatest effectiveness. The existence of such a hereto, unacceptable parameter specificity for magnetic field-organismic relations has been given strong support by the experiments of Becker (1972). Becker contends that there exists in living tissues an extremely narrow range of effective current densities in the mμ amp region. Intensities below the optimum are predictably not effective; those above are ineffectual or destructive.

Much confusion often arises by mistakenly equating two different questions: (1) can the organism detect a magnetic field, (2) can a magnetic field affect the organism? The answers to both questions are not necessarily commensurate. Various studies with birds of different species have failed to produce any consistent evidence that magnetic fields may act as conditioned stimuli (CSs) to either the delivery of positive or aversive stimuli. Grisset and De Lorge (1971) and Marr et al. (1973) have also failed to discern in several species any behavioural changes that indicate ELF magnetic fields of 45 or 75 Hz can act as conditioned stimuli (cues) to the consequent occurrence of shock or simultaneous availability of food. Scattered Soviet works on glial cell activity have been reported, suggesting changes in electrical potentials with a slight latency of a few minutes (e.g., Travkin, 1972).

On the other hand, longer term exposure to magnetic fields for continuous or repeated short durations results in a variety of behavioural and physiological changes. With static fields of intensities ranging from 10 to 5000 gauss, changes in visceral organ weights, endocrine chemicals and susceptibility to disease have been repeatedly mentioned. Reactive glial changes have been reported by Soviet scientists but not by others, (e.g., Friedman and Carey, 1969). Adult rats exposed to 0.5 Hz rotating magnetic fields (RMFs) between 0.5 to 50 gauss from 5 to 30 days show significant changes in water consumption, thyroid weights, and body weights, (Persinger et al., 1973). Removal of animals from the RMFs resulted in increased ambulation, specifically in male rats. Pulsatile fields between 130 μsec. to 3 sec. at intensities up to 300 gauss for periods of 1.5 hrs per day for 30 days to more continuous exposures have been reported to affect G-6-P activity, ATP/ADP ratios, serum glutamic oxaloacetic transaminase levels and follicle size (e.g., Porumb, 1971; Neaga and Lazar, 1972). Generally these effects take at least 2 to 5 days before they appear, thus suggesting a stress-like reaction in the animal. Two separate unpublished studies by Persinger (1967) and LaForge (1972) imply that exposure of rats to 0.2-0.5 Hz fields between learning situations may interfere with display of appropriate behaviours. Rats exposed to 8000 gauss fields after conditioned avoidance training have been reported to show a delay in criterion learning (Meszaros et al., 1971). However, these effects may be more indirectly due to competing ambulatory behaviours related to the paradigm rather than direct interference with presumed electrical correlates of memory consolidation.

The problem of detectability of magnetic fields is not unrelated to the behaviour measured. Experimental psychology underwent a period of confused comparisons between ambulatory behaviour, which is displayed in mazes and open fields, and discrete response behaviour, which is emitted in operant chambers. Experimental manipulation of stimulus patterns did not always produce comparable results in both types of behaviour. Different motoric and ancillary systems are involved with ambulatory and discrete response behaviour.

Similarly, responses associated with the autonomic nervous system are not necessarily comparable to those associated with the somatic (skeletal) nervous system. Autonomic system reactions involve smooth muscle modifications and are more susceptible to the effects of classical (respondent) conditioning. With this type of conditioning, the major stimulus (UCS, i.e., unconditioned stimulus) elicits (is followed by) a response. Repeated pairings in time and space of a previously ineffective stimulus with the UCS can result in the occurrence of a response when the previously ineffective stimulus (now the conditioned stimulus or CS) is presented by itself. An animal in this situation maintains a relatively passive role to a series of CS-UCS pattern presentations. On the other hand, the skeletal system seems more susceptible to operant conditioning. This paradigm makes the delivery of the reinforcer (a UCS or CS) more directly dependent upon the animal's behaviour, such as a lever press.

Experiments by de Lorge (1972) and Marr et al. (1973) with 7 Hz, 10 Hz, 45 Hz, 60 Hz and 70 Hz, 2-3 gauss magnetic fields almost conclusively demonstrate that the behaviour of monkeys, rats and pidgeons maintained on a variety of operant schedules does not show consistent variations when these fields are present. One could argue that the stimulus control of the reinforcement schedule would not be appreciably disrupted by such relatively weak stimuli. Behaviours maintained on some schedules of reinforcement are remarkedly immune to any large perturbations by extraneous stimulation. However, these authors have also attempted using ELF fields in a conditioned suppression paradigm as a cue to the later onset of electric shock. With normal stimuli (e.g., light), a few pairings of antecedent light followed by shock results in the suppression of ongoing food associated lever responding during the light presentation. The ELF fields used in these studies were not effective as "warnings" to aversive stimuli, since animals would continue responding during their presentations.

These results do not eliminate the possibility that whereas lever pressing (discrete) behaviours were not affected by short term ELF field presentations, other systems were responsive but not measured. Two studies which have used an autonomic response (heart rate change) as their dependent measure in a classical conditioning paradigm have produced more positive results. Reille (1968) noted that pidgeons exposed for 5 sec. to weak (0.15 gauss), 0.3 Hz, 0.5 Hz and higher frequency fields before the onset of an electric shock, began to show increases in heart rate after the presentation of the field but before the shock. Yakovleva and Medvedeva (1972) have shown similar cardiac conditioned responses to a constant magnetic field of 520 gauss presented before shock. Although recent work with biofeedback has shown that autonomically innervated organs may indeed be susceptible to operant control, the distinction between autonomically controlled and non-controlled behaviours should be realized when studies of magnetic field detectability are involved.

Organisms in different developmental stages may show differential responses to the same stimuli. Certainly the numerous experiments with the effects of ionizing radiation should emphasize the importance of this variable. During prenatal and to some extent neonatal-juvenile development the organism displays a remarkable plasticity and production. Neurons leave their proliferative matrices and travel to destinations by mechanisms which are not entirely clear. Chemical reactions which will result in the critical biochemical pools of the brain are in progress. One would suspect that such a period would show high susceptibility to even very weak forces. Many experiments by Persinger, Ossenkopp and their colleagues, (e.g., Persinger et al., 1973) have indicated that rats exposed during various stages of their prenatal development to 0.5 Hz RMFs with intensities ranging from <1 to 30 gauss, show significant changes in their behaviour relative to controls or sham exposed animals. RMF-exposed rats showed less open field activity, greater defecation, increased acquisition of a conditioned suppression paradigm, and thyroid-testicle weight changes. The effects were dose dependent and a function of developmental period at the time of exposure. Rats exposed to 0.5 Hz RMF-fields from 2 days before birth to 5 days after birth show similar behavioural-physiological differences as adults. Unpublished data collected by Persinger and Lafrenière indicate that adult changes associated with RMF-exposure during day -2 to $+5$ of development can be diminished to control values by re-exposing these subjects as adults to the same field.

In many respects, the problems and techniques which are involved with pharmacological research can be usefully applied to magnetic field research. Dose (intensity) or analog (frequency) dependent response curves are convincing data. As with drug reactions, the antecedent operations or conditions performed on the organism may determine not only the direction but extent of the effect. If the autonomic nervous system is involved primarily with biomagnetic phenomena, then exposure of the animal to extraneous conditions which stimulate this system would affect the pattern by which the magnetic field exerts its influence. Operation of a law of initial values would be a significant factor in these reactions. Age, sex and perhaps most of all the degree of autonomic lability would also be important variables.

Magnetic field ambients from both natural and man-made sources have been rarely considered in biomagnetic research. Superimposition of the experimental fields and testing area upon locally produced magnetic irregularities could have adversive effects. Tromp (1968) has measured the presence of unusual field characteristics in dwellings (with steel supports) and in areas of low soil resistivity. Interfering magnetic transients or statics can come from test equipment itself. All these factors could alter the sensitivity and response of animals exposed to experimental magnetic fields.

To any proficient scientist a repeated and measurable change of behaviour or physiology associated with the experimental presentation of magnetic fields should be sufficient stimulation to maintain further interest and investigation. The pre-isolation of a physical mechanism is not an absolute necessity to continue research. Unfortunately not all scientists agree with these statements but make the acceptability of a phenomenon contingent upon the prior existence of a "feasible model" or in terms of mechanisms known to date—even if they are not applicable. Such caution certainly inhibits the undesirable members of the academic community from publishing poorly collected or belief-influenced data, but it also delays the acknowledgement of new phenomena. Researchers who are interested in the effects of magnetic fields, specifically those associated with near-natural intensities and frequencies, will face this opposition.

Magnetic fields of different characteristics can be used as a tool to cross cut a variety of discourse levels. Magnetic field effects can be measured by molecular, physiological and behavioural indices. Perhaps the study of magnetic fields, both biogenic-like and high intensity, will allow us to understand more clearly how molecular-biochemical-physiological-electrical-behavioural sequences are normally linked. Perhaps the study of their effects will give us some insights to the complex substrates of animal behaviour. On the other hand, magnetic-associated effects may be so varied and unrelated that interdisciplinary systematic investigation will be impossible. The young researcher interested in magnetic effects should be cognizant of these possibilities and have sufficient endurance to consider the multitude of variables involved. He should realize that the study of magnetism and behaviour is a problem which cannot be contained within one discipline but requires accurate knowledge from many, many areas.

REFERENCES: see pp. 284-285.

Biological Effects of Various Physical Parameters of the Environment on Animals*

by

R. S. Pogrund

INTRODUCTION

The physiological effects of atmospheric air ionization on various organism species have been extensively investigated over the past several decades (see Chapt. 5, Sect. 1). While a substantial part of these investigations has been directed toward the biological effects of air ions contributed by such natural sources as cosmic rays, radioisotopes in the soil, lightning discharges, seasonal winds, and ultraviolet radiation, the contribution to the general air ion quality by man himself becomes a more significant concern to the specialist in occupational and environmental health. The design of experimental protocol in the well-controlled laboratory and animal exposure chambers should define and delineate as comprehensively as possible the biological effects of multiple interactions and potential synergism that exist among a myriad of physical-chemical agents and force fields. High temperature metal heat sources produce large amounts of ions which are predominantly positively charged. An open flame creates a continuous flood of ions. The strong electric fields associated with high voltages in any equipment, including R-F transmitters, such as operating radar and color television tubes, may also serve to dissociate air molecules into ions. The vehicle exhaust system of the internal combustion engine is another strong emitter of air ions in addition to the variety of noxious gases and other kinds of particulate matter.

Electric charges emanating from the earth's static electric field of about 100-200 volts per meter during fair weather conditions must be also considered. The normally negatively charged earth surface and positively charged air stream above are readily altered by adverse weather conditions and resultant oscillation of air ion concentrations and charges. Merely moving a plastic window drapery in dry air or forcing a high velocity air stream over sharp edges or through orifices will produce clouds of air ions.

A third factor in assessing the urban and especially industrial environment on biological effects is the increasing pollution of the electromagnetic spectrum by communication transmitters, whose "noise" presents an important problem area. This noise may be generated by electrical appliances, power distribution facilities, automobiles, airplanes, automatic garage door openers, and practically any other electrical equipment (Swenson and Cochran, 1973), wherein the important frequencies were measured at 74, 148, 222, and 440 MHz. Frey (1971) has reviewed the biological significance of these kinds

* See also Progress in Human Biometeorology, Part IA, Chapter 1, Section 1 and Chapter 6, Section 8, 9 and 10.

of energy, with specific reference to behavioral effects in mammalian organisms. The "hearing" of such electromagnetic noise has been described by Ingalls (1967), Sommer and von Gierke (1964) and by Frey (1962), who assigns an energy transport to such radiation and comprises a stimulus to the central nervous system, mediating certain sensory, motor, cortical, and subcortical responses. Various industrial technological processes generate this kind of radio-frequency electromagnetic energy while automating the smelting and tempering of metals, accelerating pulp drying, cutting and welding plastics (Dumanskiy, 1967). It is apparent that not only does the frequency of the electromagnetic radiation determine the biological effect but also the intensity or power of the field. On this basis, Russian investigators caution that the voltage of the electric field should not exceed 5 volts/meter, as higher levels are likely to induce a variety of neurological disorders. A sparsity of data exists with respect to experimental work on the biological effects of extremely low frequency (ELF) electromagnetic fields (45 Hz-75 Hz). This low frequency is generated by power transmission lines and has been of interest to the United States Navy (Project Sanguine) contemplating the use of these longer wavelengths to communicate with the Polaris submarine. It is clear that physiological and behavioral responses occur over a wide range of the frequency spectrum and field intensity whose total biological effect may be in turn influenced by the magnitude and existing electrical charges in air ionization.

A fourth factor is magnetic fields*, minus the electro-component, and is also mentioned for its industrial environmental significance although it is less relevant to the concept of atmospheric electricity within the present framework of discussion. However, the mobility of large and small air ions is likely to be influenced by the superimposition of magnetic fields of varying intensities. In the realm of geophysical biology, such geophysical factors as geomagnetism, telluric currents, tidal gravity, solar and lunar fluctuations have been variously documented as "Zeitgebers" that govern organic rhythms. Concrete demonstration of correlations between biological periodicities and these geophysical variants has not been generally available because of the infinite number of possible biological responses to an infinite number of combined external environmental phenomena. Moreover, the demonstration of an organism receptor for detecting magnetism or electricity has been equivocal (Weihaupt, 1964) and the exact nature of the master clock for biological rhythm remains quite uncertain (Becker, 1963). Since Becker (1963) feels that the earth's magnetic field is an important physiologic factor for living organisms, the production of magnetic fields by industrial processes becomes biologically significant, especially when the presence of a given concentration of unipolar air ions may reciprocally influence thresholds of irritability or susceptibility to environmentally induced diseases or exacerbation of existent clinical disorders. Vyalov (1967) has reviewed the evidence that implicates the generation of magnetic fields as an occupational hazard and cites the workers' exposure to magnetizing and demagnetizing equipment, ammeters, solenoids, benches equipped with magnetic plates and demagnetizers of various metallic products. Such generated magnetic fields may be direct or alternating, wherein the electromagnetic gap may contain a field intensity of up to 100,000 oersted, while a pulsing magnetic device may generate a magnetic field intensity of up to 2,500,000 oersted (Vyalov, 1967). The human central nervous

* See also Progress in Human Biometeorology, Part IA, Chapter 6, Section 14.

system has been likened to a solid-state electronic system of dimagnetic and paramagnetic protein molecules. These magnetic fields may be expected to induce a galvanomagnetic effect that reflects an interaction between the flow of direct current electricity and steady or pulsating magnetic fields. Again the concomitant effects of air ions toward either raising or lowering thresholds of neuronal irritability or action potentials within the brain stem reticular formation remain largely speculative.

ATMOSPHERIC ELECTRICITY AS A MODIFIER OF OCCUPATIONAL AND ENVIRONMENTAL HAZARDS

While the atmosphere contains air ions, electromagnetic radiation fields, static electrical fields and magnetic fields, that seasonally and geographically exist as variants, the ultimate effect on organism physiological systems must be further delineated in terms of interactions with other man-made physical-chemical substances. For example, coal miners, uranium miners, or asbestos miners are exposed to particles of varying sizes. The respirability of those particles and their ultimate influence on cellular physiology will be determined in part by size, chemical composition, concentration, particle electrical charge, presence of positive or negative air ions, presence of ionized noxious gases and electromagnetic fields, all interacting under given environmental conditions. Noxious gases in general air pollution and in specific occupational environments include sulfur dioxide, carbon monoxide, ozone, nitrogen oxide, and other gases that are ionized, and thereby, may be electrically attracted or repelled by particle charges. The threshold limiting values (TLV's) of these individual gases, as examples, in an occupational environment have little meaning within the context of these potentially interacting components.

Particles or droplets of an aerosol* are usually charged or can be induced to acquire a charge. As previously implied, these charges are influenced by electric and magnetic fields, not only when they are fully grown but also during the process of their formation or disappearance. Thus their size, number, trajectories, form and site of deposition within the organism can vary within wide limits. Each particle, therefore, represents a potential condensation nucleus whose particular charge enhances the effectiveness of particle growth due to chemical reaction with gases.

Man from earliest times must have inhaled air ions with their specific electrical charges carried on gaseous molecules and particles in the atmosphere, which seem to have clinical effects. Smoke from fires, bacteria and virus-containing droplets from other men and animals, molds, spores, pollens and animal danders from his husbandry might have all produced disease. It cannot be assumed that mineral industrial aerosols such as those released in our present age were without effect on primitive man. But because of the steadily increasing air pollution in the urban community where the air quality outdoors also influences the air quality indoors, including hospitals, residences, and the workplace (Henderson et al., 1973), the nature of air ions takes on a very special significance as already mentioned. Kato (1964) observed a preponderance of large

* See also Progress in Human Biometeorology, Part IA, Chapter 6, Section 10.

positively charged ions in an area of high industrial density while the converse was measured in clean country air, that of a higher concentration of smaller negatively charged air ions to the extent of seven to twenty times the concentration found in the vicinity of industrial areas. While the quantity of carbon dioxide and dust particles in the air was found to be directly proportional to each other, the concentrations of dust and air ions were inversely proportional. The mere release of an exhaust gas, such as carbon monoxide, into a room for ten minutes reduced the ion concentration from 1110/cc to 120/cc. The presence of people in a room reduces the number of air ions; passage of air across polyethylene and vinyl filters or polyurethane and glass fiber filters can reduce the air ion concentration anywhere from one-tenth to one-third of the air ion concentration present at the inlet.

THE BIOMEDICAL COMPONENTS

Kornblueh's (1971) preliminary observation that air ions might affect human comfort may have far reaching significance to Occupational and Environmental Health specialists, and in fact, in the entire area of Public Health. For if there is really some optimal concentration of air ions of specific size range and polarity ratio, then air ions that are less than optimal by the same criteria would be expected to conversely exert varying degrees of human discomfort. Corrado and Beckett (1960) have reviewed the air ion considerations for human comfort and/or discomfort to include the association of air ions with gaseous molecules and particulate matter. As such, small ions are associated with gaseous molecule clusters, averaging 1 millimicron in size, while

Table 1. Reported physiological effects of positive ions

1. Inhibition of growth of tissue cell cultures	7. Increased skin temperature
2. Increased respiratory rate	8. Depressed rate of ciliary activity
3. Increased basal metabolism	9. Increased muscle chronaxie
4. Increased blood pressure	10. Altered alpha wave of the EEG
5. Produces headache, fatigue, nausea	11. Reduced succinic oxidase activity in the adrenals
6. Produces nasal obstruction, sore throat, dizziness	12. Increased the blood level of 5-hydroxytryptamine

NOTE BY THE EDITOR-IN-CHIEF: Most of the findings require considerably more research and further confirmation before they can be generally accepted and applied clinically. Readers are recommended to study the paper by I. ANDERSON: Effects of Natural and Artificial Generated Air Ions on Mammals. In: Biometeorology, 5, II. Proc. 6th Int. Biometeor. Congr. Swets and Zeitlinger Publ. Co., Amsterdam: 229-238 (1972).

the larger air ions are those coalesced on particles of 1 micron in diameter. The small ions are highly mobile in an electrical field while the large air ion is considerably less mobile. The nature of these gases and particles in association with these defined air ion characteristics is the matter of primary concern in the field of Occupational and Environmental Health.

While these public health aspects are emphasized in this Section, the psychophysiological effects of air ions, per se, and independent of coexisting physical-chemical substances are basic to organism responsiveness. In this regard, investigators have used the central nervous system and a variety of behavioral aspects as the prime target organ system (Oliverau, 1973; McDonald et al., 1967; Nazzaro et al., 1967; Frey and Granda, 1961; Krueger, 1968). These investigations generally consider both humoral and nervous factors to account for the reported favourable effects of negatively charged small ions and the usually unfavourable influence of positively charged air ions as well as the ratio of one charge over the other. For further details see Chapter 5, Section 1 of this volume by A. P. Krueger. A summary of reported physiological effects of positive and negative ions is given in Table 1 and 2.

Table 2. Reported physiological effects of negative ions

1. Decreased respiratory rate	9. Decreased eosinophilia and lymphocyte count
2. Decreased basal metabolic rate	10. Increased CO_2 combining power of plasma
3. Decreased blood pressure	11. Decreased blood sedimentation rate
4. Produces a feeling of well-being	12. Decreased muscle chronaxie
5. Increased vital capacity	13. Increased ciliary activity
6. Decreased skin temperature	14. Increased frequency of mitosis
7. Acceleration of the conversion of succinate to fumerate	15. Increased resistance to infection
8. Stimulation of cytochrome oxidase activity	16. Suggested as therapy in chronic rhinitis sinusitis, migraine, insomnia, tuberculosis, wound and burn healing, asthma, hay fever, emphysema, bronchitis, conjunctivitis, chlorine gas poisoning

NOTE BY THE EDITOR-IN-CHIEF: Most of the findings require considerably more research and further confirmation before they can be generally accepted and applied clinically. Readers are recommended to study the paper by I. ANDERSON: Effects of Natural and Artificial Generated Air Ions on Mammals. In: Biometeorology, 5, II. Proc. 6th Int. Biometeor. Congr. Swets and Zeitlinger Publ. Co., Amsterdam: 229-238 (1972).

As previously indicated, air ion concentrations, polarity, size, and mobility are influenced by outdoor and indoor air quality. The indoor aspects include human habitation in the residence, workplace, automobile, schoolbus, hospital, train or boat. Human habitation in orbital space capsules or in submarines represents extreme departures from sea level in engineered closed cycle environmental control systems whose air ion composition has been of concern to those guarding crew safety, survival and psychomotor performance. Limitations of space preclude any attempt to deal in detail with the fate and effect of all classes of aerosols in these controlled environments or uncontrolled environments in which aggregates and water soluble particles are prevalent.

In these respects, the target organ of concern is the lungs. The net adverse respiratory effects include increased airway resistance via bronchospasm, depressed ciliary and mucus transport and phagocytosis whose total activity is compromised when such particles as asbestos, coal mine dust, lead, cadmium*, or a variety of carcinogenic polycyclic hydrocarbons are inhaled. The contribution of such gases as sulfur dioxide, nitrogen dioxide, and ozone to these pathologies is well documented*.

As such, carbon monoxide within the total milieu of atmospheric pollutants can not only contribute to the impaired diffusion of oxygen across the alveolar membrane but also actively combines with hemoglobin with the formation of carboxyhemoglobin. The total hypoxemia is the resultant of the varying degrees of interactions as described and the extent of which leads to the etiology of cardiovascular disease that includes the agglutination of platelets, atherosclerosis, and myocardial dysfunction. Rapid industrialization, urbanization and technological innovativeness make the clear cut separation of air ions from physical-chemical products of combustion regarding physiological effects more academic than meaningful in the real world. While the inhalation of positive air ions may stimulate sympathetic innervation and negatively charged air ions may act conversely via parasympathetic innervation, the real concern is the influence that these ions may have on either accelerating or ameliorating the adverse physiological effects of these products of current civilization.

CONCLUSIONS

One of the more persistent frustrations experienced by the scientific investigator of health and disease stems from the nagging awareness that a vast array of forces affects bodily function and that he can study the influence of only a few of them at a time. In part, this limited approach is necessitated by the sheer complexity of the phenomena that have been described; in part it is attributable to the diversity of disciplines involved in investigating these biometeorological components and to the difficulty researchers have in transcending methodological and conceptual barriers. Space does not permit to discuss in detail the methods and equipment which have been developed to study these complex problems. Readers are referred to the investigations by Pogrund (1969), Pogrund and Thrasher (1970), Weihe (1974) and others.

REFERENCES: see pp. 285-286.

* See Chapter 5, Sections 2 and 3.

CHAPTER 6

EFFECTS OF GRAVITATIONAL FIELDS ON ANIMALS

SECTION 1: Effects of Altered Gravitational Fields on Laboratory Animals (J. OYAMA)

Chapter 6

Effects of Altered Gravitational Fields on Laboratory Animals*

by

J. Oyama

INTRODUCTION

Gravity is one of the most important environmental factors affecting the structure, function and behavior of animals. Its directive influence in evolution is exemplified by the highly developed anti-gravity musculoskeletal system of terrestrial vertebrate animals. Experimental studies on gravity, however, have been quite limited due to its all pervasive nature and virtual constancy on earth. Early gravitational studies on animals were mainly concerned with physiological responses to changes in the applied direction of the ambient field, e.g., prone versus upright position. Later with the development of high-speed aircraft during World War II, considerable attention was directed to the physiologic problems of pilots during high acceleration maneuvers. In support of these studies, extensive use was made of animals to serve as human analogues. Exposure to high acceleration forces is comparable to being exposed to an increased gravitational field according to Einstein's Principle of Equivalence which asserts that the inertial reaction derived from accelerated motion is physically equivalent and indistinguishable from the inertial reaction derived from a gravitational field of equal magnitude. Accordingly, the terms: hypo-G and hyper-G, will be used in this review to designate accelerations which simulate gravitational fields that are less or greater than normal earth's gravity.

In 1957 the Space Age was ushured in with the launching of Sputnik I, the first successful orbiting earth satellite. Since then man has been able to "escape" from earth's gravitational influence and as a result there has been a concerted effort to investigate the effects of the space environment on basic life processes. It has generated an intense interest in weightlessness as well as in the entire gravitational spectrum including fields above and below normal earth's gravity. The objectives of this review are to describe some of the major physiologic changes effected in laboratory animals subjected to hyper-G and hypo-G, some of the basic controlling mechanisms, and areas of potential importance for future research. A review of the basic principles in gravitational biology and of animal studies has been made recently by Smith (1974) and is recommended for its timeliness and coverage of the field.

PHYSICAL ASPECTS OF ALTERED-G EXPOSURES

The reaction or resistance to an abnormal G-field manifests itself in the animal by

* See also Progress in Human Biometeorology, Part IA, Chapter 6, Section 15, pp. 400-408, paper by J. F. Saunders.

the following physical effects: (a) changes in hydrostatic pressure gradients, (b) redistribution of blood and other body fluids, (c) changes in weight bearing by bone, muscle and other supporting tissues, (d) displacement of mobile organs, (e) distortion of organs and tissues through compression, distension, stretching etc. These physical changes are perceived, integrated and transduced by the nervous and endocrine systems. The subsequent response or change effected by the animal depends on: (a) the G-intensity, (b) the rate of onset, (c) the duration of exposure, (d) the direction of the G. A standard triaxial directional terminology system has come into general use (Pesman, 1971). The spinal column is taken as the Z axis with the X and Y axis being mutually perpendicular to it. The X axis runs front to back while the Y axis runs laterally, right to left. Only the directional symbols and descriptive terms used in this review are presented below:

Direction of Accelerative Force (relative to spine)	Kinetic Reaction or Organ Displacement	Symbol
Forward Transverse (back to chest)	Towards spine	$+G_X$
Backward Transverse (chest to back)	Away from spine	$-G_X$
Positive Headward	Tailward	$+G_Z$
Negative Tailward	Headward	$-G_Z$

Based on this terminology, the normal gravitational orientation for a biped animal would be $+G_Z$ and for the quadruped animal $-G_X$. It should be noted that no directional orientation exists in weightlessness.

TOLERANCE AND SURVIVABILITY TO ALTERED-G

Results from the various space flights and particularly from the recent record-setting 84 day Skylab III mission support the conclusion that weightlessness is well tolerated by man and does not cause any serious impairment or deterioration in physiologic function. Animal space flights, albeit of shorter duration, generally support the same conclusion. The only exception was the premature death of the monkey in the 8 day space flight (Biosatellite III) which was probably due to factors other than weightlessness (Meehan, 1971). Similar conclusions can undoubtedly be drawn with respect to subnormal gravitational fields such as encountered on the lunar surface.

Results from exposures to hyper-G indicate generally stressful responses which can cause serious impairment of normal functions or death if tolerance limits are exceeded. Various tolerance indices or endpoints have been used in animals including: mortality (Britton et al., 1946); survival time (Polis, 1961); decrease in heart rate (Sipple and Polis, 1959); decrease in blood pressure (Burton, 1973); and lowering of body temperature (Oyama et al., 1971). In addition to the G-intensity and duration, the direction of the acceleration has an important effect on tolerance. In general, transverse accelerations are better tolerated than either the headward or tailward accelerations with the headward, in turn, better than the tailward.

Tolerances of animals generally decrease as one proceeds up the phylogenetic scale and as the body size of the species increases (Smith, 1971). Sex and age appear to have variable effects on tolerance depending on the animal species. Britton et al. (1946) in a very extensive series of studies on a variety of animal species found females to have a higher tolerance than males. Reeves (1961, 1963) found no differences in tolerance due to sex but did find a progressive decline with age in rats. Wunder (1962), on the other hand, found the tolerance to increase with age in mice subjected to chronic hyper-G exposures.

Comparisons in tolerances between biped and quadruped animals are probably invalid since a common and comparable directional orientation does not exist. For example, if comparisons were made between the two groups in their normal gravitational orientation, it would be between biped animals subjected to headward ($+G_z$) acceleration and quadruped animals subjected to transverse ($-G_x$) acceleration. The quadruped animals would be expected to have a higher tolerance since the transverse acceleration is better tolerated than the headward acceleration.

A number of techniques has been found which enhance the tolerance and survivability of animals to hyper-G. Water immersion can increase the G-tolerance significantly by balancing the increased hydrostatic pressures generated (Magaria et al., 1958). Increased protection has been reported in animals given various anti-hypoxic drugs (Vinogradov and Pastushenkov, 1969); neurotropic drugs (Belay et al., 1970); Lucidril (Polis, 1962); and endotoxins which stimulate the reticulo-endothelia system (Stiehm, 1961). Repeated low-G exposures can also increase the tolerance and survivability of animals to a subsequent high-G field (Burton and Smith, 1972). Hypothermia induced by external cooling produces either an increase or decrease in G-tolerance in rats depending on the G-intensity (Stiehm, 1962).

ACUTE RESPONSES TO HYPER-G EXPOSURES

Selected examples of major physiologic system responses to short-duration hyper-G exposures are shown for various animal species in Table 1. The changes should not be taken as representing general responses but only for the conditions specified. In the very dynamic cardiovascular system, reversal of the initial response can often occur with time or with an increase or decrease in G-intensity. The G onset rate is not listed in the table due unfortunately to its frequent omission in published reports. In many of the studies the time to reach the designated G-intensity is considerably less than one minute. References are included so that the reader can obtain more detailed information on the conditions employed and other physiologic measurements recorded.

NERVOUS SYSTEM. There have been only a few studies reported on the effects of hyper-G exposures on brain function in animals. Britton et al. (1946) found the normal EEG patterns of cats, dogs, and monkeys interrupted by high voltage, low frequency "delta" waves, while Adey (1964) found a general flattening of the wave patterns in monkeys. These changes have been interpreted as a general suppression of cortical activity. The technical problems in conducting EEG measurements in animals

Table 1. Acute responses to hyper-G

SYSTEM/FUNCTION	G	DURATION	RESPONSE	SPECIES	REFERENCE
NERVOUS					
EEG	$+5G_z$	30″	Flattening of pattern	Cat, monkey	Adey (1964)
Sympathoadrenal	$+4G_z$	5′	Increase in blood adrenalin	Rat, dog	Barer (1958)
ENDOCRINE					
Hypothalamic-pituitary	$-10G_x$	4′	Increase in ADH secretion	Rabbit	Andrianova (1968)
Pituitary-adrenal	$-5G_x$	30′	Increase in corticosteroids	Mouse, rat	Oyama (1964, 1967)
CARDIOVASCULAR					
Heart Rate	$+4G_x$	5″-10′	Increase then decrease	Cat, dog, monkey	Britton (1947)
Arterial Pressure	$+6G_z$	25″	Decrease (carotid)	Monkey	Lawton (1958)
Cardiac Output	$+5G_x$	3′	Decrease	Dog	Sandler (1966)
Blood Flow	$-8G_x$	10′	Decrease (brain)	Rabbit	Preobrazhenskaya (1970)
PULMONARY					
Respiration Rate	$+6G_z$	10′	Decrease	Rat, dog, monkey	Britton (1947)
O_2 Consumption	$-4G_x$	1 hr	Decrease	Rat	Oyama (1973)
RENAL					
Urine Output	$-4G_x$	24 hr	Decrease	Mouse	Wunder (1971)
Urine, Na, K	$-7G_x$	1 hr	Decrease output	Rabbit	Bricker (1967)
Renin-angiotensin	$-4G_x$	3 hr	Increase activity	Rat	Pankova (1972)
THERMOREGULATION					
Body Temperature	$-2G_x$	1 hr	Decrease	Rat	Oyama (1971)
Heat Production	$-10G_x$	6 hr	Decrease	Rat	Gyurdzhian (1966)
METABOLISM					
Carbohydrate	$-5G_x$	1-24 hr	Increase glucose, liver glycogen synthesis	Mouse, rat	Oyama (1964, 1967)
Fat	$-5G_x$	1-24 hr	Increase fatty acid synthesis, NEFA (blood)	Rat	Feller (1968)
Protein	$-5G_x$	1-24 hr	Increase synthesis (liver)	Rat	Leon (1968)

during acceleration have undoubtedly limited the number of studies of this system. In contrast to the depressed cortical response, the sympathetic nervous system is stimulated as evidenced from a number of characteristic responses including: a rapid rise in blood adrenalin (Barer, 1958); increase in heart rate (Britton et al., 1947); increase in blood glucose (Oyama and Platt, 1964); increase in muscle glycogen breakdown (Harrison, 1970); decrease in secretory activity and motility of the gastrointestinal tract (Surorov, 1958); and vasoconstriction or dilitation of the vasculature of specific organs and tissues.

ENDOCRINE SYSTEM. Like other stressors, hyper-G stimulates the hypothalamic-pituitary-adrenal system effecting a rise in blood ACTH and adrenal corticosteroid levels (Oyama and Daligcon, 1967). The degree of stimulation has also been shown to be proportional to the G-intensity and duration of the exposure as well as age of the animal. Adrianova (1968) found an increase in anti-diurectic hormone (ADH) secretion in hyper-G stressed rabbits indicating stimulation of the neurohypophysial system. Very little work has been reported on other hormonal changes in animals during G-stress. With the development of very sensitive radioimmunoassay methods, it is anticipated that more information will be available on such important hormones as thyroid stimulating hormone (TSH), growth hormone, and insulin in the near future.

CARDIOVASCULAR SYSTEM. In view of its critical importance, this system has been more extensively studied than any other physiologic system under hyper-G. As a result of hydrostatic pressure changes and the general movement of blood in the direction of the inertial force, systemic as well as regional blood pressure and blood flow patterns are affected along with heart rate and cardiac output (cf. Table 1). These changes are highly dependent on the direction of the applied G-force. In the headward ($+G_z$) acceleration, blood may be diverted from the head to induce varying levels of cerebral hypoxia which produce effects similar to hypobaric hypoxia (Fluckiger, 1956). One of the most sensitive indices of reduced cerebral blood flow is loss of peripheral vision ("grey-out") or central vision ("black-out"). They are widely used as important indices of acceleration tolerance in aviation physiology (Howard, 1965). Burton (1973) has used corresponding blood pressure changes as tolerance indices in swine. Kupriyanov (1971) studied the effects of headward acceleration ($+G_z$) on the morphology of the vasculature in rats and dogs. Changes are first observed in the capillaries and venules. The vasculature distal to the head show hemostasis, dilitation of the venules, tortuosity, and hemorrhage.

RESPIRATORY SYSTEM. Respiration rate and pulmonary ventilation are highly dependent on the intensity, duration, and direction of the applied acceleration force. Britton et al. (1947) found the respiration rate to be depressed by headward acceleration in rats, cats, dogs, and monkeys. After exposure there was a brief increase in the rate indicating that animals were hypoventilating during acceleration. Barr et al. (1948) have attributed the decrease in ventilation rate to an inhibition of the pulmonary stretch receptors. Sandler (1966) noted an increase in respiration rate in dogs subjected to transverse accelerations. He attributed this increase to a compensatory reaction by the dog to a reduced lung volume caused by compressive loading of the chest. Rats exposed to $-4G_x$ for periods of more than one hour show a decrease in oxygen consumption and production of carbon dioxide (Oyama and Chan, 1973).

RENAL FUNCTION. Renal function tests in both humans and animals generally support the conclusion that headward ($+G_z$) accelerations reduce urine output. The effect of transverse accelerations appear less conclusive. Bengele (1969) found an increase in urine output in centrifuged rats which differs from the results of other workers (Wunder et al., 1971). Changes in urinary excretion rates can be attributed to hemodynamic changes which alter the glomerulo filtration rate as well as to alterations in ADH secretion. The extent to which these two factors participate under hyper-G exposures has not been determined but Bricker (1967) suggests that the immediate decrease in urine output is due to hemodynamic factors rather than to ADH or other hormonal factors. Pankova (1972) has shown that hyper-G exposures of rats activates the renin-angiotensin system which mediates the secretion of aldosterone by the adrenal cortex. Although the role of aldosterone in regulating sodium reabsorption by the kidney is well established, the specific action of this hormone during hyper-G exposures has not been determined.

GASTROINTESTINAL SYSTEM. Most of the animal studies dealing with hyper-G effects on this system have been conducted by the Russian workers. The general response effected by hyper-G exposure is a decrease in secretory activity and motility of this system (Surorov, 1958). Smirnov (1973) in more recent studies observed the intestinal absorption of glucose was enhanced after exposure of rats to hyper-G. Skinner and Camp (1969) found the gastroesophageal reflux test inmonkeys under tailward acceleration ($-G_z$) to be a valid means of predicting the competency of the cardia under gravitational stress.

THERMOREGULATION. Mice, rats and other small animals show a rapid and substantial fall in their deep body temperature ranging up to 5-6° C when exposed to hyper-G. Smaller temperature decreases occur in larger animals. In rats, the decrease in temperature has been shown to be directly proportional to the intensity of the acceleration imposed in the headward ($+G_z$) and transverse ($-G_x$) directions but not in the tailward ($-G_z$) direction (Saito et al., 1965). Continuous measurements of deep body temperature by implant biotelemetry in rats have shown that temperature changes can be used as sensitive and reliable indices of the severity of the stress and the time course of recovery and adaptation to hyper-G (Oyama et al., 1971). Temperature decreases have been measured in the rectum, brain, liver and muscles of rats exposed to hyper-G (Matsynin, 1964). The decrease in body temperature is accompanied by a decrease in heat production as determined by direct calorimetry (Matsynin, 1964) and by indirect measurements of oxygen consumption rates (Gyurdzhian, 1966; Oyama and Chan, 1973). There is no information on how the thermoregulatory center is affected by hyper-G in spite of the demonstrated reduction in cerebral blood flow and the induced state of hypoxia. Savin (1966) has shown a correlation of reduced oxygen consumption or heat production with a lowering of the oxygen tension in the brains of centrifuged cats and rabbits. Recent studies by Oyama (1974) have shown that the hypothermia induced in rats by hyper-G exposures can be altered by drugs which affect brain levels of norepinephrine and serotonin.

METABOLISM. Hyper-G effects on metabolic processes have been studied in animals through changes in blood and tissue levels of key intermediate compounds and enzymes in the intact animal as well as in isolated tissue preparations. All animals show a rise in blood glucose with hyper-G exposure. In mice, the rise has been shown to be proportional to the G-intensity (Oyama and Platt, 1964). A phasic response pattern in both glucose and corticosterone levels in the blood of rats exposed continuously for 24 hours has been reported by Feller and Neville (1966) which could be attributed, in part, to the action of insulin. A rather surprising finding is the increase in liver glycogen in hyper-G exposed animals since stress activation of the sympathetic nervous system is known to increase glycogenolysis (Oyama and Daligcon, 1967). In some cases transient decreases in liver glycogen are seen but if the hyper-G exposure exceeds 0.5 hr, there is a net increase in glycogen. This suggests that the initial glycogenolytic action of epinephrine or other hormones is eventually overwhelmed by the subsequent action of adrenal corticosteroids which stimulate glycogenesis. Changes in lipid metabolism in hyper-G rats have been studied in detail by Feller et al. (1968) and Neville (1969). They observed sustained increases in plasma non-esterified fatty acids (NEFA) and triglyceride levels for periods up to 24 hrs in hyper-G exposed rats. During this same time period, the rates of fatty acid synthesis into adipose tissue and plasma lipids were also increased indicating that both fat synthesis and breakdown are accelerated by hyper-G. Oxidative metabolic processes measured by CO_2 production rates were unaffected. Leon and Chakerian (1968) found the rate of liver protein synthesis to be increased as much as 157% in the cell sap fraction of hyper-G rats compared to controls. The increased synthesis depends on the availability of insulin in the stressed rats. The increase in liver protein synthesis has been substantiated by Abakumova (1973) who also found the rate of nuclear and cytoplasmic RNA synthesis to be increased.

MORPHOLOGY. Gross morphologic and anatomic changes have been determined by a quick-freeze technique in rats exposed to hyper-G (Gell and Cranmore, 1956). They showed by this in situ fixation technique that elongation and torsion of the lungs and heart occurred in the $+G_z$ direction while compression of both organs occurred in the $-G_z$ direction. Histologic studies of various elements of the hypothalamic-pituitary-adrenal system of animals exposed to hyper-G show increase secretory activity. Podymov (1970) found an increase in the number of secretory granules in the neurons of the supra-optic nucleus which subsequently disappeared indicating an initial stimulation followed by exhaustion. Savina (1972) studied the progressive changes in the morphology of the adrenal cortex of rats subjected to $-4G_x$ for periods of 3-24 hrs. He observed marked decreases in lipid and ascorbic acid concentrations and increases in acidic phosphatase activity. Morphologic studies of the vestibular system after exposure to moderately high G-intensities have revealed no abnormal changes. At very high intensities of $-60G_x$ or higher, Igarashi and Nagaba (1967) noted the otoconia displaced from the saccular macula in squirrel monkeys; but no changes were detectable in the semicircular canal cristae or in the organ of Corti. The displacement resulted in an ataxic posture and gait and spontaneous eye movements.

NEUROENDOCRINE CONTROL MECHANISMS

Figure 1 illustrates some of the effector pathways of the neuroendocrine system which

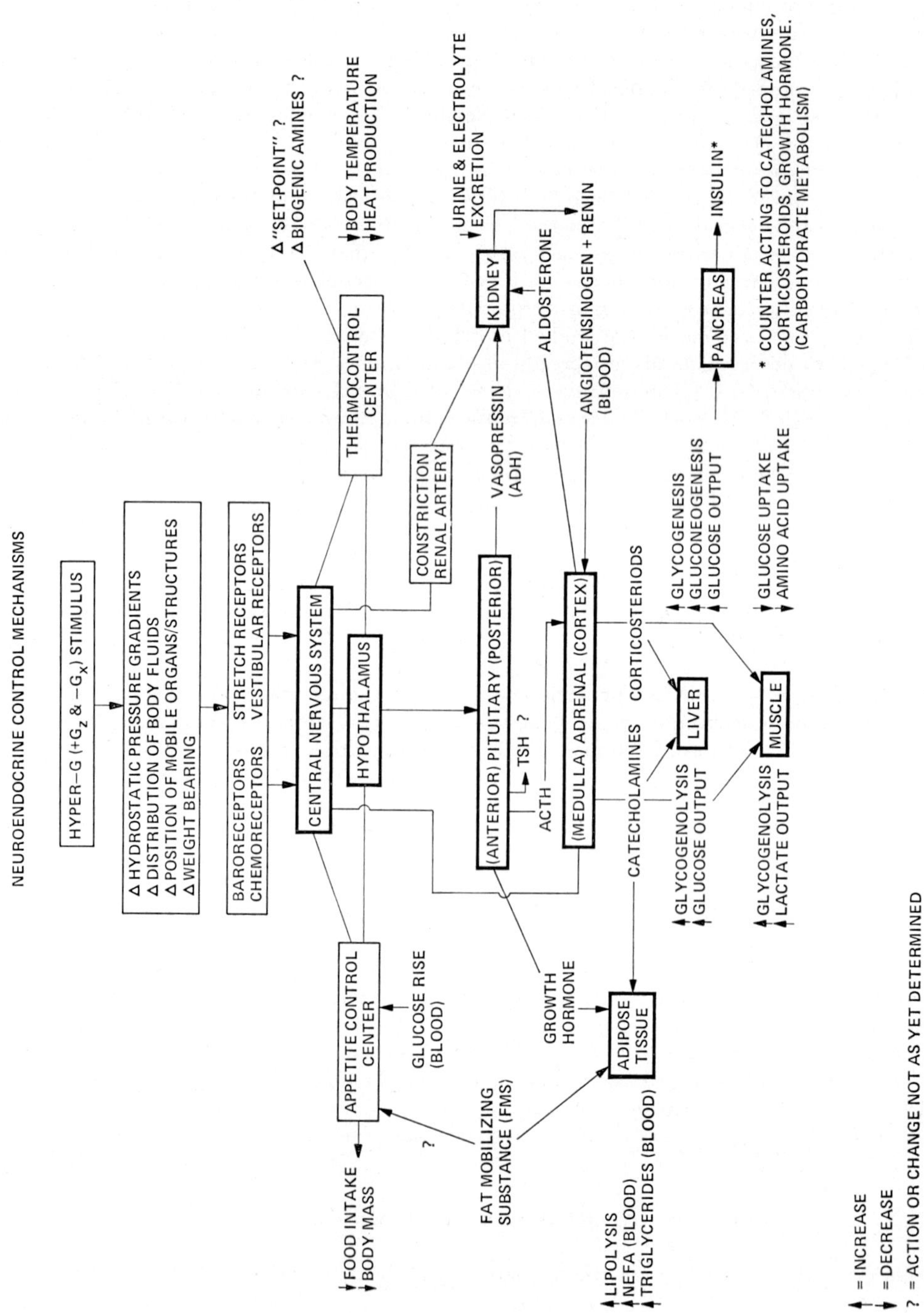

Fig. 1. Hyper-G and neuroendocrine coutrol mechanisms.

are activated by hyper-G. Feedback control systems are omitted for the sake of simplicity and because in a number of the systems they have not been fully resolved. The figure includes many of the responses listed in Table 1 and will not be further elaborated on here. There appears to be little doubt that central control of thermoregulation and appetite are both affected by hyper-G, but there have been few investigations into the nature of the changes and the mechanisms involved. Smith has shown (pers. comm.) that hyper-G exposures stimulate the production of FMS, fat mobilizing substance, which is excreted in the urine. When administered to normal rabbits, FMS has a profound appetite suppressing action. It remains to be determined whether FMS is physiologically active in the hyper-G exposed animal and if so whether its action is directed at the central control system. Since FMS also has a lipolytic action, it may also participate in the increased rate of lipolysis shown by hyper-G animals.

The neuroendocrine system regulation of intermediary metabolism is directed towards ensuring an adequate blood level of glucose for the brain and other vital organs such as the heart. This is achieved by increasing the rate of glucose synthesis by the liver as well as decreasing its utilization rate by other tissues than the brain and heart directly or through indirect sparing action of accelerated fat metabolism. The rise in blood glucose undoubtedly increases the secretion of insulin by the pancreas; however, the fact that in rats blood glucose levels remain abnormally high for at least the first 24 hours of hyper-G exposure (Feller and Neville, 1966) indicates that its action is more than counter balanced by hyperglycemia inducing hormones such as growth hormone, epinephrine and the adrenal corticosteroids. In view of its rather profound effect on metabolism, it is surprising that very little work, if any, has been done on the thyroid system during acute hyper-G exposures.

LONG-TERM HYPER-G EXPOSURE EFFECTS

Table 2 lists some of the major changes which have been reported in various animals exposed to hyper-G conditions by chronic centrifugation for periods ranging from 4-52 weeks, a time period sufficiently long to ensure physiologic adaptation. In all these studies, animals were maintained in pivitol cages so that they were subjected to the resultant of the centrifugal and gravitational forces in the normal or vertical direction with respect to the cage floor. Comparisons are made with noncentrifuged control animals maintained under essentially the same conditions except for the G exposure.

GROWTH. One of the most consistent and reproducible effects common to all animal species so far studied is a generalized suppression of their normal growth rate. All show a suppression which is rectilinearly related to the G-intensity (Smith and Burton, 1967). One interesting exception to this general response has been reported by Dodge and Wunder (1963). They observed an accelerated growth rate in turtles exposed to $-5G_z$ which was attributed to a stimulated appetite. At higher G-levels the turtles experienced the same general suppression in growth as other animal species. When first exposed to hyper-G, all animals show a loss in body mass. In rats, the loss can range from 10-15% during the first 24 hrs of centrifugation with additional losses occurring over the next several days (Oyama and Platt, 1965). Recovery of the initial body mass loss is usually achieved within the first week of chronic centrifugation in

Table 2. LONG-TERM HYPER-G EXPOSURE EFFECTS

SYSTEM/FUNCTION	G (range)	DURATION (weeks)	CHANGE FROM NORMAL G	SPECIES	REFERENCE
GROWTH	2-7	10	Decrease	Mouse	Wunder (1961)
	5-6	11	Decrease	Hamster	Wunder (1966)
	2.5-4.7	18	Decrease	Rat	Oyama (1965)
	1.5-4.0	24	Decrease	Chicken	Smith (1959)
BODY FAT	4.7	4-8	Decrease	Mouse	Keil (1969)
	2.8-4.1	8	Decrease	Rat	Pitts (1972)
	1.5-3.0	10-24	Decrease	Chicken	Smith (1963)
	1.5-2.5	4-28	Increase	Monkey	Smith (1973)
MUSCLE	4	8	Decrease (heart, diaphragm)	Mouse	Bird (1963)
	1.5-3.0	12-52	Increase (adductor)	Chicken	Burton (1967)
BONE	4	8	Increase (femur/body mass)	Mouse	Wunder (1960)
	4.7	52	Increase (,,)	Rat	Oyama (1967)
	1.5-3.0	11	Increase (,,)	Chicken	Smith (1963)
GLUCOSE METABOLISM	2.7-4.1	4-12	Increase (diaphragm muscle)	Rat	Daligcon (1970)
	1.7-3.0	24	Increase (whole body)	Chicken	Evans (1970)
FAT METABOLISM	4.7	52	Increase non-saponifiable fat	Rat	Feller (1965)
	1.8-3.0	12-24	Decrease fatty acid uptake	Chicken	Evans (1969)
RESPIRATORY METABOLISM	4.1	12-52	Increase O_2 consumption	Rat	Oyama (1973)

young growing rats while mature rats may never fully recover their body mass loss particularly at the higher G-levels.

BODY COMPOSITION. The most apparent change in body composition in hyper-G adapted animals is in their body fat. Marked losses in fat have been found in mice, rats, hamsters and chickens with the amount of loss increasing with the G-intensity. One striking exception has been reported recently by Smith et al. (1974) who found a significant increase in the total body fat of chronically centrifuged monkeys. They hypothesized that this unexpected increase could be due to differences in the weight distribution on the brain and cord of monkeys compared to other animal species. Oyama and Zeitman (1967) found marked decreases in lipid content of the kidney and liver in chronically centrifuged rats while plasma lipids showed relatively minor changes. The fatty acid distribution pattern in the lipids of the kidney and adipose tissues were altered significantly, but the ratio of saturated to unsaturated fatty acids remained the same.

MUSCLE. Muscular development under long-duration hyper-G exposures has been studied in the mouse, rat, and chicken. Comparison of absolute muscle masses among hyper-G adapted animals with corresponding normal gravity controls almost invariably shows the hyper-G animals to be smaller. This is due to the smaller body size of the hyper-G adapted animals. In order to normalize the body size differences, the muscle data are usually expressed in relative terms, i.e., muscle mass/body mass. Bird et al. (1963) found both the absolute and relative mass of muscle of chronically centrifuged mice to be smaller than controls. In rats, Pitts al. et (1972) reported a decrease in absolute muscle mass but the relative mass was unchanged. Burton et al. (1967) found the adductor muscle (hip extensor) increased absolutely as well as relatively in the chronically centrifuged chicken while opposite changes were found in the sartorius muscle (hip flexor). These different effects on muscle mass may reflect different levels of physical activity as well as mass differences between the various animal species.

BONE. Because the skeletal system bears most of the weight of the animal, changes might be expected in the major load bearing bones such as the tibia and femur. No absolute increases in bone mass have been reported so far in the chronically centrifuged mouse, rat or chicken although the relative mass is increased significantly in all three species as shown in Table 2. Jankovich (1971) found no differences in femur bone in chronically centrifuged rats based on densiometric measurements. Fosse (1971) reported essentially the same results in centrifuged male mice but found decreases in the density of the femur bone in female mice. The cross sectional dimensions of the femur assume a more circular form in chronically centrifuged mice (Wunder et al., 1960) and in the rat (Amtmann and Oyama, 1973). Amtmann suggests that this dimensional change may be due to immobilization rather than to increased weight bearing as first proposed by Wunder. The effect of calcium metabolism on bone development in rats during chronic hyper-G exposures has been investigated recently by Keil (1973). He found no significant changes in total calcium balance due to hyper-G exposure or any increase in the rate of calcium incorporation into the femur bone of young growing rats during chronic centrifugation. These findings along with those previously discussed support the conclusion that with small animals hyper-G exposures do not stimulate bone growth or increase its strength. Differences between hyper-G and normal gravity animals probably reflect retardation of non-skeletal tissue and organ growth.

BLOOD. The initial rise in blood glucose, fatty acids and adrenal corticosteroids levels are returned to pre-stress levels after several weeks of chronic centrifugation indicating the reestablishment of a new, steady state condition and physiologic adaptation. Other blood constituents show permanent changes of significant magnitude. Burton and Smith (1969) found increases in plasma volume, protein concentration, RBC and hematocrit in chickens after several months of centrifugation; opposite changes were found in rats (Oyama and Platt, 1965). Keil (1973) found a decrease in plasma calcium levels after 4 weeks of centrifugation of rats and Oyama and Zeitman (1967) reported the same changes in rats after one year of centrifugation. The lymphocyte count which falls markedly during hyper-G stress in chickens is restored to near normal levels after adaptation and has been used as a stress and adaptation index (Burton and Smith, 1972).

METABOLISM. Animals adapted to hyper-G consume more food and expend more energy than when under normal gravity (Smith, 1971). This increase in metabolism has been demonstrated by whole animal metabolic rate measurements during centrifugation as well as by measuring the metabolism of isolated tissue preparations from centrifuged animals. Resting metabolic rates of hyper-G adapted rats are more than 25% higher than controls (Oyama and Chan, 1973). The half-life of administered glucose is reduced by about 40% in chronically centrifuged chickens with a corresponding increase in glucose utilization rate according to Evans and Boda (1970). In isolated tissue preparations from chronically centrifuged rats, Feller and Neville (1965) determined the rate of lipid synthesis from acetate to increase more than two fold in the liver. Diaphragm muscle preparations from 2-20 weeks chronically centrifuged rats show as much as a 35% increase in glucose uptake compared to tissues from control animals (Daligcon and Oyama, 1970).

REPRODUCTION. Mice and rats can mate and successfully reproduce under hyper-G. Oyama and Platt (1967) observed that although rats can conceive and deliver viable pups at intensities as high as 3.5G, the newborn animals do not survive unless the intensity is less than 2.5G. Mice can reproduce with little difficulty at 2.5G. In the author's laboratory, rats are routinely reproduced at 2.3G under continuous centrifugation with an occasional loss of litters. No differences have been noted in the estrus cycle or gestation period in rats adapted to hyper-G. Newborn animals appear normal in all respects except that they do not show the head nystagmus response which is generally seen in normal gravity reared animals when initially subjected to angular accelerations.

FACILITIES FOR HYPER-G STUDIES ON ANIMALS

Various devices are used for hyper-G studies including catapults, vertical accelerators and centrifuges. Only the centrifuge provides the capability of short and indefinitely long exposures to hyper-G by virtue of its use of centripetal acceleration rather than linear acceleration. Many of the animal studies have been run on human centrifuges. A descriptive catalog of major human centrifuges in the U.S. and Canada has been compiled by Gell (1971) and in Europe by Scano (1971). Centrifuges built exclusively for acute and chronic hyper-G studies on animals and in current operation are located at the University of California, Davis; University of Iowa, Iowa City; and at Ames Research Center, Moffett Field, California. Description of some of the animal centrifuges and principles of chronic centrifugation are reviewed by Wunder (1965, 1971). Centrifuges nominally require daily or twice weekly stoppages for servicing of animals and removal of excreta. The Ames Research Center animal centrifuge is unique for its animal holding capacity and capability of being run continuously without any interruption for many months. An automated waste disposal system is installed in all cages along with an automated drinking water system. Food is delivered from a stationary platform located above the center of rotation through a tubing system which can telescope whenever the cages are moved along one of the 10 radial tracks. Physiologic measurements can be made continuously with a biotelemetry system mounted on each cage. Deep body temperature and heart rate measurements have been recorded in rats, rabbits, and dogs for periods up to 3 months on the centrifuge. An on-board T.V. camera permits continuous viewing and recording of physical activity of animals during centrifugation from an adjacent control room. The maximum operating speed is 20 RPM and at the maximum effective radius of 20 ft; the load that can be imposed is approximately 3G.

WEIGHTLESSNESS AND HYPO-G SIMULATIONS

Table 3 lists some of the methods which have been employed to simulate reduced gravity along with methods to achieve brief and long-term weightsness. A comprehensive review of the methods and physiologic consequences of hypodynamia and weightlessness has been compiled and edited by McCally (1968). In general, simulation methods have not been used very much on animals in comparison to human simulation studies because of the inherent problems in maintaining animals under such conditions for any length of time. Meyers and Godly (1967) studied the effects of total water immersion for two hours on chloralosed dogs and found increase in diuresis and cardiac output. Popovic (1973) measured circulatory parameters in hypothermic hamsters totally submersed in fluorocarbon for periods up to 24 hrs and have them fully recover. Using the water displacement method, Vose and Smith (1968) bouyed up dogs by water filled plastic containers and measured the progressive loss of calcium from the distal radius by radiographic bone densiometry. Hypokinesia has been extensively employed on animals, but the difficulty in delineating psychologic stress factors induced by restraint or immobilization from physiologic effects must be considered in using this method as a simulation of hypo-G. Besch et al. (1972) reported a 63% loss in body mass in totally restrained chickens after 9 days along with hypertrophy of certain organs such as the heart, lungs, liver and adrenals. Baybara (1971) found progressive degeneration of the bronchial vasculature in restrained rabbits between 2-12 weeks of exposure. In the de-acceleration method, animals are first adapted to hyper-G then subsequently studied upon return to a lower G level or to normal gravity.

Table 3. Weightlessness and hypo-G simulations

METHOD	DURATION	SPECIES	FUNCTION/SYSTEM	REFERENCE
Water Immersion	2 hr	Dog	Cardiovascular-Renal	Myers (1967)
Fluorocarbon Immersion	24 hr	Hamster	Cardiovascular	Popovic (1973)
Water Displacement	32 day	Dog	Skeletal	Vose (1968)
Hypokinesia (restraint)	9 day	Chicken	Organ and Body Mass	Besch (1972)
,, ,,	14 week	Rabbit	Vascular Morphology	Baybara (1971)
De-acceleration	1-7 day	Rat	Food Intake-Body Mass	Oyama (1964)
,,	43 day	Rat	Body Composition	Pitts (1972)
,,	1-4 day	Rat	Liver Regeneration	Feller (1967)
Parabolic Flight (zero-G)	7 second	Turtle	Eating Behavior	Von Beckh (1954)
,,	30 second	Monkey	Behavior	Thach (1968)
,,	30 second	Goldfish	Vestibular-Behavior	Von Baumgarten (1972)
Balistic Flight	2-3 min	Monkey, Mice	Cardiovascular-Behavior	Henry (1952)
Orbital Flight				
COSMOS-110	22 day	Dog	Cardiovascular-Pulmonary	Nefedov (1968)
Zond-7	6 day	Turtle	Morphology-Biochemistry	Sutulov (1971)
Biosatellite III	8 day	Monkey	EEG-Cardiovascular	Adey (1971)
OFO-A	6 day	Frog	Otolith Function	Qualtierotti (1972)
Apollo 17	12 day	Pocket Mouse	Radiation-Brain Morphology	Haymaker (1975)
Skylab II	59 day	Minnow Fish	Swimming Behavior	Von Baumgarten (1975)
Cosmos-605	22 day	Rat	Physiology-Morphology Biochemistry	Ilyin (1975)

The method is analogous to normal gravity reared animals being exposed to the lower gravity of the moon. De-accelerated rats show a marked and rapid increase in food and water consumption and body mass within the first 24 hrs upon return to normal gravity from 4.7G (Oyama and Platt, 1964). Pitts et al. (1972) measured the body composition of rats 43 days after de-acceleration from 60 days of chronic centrifugation at 2.8 and 4.1G. Although the body mass of the centrifuged rats did not reach control levels even after 43 days, no significant differences were found in body composition. Feller et al. (1967) found the mitotic activity in partially hepatectomized rats to be enhanced by de-acceleration; increased-G exposures on the other hand decreased liver mitotic activity. Parabolic and ballistic flights only provide at most a few minutes of weightlessness. Because hyper-G pulses are generated immediately before and after the weightlessness phase of the flight, physiologic data may be difficult to assess. Most of the studies have dealt with behavioral responses to weightlessness. Orbital flights have provided the only feasible means of exposing animals to weightlessness for any extended period of time. In general, the results obtained on animals are supportive of the human flight results. The original reports should be consulted for detailed accounts of the physiologic responses and changes effected.

CONCLUSIONS

The available data support the conclusion that within certain limits laboratory animals are fully capable of coping with and physiologically adapting to gravitational fields which are either lower or higher than normal earth's gravity including weightlessness.

In adapting to hyper-G, animals undergo a wide range of functional changes indicative of a state of generalized stress. The nature and extent of the changes effected vary considerably depending on various intrinsic factors (species, age, sex, body size, shape etc.) as well as on extrinsic factors (G-intensity, onset rate, duration, direction etc.) Many of the observed changes are mediated through the sympathetic nervous system and the hypothalamic-pituitary-adrenal system. Comparatively little is known, however, as to how the nervous and endocrine systems integrate their functions to effect a restoration of homeostasis and physiological adaptation.

Based mainly on human space flight studies along with the few animal experiments, weightlessness appears to be well tolerated with little evidence that it is stressful. Physiologic changes have been observed during weightlessness notably in body mass, cardiovascular/hemodynamic, musculoskeletal, endocrine/electrolyte, and vestibular systems. The vestibular system appears to be the most affected during the initial period of exposure to weightlessness and can induce some adverse responses.

Results from long-duration hyper-G exposure studies indicate that the major adaptive changes in animals to altered-G will be affected in their growth and development and in their metabolism. It is anticipated that these and other important physiological parameters will be studied in animals exposed to long-duration space flight weightlessness in the near future. The development of orbiting space laboratories for life science research within the next decade should lead to a more comprehensive understanding of the basic role of gravitational force on the structure-function-behavior of animals including man.

REFERENCES: see pp. 287-291.

CHAPTER 7

EFFECTS OF CHEMICAL AIR POLLUTANTS ON ANIMALS

SECTION 1: Effects of Various Air Pollutants on Domestic Animals (F. L. HAYS)

SECTION 2: Effects of Lead on the Growth of the Laboratory Rat (M. A. LESSLER and G. L. WRIGHT)

SECTION 3: Effects of Ozone, Carbon Monoxide, Cadmium and Nitrogen Dioxide on the Cardiovascular System of Animals (Alice Y. S. P'AN)

SECTION 4: Effects of Sulphur Dioxide and Related Compounds on Monkeys (Y. ALARIE)

SECTION 5: Effects of Pesticides on Animals (S. C. BHATIA)

Chapter 7, Section 1

Effects of Various Air Pollutants on Domestic Animals

by

F. L. Hays

1. INTRODUCTION

Man's success in populating and industrializing the world has been burdened with massive amounts of wastes, which have either made life unpleasant or endangered the health and existance of man as well as the plants and animals on which he depends for food and enjoyment. This is pollution. An air pollutant is some impurity discharged into the air. The air pollutant may simply be a nuisance or it may have increased the toxic load on the environment to the extent that in many areas the health and productivity of farm animals are affected. At the present air pollutants commonly tolerated by, or toxic to farm animals include smoke, automobile exhausts, various types of dusts, wastes from the oil or natural gas industry, heavy metals, respiratory gases and gases from excreta, fluorides, hydrocarbons and others. In this discussion only the toxic problems of the pollutants will be examined.

Industry has been obligated to conform to health practices for workers to maintain productivity. Health regulating groups have recognized maximal tolerance levels of toxicants for specified periods in working man while information about the tolerance levels for farm animals is limited.

Threshold Limit Values (TLV) are recommended in the U.S.A. by the Committee on Threshold Limits associated with the American Conference of Governmental Industrial Hygienists. These values are based on the best information available to the committee from industrial experience and from experimental studies designed to give the best criteria of either discomfort or toxic effects for workers. They are designed for use in the field of industrial hygiene and are reviewed yearly by the committee applying changes, revisions or additions as more information becomes available. Other countries have similar types of safety recommendations for workers, for example in Russia the MAC (Maximum Allowable Concentration) and in Germany the MAK (Maximum work-side concentration of gases, vapors, dusts, fumes and mists, the close observation of which may prevent damage to health even in the case of long-term, daily, eight-hour work periods). All of these values are used to predict healthy environments for workers; nonetheless, there are dangers since deviations of 50 to 100% do not have much significance since individual biological variability can be of that order of magnitude. Safety levels recommended by different groups may vary by a factor of 100.

The London smog in 1952 (Hudson, 1954) and the Poza Rica natural gas disaster in Mexico in 1950 (McCabe and Clayton, 1952) demonstrated a variety of toxic effects of air pollutants. These suggested that anatomical and physiological peculiarities of the various species had a modifying role in tolerance to air pollution. These and similar reports indicate the need for well controlled experiments in order to evaluate specific toxicological environments which can be used for evaluation of safety standards for farm animals.

2. SOURCES OF AIR POLLUTION

The variable sources of air pollutants can be generalized into three categories: natural-made, which include volcanic gases, marsh gases, smoke, dusts, pollen, viruses, fungi and bacteria; man-made, which include industrial pollution, silo gases, automobile, airplane and rocket exhausts, pesticides and radioactive contaminants and; animal-made, which include gases and micro-organisms that result, directly or indirectly from metabolic processes (i.e., respiratory gas and gases and micro-organisms from excreta).

Air pollutants toxic to farm animals could perhaps be thought of as those toxicants which irritate the eyes, skin or other periphery of the animal's body or which enter the animal to produce undesirable systemic effects. In addition, toxicants from the air may settle on feed or water such that the animal is exposed to an immediate atmosphere of the contaminant together with its ingestion. This occurs with fluoride and heavy metal emissions. Consequently, field surveys for air levels of many toxicants are meaningless without this additional information about the concentrations of these particular materials in forage or other feed which is consumed by these animals.

3. METHODS OF STUDY

Usually information about the toxic effects of air pollutants in domestic animals comes from two main sources: field surveys or observations and laboratory trials. Deaths, decreased growth or production, reproduction difficulties, increased incidence of disease and pathological signs of dead or sick animals are usually reported when suspected contaminants have polluted farm animals after industrial or farm accidents and after long-term exposures near specified sources. In most cases the concentrations or time intervals have not been measured and the toxicant thought to be responsible has been identified from previous experiences. All the same these reports are valuable since they are the result of actual exposure where interspecies comparisons may have been made but they are also, in many cases, difficult for interpretation because of the multitude of factors involved.

Laboratory trials utilize exposure chambers where precise levels of one or several pollutants (singly or in combination) can be introduced to the animals under otherwise controlled conditions. These pollutants can be assayed and their biological effects studied. These facilities, especially for large farm animals, are very expensive to construct and operate but they are the only choice for well controlled experimental work.

Many toxicological studies have involved rodents, cats, dogs, rabbits, monkeys and man or other species with little direct evidence for the toxic effects in farm animals. On one hand, this type of information is helpful for estimating tolerance levels or toxic effects in farm animals since little other information may be available. However, interspecies comparisons may be difficult to interpret because of metabolic size of the animals, methods of uptake of the toxicants by the animals, detoxification and excretion or other biological variabilities. Mice, for instance, exposed to an environment of 20 ppm hydrogen sulfide demonstrated thermoregulatory problems where as this concentration is close to the Threshold Limit Value (TLV) accepted for man (10 ppm). Dairy cows exposed to this H_2S environment for 2 weeks were not affected to any great extent (Hays, 1972).

4. ACTION OF AIR POLLUTANTS

A. SHORT AND LONG TERM EFFECTS

Toxicity of air pollutants in animals is referred to as "acute" or short-term exposures which is the result of inhalation of large quantities of toxicant over a short period and "chronic" or long-term exposure of a level considered sublethal for the short-term which results in pathological functions or death after a considerable length of time. Sub-acute intoxications have identified low levels of a toxicant which primarily involves irritation such as keratoconjunctivitis, pharyngitis and bronchitis with photophobia, pain and other eye disturbances. The effects of the long-term exposure can be complicated by modifying factors as shown in Table 1. The detrimental effect of the air pollutant may not be due primarily to the action of the toxicant on some organs or tissues but secondarily to starvation or disease. These factors complicate air pollution research.

Table 1. Factors which influence the toxicity of air pollutants in domestic animals

1. Type and concentration of pollutant
2. Length of time of exposure
3. Solubility of pollutant
4. Additional pollutant interaction
5. Climatic effects (temperature; humidity; radiation; prevailing wind; rainfall; season)
6. Biological effects (species of animal; age of animal; individual biological response (genetic); nutritional status; adaptative response; management and activity of animal)
7. Topography
8. Distance from contamination sources
9. Other stress conditions

B. SPECIFIC PROPERTIES OF POLLUTANTS IN VIVO

Physical or chemical behavior of pollutants in vivo are of paramount importance in determining possible toxicological effects. Increased solubility increases the amount of material that enters the systemic and tissue levels of the animals. Aviado and Salem (1968) suggested that inhaled pollutants such as sulfur dioxide and trioxide, nitrogen dioxide, ozone, hydrocarbons, carbon monoxide, aldehydes and particulates in the air triggered responses in the airways and bronchial, pulmonary and systemic circulation which were qualitatively similar. Depending on the level and the time of exposure, SO_2 and SO_3

generally exerted their primary effects on the airways resulting in mucosal irritation, bronchioconstriction and coughing. NO_2 and O_3 exerted their primary effects on the airways resulting in thickening of the alveolar wall and pulmonary edema. Hydrocarbons predominately affected bronchial, pulmonary and systemic blood vessels.

Transfer of any gas across the alveolar capillary membrane is a function of the rate gas enters the plasma from the alveoli, the rate at which gas is removed from the plasma by both the circulation of the blood and reaction in the blood and the rate at which gas is added to the alveolar capillary plasma by breakdown of the formed compound and the entry of more gas in solution in the mixed plasma (Cotes, 1968).

C. CONSEQUENCES OF ANIMAL SPECIES AND SIZE

Differential species sensitivites to given concentrations of air pollutants have long been recognized to be partially due to animal size. Small animals such as birds were at one time used to indicate dangerous levels of mine gas. Figure 1 demonstrates the body size and time effects of three species subjected to carbon monoxide. Nonetheless, differential interspecies sensitivities to certain toxicants may cause animals to deviate from this generality in the short-term and long-term exposures due to secondary effects.

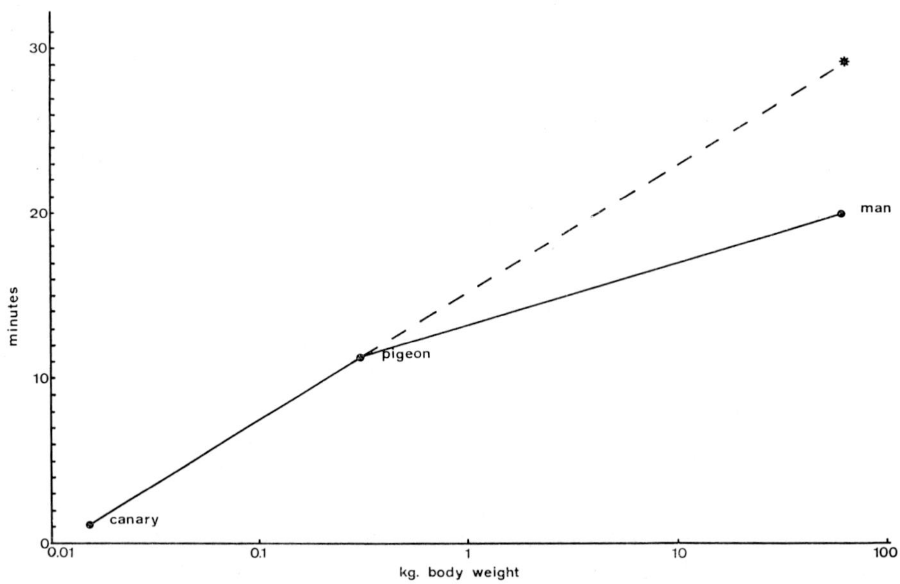

Fig. 1. Time necessary for distress in 3 species with exposure to atmospheric carbon monoxide (25 ppm). The birds were unable to maintain their equilibrium and the investigator suffered a slight headache and then became ill. A predicted time (by extrapolation compared to body weight) for man (*) is 9 min greater than the actual time recorded. This could be due possibly to the different index of distress used (Modified from Burrell, 1914).

The ratio of lung volume or minute ventilation and body weight do not change with increasing animal size where as the internal diffusing surface area (alveolar surface area) is proportional to the metabolic rate (Tenny, 1970). If all bronchiolar and pulmonary tissue for all animals were of equal sensitivity, then exposure to the same atmosphere of an air pollutant would undoubtedly affect a greater proportionate mass of tissues of the smaller animals compared to the larger animals in a equal period and thus increase that amount of pollutant which is available to diffuse into the systemic circulation. The ventilatory dead space compared to the lung volume is constant for all sizes of animals but as size decreases the respiratory frequency usually increases. This suggests that the rate of air movement into the airways is faster in smaller animals and that air would remain in contact with the tracheal tissues for a relatively shorter period and possibly deliver more of the air pollutant to the lower alveolar tissues. Toxic materials such as SO_2 have been isolated in dogs more in the upper respiratory tract than in the alveolar regions and from here enter the circulation (Frank et al., 1969). This is partly due to the concentration of SO_2 in the air, partly to the rate of air movement through the airways and partly to the surface area of the airways. Smaller animals breathing this same atmosphere could have more of the lower airways irritated. The inverse also holds since larger animals such as adult cattle would not necessarily be breathing air with such a great velocity and therefore not introducing so much SO_2 to the lower airways, over a given period. Differential ventilatory patterns of different species or during different stages of pollution induction or varying climatic interactions (i.e., such as heat stress) can affect the signs of exposure to given concentrations of air pollutants. Rapid, shallow breathing favors upper respiratory tract and anatomical dead space deposition of toxicants while slow, deep respiration tends to deposit gases, vapors and particles into the alveolar regions.

5. RESPONSES OF THE BODY TO AIR POLLUTANTS

A. TOLERANCE TO AIR POLLUTANTS

The principal centers of the body to register any meteorological stimuli are the eyes, skin, nose throat, lungs and nervous system. Many air pollutants instill acute inflammation and pulmonary edema not through neural or hormonal effects but by direct action, for example with sulfhydryl(SH^-) groups of proteins. It has been speculated that sulfhydryl moieties such as glutathione (GSH) or other "free radical scavengers" can maintain the cellular integrity of the respiratory tract (Stokinger, 1965). During exposure to irritants GSH levels in the body are increased through activity of the Pentose Phosphate Pathway. The increased GSH diminishes the degree of injury and tolerance ensues. This tolerance, nonetheless, has only been demonstrated for acute exposures to air pollutants. Many irritants afford a cross-tolerance effect to other irritants (Morrow, 1964). This is observed by induction of mucosal membranes to certain irritants followed by exposure to lethal levels of other pollutants. Edema, which can contribute to the tolerance effect in the respiratory tract, is not necessary for tolerance.

B. ADAPTATION TO AIR POLLUTANTS

The action of the hypothalamic-pituitary-adrenal axis adapts the body to secondary effects of air pollutants which influence many physiological functions. Corticoids mobilize amino acids for repair of damaged tissues. They also play a role in carbohydrate, protein, fat and purine metabolism as well as in diversified functional capacities of many systems (Goodman and Gilman, 1970).

C. ELIMINATION AND DETOXIFICATION OF AIR POLLUTANTS

The effect of any toxicant in the body at any specific cellular site which advertently affects the entire body depends on the amount which is absorbed or taken in by the body and the amount which is metabolized, conjugated, stored and excreted. This is also modified by adaptative processes with time. Signs and symptoms appear when the differential amount of pollutant delivered at the cellular level is high enough to be toxic. If detoxification, storage and excretion mechanisms are saturated, the poisonous signs

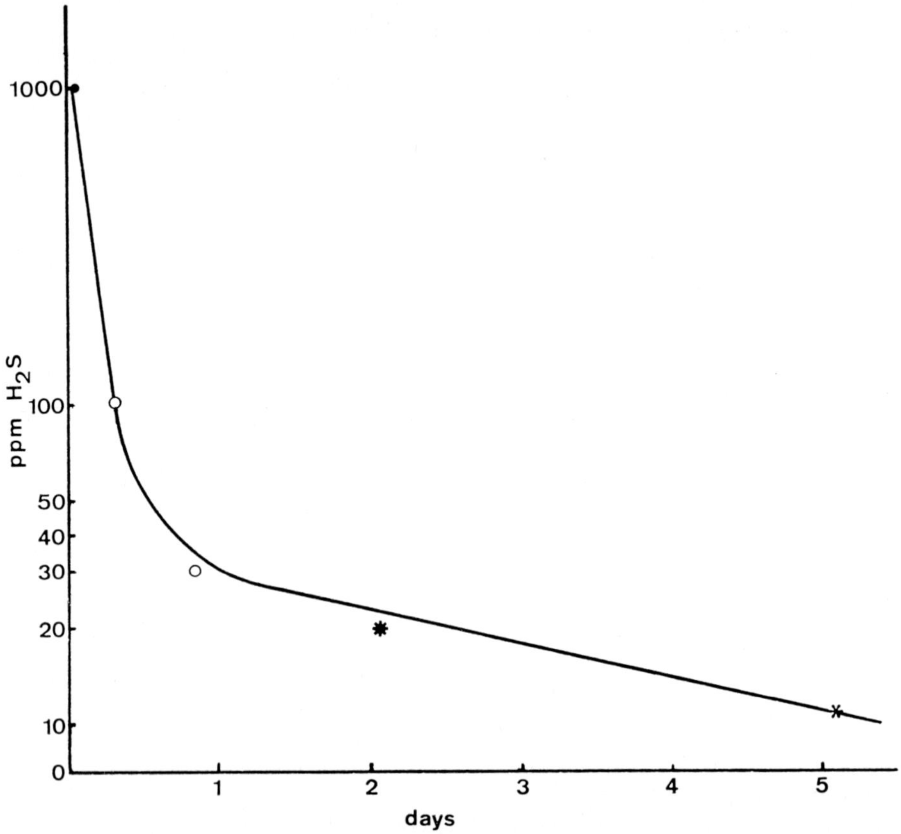

Fig. 2. Tolerance of mice exposed to atmospheric hydrogen sulfide. Circles indicate the level and time of death of 50% of the exposed mice. * is exposure to 20 ppm with no deaths but with hypothermia of 3.5 C less that normal and * is exposure to 10 ppm with no hypothermia. A minimum level of H_2S between 10 and 20 ppm is necessary for complete saturation of degradation and elimination processes which then expresses toxic effects (The closed circle value from Weedon et al., 1940. Other values from Hays, 1972).

may appear. If the animal is able to maintain those elimination processes at a level greater than intake, then signs should not appear. Figure 2 presents toxic affects of atmospheric hydrogen sulfide in mice. A level of H_2S between 10 and 20 parts per million (ppm) is lethal for mice. Exposure to this maximum value represents the differential between uptake and elimination which gradually saturates cellular mechanisms. Mice exposed to lower levels tolerate that environment since the elimination processes are not saturated or exhausted.

Clearance mechanisms in the airways are relatively unimportant with exposure to soluble compounds but are most important with insoluble particles and gases. Mucociliary escalation is a efficient mechanism for clearing deposited particles from the lungs. This mechanism transports particles deposited on the surface of the upper respiratory tract as well as those particles carried to the escalator from the non-ciliated lung surfaces to the oral pharynx and the to the gastrointestinal tract (GIT).

D. PHYSIOLOGICAL EFFECTS OF DIFFERENT BODY SYSTEMS AFTER EXPOSURE TO AIR POLLUTANTS

Although exposure to specific toxicants in the air can bring about specific signs and symptoms, many of these are non-specific effects. Table 2 presents physiological effects in different farm animals for different body systems upon exposure to some air pollutants.

E. SPECIFIC TOXIC EFFECTS OF FOUR COMMON AIR POLLUTANTS

1. Mercury

A toxic condition seen frequently in cattle and dogs is poisoning by mercury compounds. The elemental, liquid form of mercury is not poisonous whereas the alkyl, methyl and ethyl compounds are potentially dangerous (Goldwater, 1971). Organic mercurial antiseptics, fungicides, diuretics and insecticides are the common sources of poisoning in farm animals while mercury compounds in the air from industry are only a local problem. Since mercury is a cumulative poison acute and chronic toxic effects have been reported. Like other irritants, mercury ions attach to the Sulfhydryl groups of protein and destroy lung tissue, injure the gastrointestinal tract causing bloody diarrhea and damage kidney tubules and glomeruli which results first by polyuria and then anuria. Chills, fever, increased respiration rates and increased heart rates have been reported (Siegmund, 1967). Chronic mercury exposure can be characterized by diarrhea, weakness, paralysis, blindness, muscular tremors and incoordination. Mental retardation, abnormal effects of chromosomes and the fetus in utero and cerebral palsy have been reported in cases of mercury poisoning. Exposure to low levels of mercury compounds for long periods demonstrates no signs of adverse effects indicating possible increased tolerance to the compounds. Although determination of blood or urine levels is of little use for early diagnosis to acute exposure of mercurial compounds, the highest levels in the blood for normal people is 30 to 50 parts per billion (ppb). Nonetheless, levels of 50 parts per million (ppm) have been found in fish.

2. Lead

Farm animals in the vicinity of lead smelters have been subjected to the turmoils of

Table 2. Physiological effects in different species when exposed to atmospheric pollutants*

PHYSIOLOGICAL SITE OR RESPONSE	CATTLE	SWINE	SHEEP AND GOATS	POULTRY	HORSES	OTHERS
Decreased Growth	Sm Ar Cd D (flyash) F	F	Sm	Sm NH_3 CO D F Ions F		Be: a, b, e, i CO: a F: a Pb: i
Decreased Production	↓ milk: Sm Ar Cd F Pb Mo D (cement)		↓ wool quality: Sm; loss of wool: F	↓ eggs: Sm NH_3 D F		
Decreased Food Intake	Cd F Pb Hg NO_2 Sm	Pb F	F Pb Hg Sm	NH_3 CO		Auto: e
Body Temperature Change	↑: Sm NO_2 Pb		↑: Sm Pb H_2S	↓: CO		↓: CO; a ↓ H_2S; c
Activity Change	lameness: F	stiff gait: F	depression: Pb	incoord. gait: CO		stiffness: F; a
	weakness: Pb		lameness: F	hyperexcit.: HC		dullness: Auto; c, e
	pacing gait: Mo					disturbed postural reflexes: CO; e
	ataxia: NO_2 Ar					unsteady gait: NO_x^-; c
						unsteadiness: Gasoline; a, c, f, g, h
						hyperexcit.: HC; a, b
						↓ reflex activity: Cd; a
Reproduction	abortion: NO_2 D (flyash) Ar ↓ efficiency: F sterility of bulls: Mo		abortion: Sm Pb	catarrh in ovaries: Sm affects offspring: Pb ↓ egg numbers: F		testicle changes: F; e

Increased Susceptibility to Disease	Sm NH_3		Sm	Sm NH_3 D	Pb pneumonia	↓ Immuno. response: F; a, f, k H_2S; a, b, c, f, k SO_2; a ↓ resistance: NO_x^-; a, b, c, f, k
Death	NH_3 CO F Pb Mo NO_2 HC H_2S SO_2 (fog)	NH_3 Cd CO NO_2 HC H_2S	Pb NO_2 HC H_2S Ar CO	CO NO_2 HC H_2S Sm	Pb HC Ar F	Sm: a NH_3: c, k Auto: a, c, g F: a, e NO_2: a, b, c HC: a, b, c, e, Cd: a, b H_2S: a, b, c, e, f, g, h, i, k
Eyes	inflamed: Ar blindness: Pb irritant: H_2S	irritant: SO_2	conjunctivitis: F, H_2S blindness: H_2S	conjunctivitis: NH_3 HC	blindness: Pb	conjunctivitis: Auto; e F; a, e HC; a, b H_2S; f
Skin	irritant: Ar coarse hides: F	icteric: Cd		irritant: HC	lost hair, ulcers: Ar	scleritis: Auto; e F; a, e HC; a, b lost hair: Ar; a

* Most examples are reported from the literature where exposure to toxicants are found in the atmosphere or at levels usually experienced by these animals in the atmosphere. No differentiation has been made between short-term or long-term effects.
This table is compiled from various sources but mainly from Lillie, 1970.
Symbols: Materials are represented by their chemical formula where appropriates. Others include Smoke (Sm), Dust (D), Auto Exhausts (Auto), Nitrogen oxides (NO_x), Hydrocarbons (HC), Aluminium dust (D(Al)).
Responses include increase (↑), decrease (↓), calcification (calcif.), congestion (congest.), bronchiole (bronch.), pulmonary (pul.), respiratory (resp.), alveolar (alv.), center (cent.), unconsciousness (uncons.), respiration rate (RR), heart rate (HR), immunological (immuno.), hyperexcitable (hyperexcit.), carboxyhemoglobin (COHb), hemoglobin (Hb), methemoglobin (metHb), change (△).
Other animals include: a- rabbit, b- rat, c- mouse, e- dog, f- man, g- guinea pig, h- monkey, i- cat, k- fowl.

Table 2. Cont.

Table 2. Cont.

PHYSIOLOGICAL SITE OR RESPONSE	CATTLE	SWINE	SHEEP AND GOATS	POULTRY	HORSES	OTHERS
Teeth and Bones:	chalky, white, thick bones, dental mottling calcif. ligaments, osteomalacia: F fragile bones: Pb	soft teeth & bones: F	dental lesions, exostoses: F fragile bones: Pb	calcif. of bones: F	bone fractures, lameness: F	ricketts: F; a, e
Lung and Resp. Tract	edema: Hg NO_2 hyperemia: emphysema: NO_2 inflamed: CO, Ar	inflamed: CO emphysema: SO_2	edema: Cd Hg O_3	irritant: HC inflamed: CO catarrh in trachea & lung: Sm edema, congestion: NH_3	emphysema: F Pb bronch. paralysis: Pb	edema: O_3; b, e Cd; a, b, e NO_2; a, f SO_2; a, b, c, g Auto; a, b, g, e Pul. congestion: NO_2; a, f O_3; f bronchitis: Sm; a, b F; a, e O_3; b, c, g irritant: D; a, b, k O_3; e H_2S; a, f NO_2; b resp. tract damage: Cd; a, b, e HC; a, b, c, g Be; a, b, e, i SO_2; a, b, c, g NH_3; a, g, i inflamed: Ar; a NH_3; a, g, i emphysema: D(Al); a, b
Blood	anemia: F ↓ Hb, ↓ RBC: F Pb Mo	hemorrhage: SO_2 ↓ Hb: NH_3 anemia: Cd		↓ Hb, ↑ pH: NH_3 COHb: CO ↑ coagula-	anemia: Pb	anemia: Be; a, b, e, i ↓ RBC: F; a, e ↑ metHb: NO_2; c resp. alkalosis: CO; e

	↓ esino philes: F ↓ thrombo-cytes: Pb ↑ metHb: NO_2			tion-time: F		↑ RBC; ↑ COHb: CO; e
Nervous System	↓ resp. center: H_2S degenerated brain, spinal cord: Hg delirium: Pb F Ar	convulsions: NH_3 H_2S	↓ cholin-esterase: HC	convulsions: HC spasms, coma: CO	twitching, cerebral effects: Ar delirium, coma, convulsions: Pb	paralysis: Pb; i convulsions: HC; a, b CO; a tetanic spasms, unconscious: H_2S; a ↓ resp. centers: H_2S; e △ in cortex & white matter: CO; a, e
Respiration Rate	dyspnea: Pb Hg NO_2 increased: NO_2 F	dyspnea: Pb	dyspnea: Hg	dyspnea: CO decreased: NH_3 increased & decreased: HC	dyspnea: Ar F increased: Pb	increased: CO; a Cd; e dyspnea: F; f
Heart & Circulation	↑ heart rate: NO_2 F Pb, Sm vasodilation: NO_2 hemorrhage: Ar degeneration: Hg pericarditis: NH_3	↑ heart size: Cd degeneration & hemorrhage: Pb	↑ heart rate: Pb Sm hydropericarditis: Hg	↑ heart rate: HC vasodilation: CO	↑ heart rate: Ar	↑ heart rate, ↑ blood pressure: CO; e ↓ heart rate: Cd; e △EKG: CO; a, e hemorrhage: D(Al); a, b CO; a, e heart congestion, pulmonary hemorrhage: F; a pericarditis: NH_3; a, g, i degeneration: D(Al); a, b
Ears						deafness: CO; e, i Hg; g

PHYSIOLOGICAL SITE OR RESPONSE	CATTLE	SWINE	SHEEP & GOATS	POULTRY	HORSES	OTHERS
Gastro-intestinal Tract	diarrhea: Ar F Mo gastroenteritis: Hg hyperemia of Rumen: Pb hyperemia: Sm	ulcer: Cd hemorrhage: Pb	diarrhea: F hyperemia: Sm ulcer: Cd	↑ wall thickness: F catarrh in GIT: Sm	diarrhea, gastritis: Pb	inflamed: Ar; a Cd; a, b
Liver	degeneration: Pb Hg F necrosis: NO_2 Sm fatty liver: Mo ↑ liver size: Sm	fatty liver: Cd pale, covered with necrotic spots: Pb	necrosis, ↑ size: Sm		fatty degeneration: F	degeneration: Pb; g, k D(Al); a, b hyperemia: Gasoline; a, e, g, h congestion: Auto; e F; a necrosis: Mo; g
Muscle	atrophy, paralysis: Ar lameness: F weakness: NO_2 spasms, tetany: Pb				partial paralysis: Ar weakness, stiffness: Pb	

lead poisoning. Forage sprayed with lead arsenate, discarded storage batteries and paint containing lead are other common sources of poisoning. It is of little importance whether lead is inhaled or ingested since circulating blood lead levels are of immediate importance. Lead inhibits enzymes that are dependent on sulfhydryl making them unavailable. In addition lead is also deposited among the bone salts which indicates possible interactions with carboxyl, phosphoryl and other ionic groups (Chisolm, 1971).

Inorganic lead absorption by the skin is relatively unimportant while the greatest level of access to the body is through the gastrointestinal tract and by way of the airways and lungs. Acute lead poisoning is characterized by depression, staggering, blindness, muscular twitching and "snapping" of the eyelids. Convulsive seizures are common with elevated body temperature and the pulse is fast and weak. Respiration is usually rapid, shallow and labored. Abdominal pain may be evident with diarrhea or constipation. Acute hemolytic crisis is sometimes observed which results in hemoglobinuria and anemia. In clinical situations the toxic effects of acute poisoning are not well understood but observations of edema and direct injury to nerve cells have been noted. Blood vessels in the brain have an increased permeability with fluid leakage which can destroy brain tissue.

A distinct syndrome of chronic lead poisoning has never been sufficiently documented for cattle. Usually long-term intake results in a sudden development of the acute syndrome. Three distinct types of chronic poisoning have been described (a) gastrointestinal, (b) neuromuscular and (c) CNS syndromes. (Goodman and Gilman, 1970). Certain hematological signs may also be associated with these syndromes. In lead poisoning δ-aminolevulinic acid (δ-ALA) dehydratase activity which is dependent on free sulfhydryl ions is reduced which in turn leads to an increase of δ-ALA in the blood and urine and decreased formation of porphyrobilinogen. Excess δ-ALA is also converted to coproporphyrin III which is also found in the blood and urine. Mild symptoms are usually associated with 60 to 80 μg/100 ml of blood. Blood levels in excess of 100 μg/100 ml indicate relatively heavy exposure.

3. Nitrogen Oxides

Nitrogen oxides can be released from fermented plants (silage), automobile exhausts and industrial emissions. Normally nitrates are utilized in plants by conversion to proteins in the presence of sunlight. However, nitrates can accumulate under intense heat in the plants to produce nitrous acid (HNO), then nitric acid (HNO_3), nitrogen oxide (NO), nitrogen dioxide (NO_2) and nitrogen tetroxide (N_2O_4) (Lillie, 1970). Pharmacologically, the nitrate ion (NO_3^-) is inert but inhaled or ingested nitrate can be transformed into nitrite (NO_2^-) in the body; and it is the latter which is highly toxic to farm animals and other species. Toxic levels of nitrate can be found in common pasture grasses during rapid growth. Many of the cereal grains and crops grown on summer-fallow are high in nitrate and it usually increases in feed after rains or excessive dampness. The nitrite ion absorbed by the blood converts hemoglobin to methemoglobin which changes the blood to a chocolate brown color. Although the exact mechanism is not understood the basic pharmacological action of nitrite is to relax smooth muscle (Goodman and Gilman, 1970). Smooth muscle of the vascular system dialates which decreases venous return and pressure. Fainting can occur because of blood pooling in post-capillary vessels and the direct cause of death is the result of hypoxia. Signs of nitrite poisoning include dyspnea, with progressive severity in respiration, rapid, weak heart rate, decreased body temperature, muscular weakness, ataxia and cyanosis. Reports of chronic exposure to nitrogen oxide have indicated a greater incidence of abortion in farm animals. A maximum safe concentration of 25 ppm has been suggested for farm animals. The TLV for man was recommended to be 5 ppm.

4. Hydrogen Sulfide

Hydrogen sulfide is found widely distributed in nature as a product of bacterial decomposition of sulfur containing protein matter. Large quantities occur in certain petroleum and natural gas deposits, in volcanic gases and in sulfur springs. Hydrogen sulfide, ammonia, methane and carbon dioxide are some of the major gases from excreta which can evolve from anaerobic fermentation of animal wastes, from urine and from respiration. Hydrogen sulfide poisoning has been categorized as acute which refers to the systemic poisoning due to inhalation that has been of rapid onset where the central nervous system effects have predominated. This is usually marked by respiratory paralysis. Sub-acute exposure has been used to describe cases where local irritation has been the predominant factor. Long-term or chronic exposures have been ambiguous in the literature although serious effects of low levels for long periods have been reported in dairy cows (Hays, 1972). Irritation and inflammation of the conjunctivial and corneal tissues is a condition referred to as "gas eye" and is observed in about 1 hour at 50 ppm. At concentration from 50 to 100 ppm mucous membranes of the respiratory tract are irritated causing pulmonary edema. Concentrations of a least 1000 ppm can produce respiratory failure and asphyxia. It is accepted that H_2S inhibits enzyme systems such as catalases, peroxidases, DOPA oxidase, succinic dehydrogenase, carbonic anhydrase, dipeptidase and benzamidase. It is speculated that the activity in the respiratory center is inhibition of the iron containing cytochrome oxidase in the mitochondria. In general the mechanism of H_2S toxicity is based on its ability to bind with alkali metals which play an important role as activators of some enzymes. It is of interest that although large amounts of the toxicant is detoxified in the liver, large amounts have been reported to be detoxified in the erythrocyte (Evans, 1967). The TLV for man has been suggested at 10 ppm.

F. GENERAL ACTIONS OF AIR POLLUTANTS

Although the effects of specific air pollutants have been described many of these compounds induce non-specific effects in many animals. These generalized activities such as acute irritation or the chronic effects in other body systems are shown in relation to mechanisms of tolerance or adaptation (Fig. 3). Short-term exposure causing irritation can decrease appetite which ultimately decreases productivity. This can be classified as a primary effect of the pollutant. Long-term exposure affecting productivity due to unbalanced systems can be classified as secondary effects or effects not directly due to the toxicant.

G. TOLERANCE LEVELS OF SELECTED AIR POLLUTANTS FOR DOMESTIC ANIMALS

Table 3 attempts to indicate tolerance levels of some of the more common air pollutants for domestic animals compared to the TLV or MAK recommended for man. Much

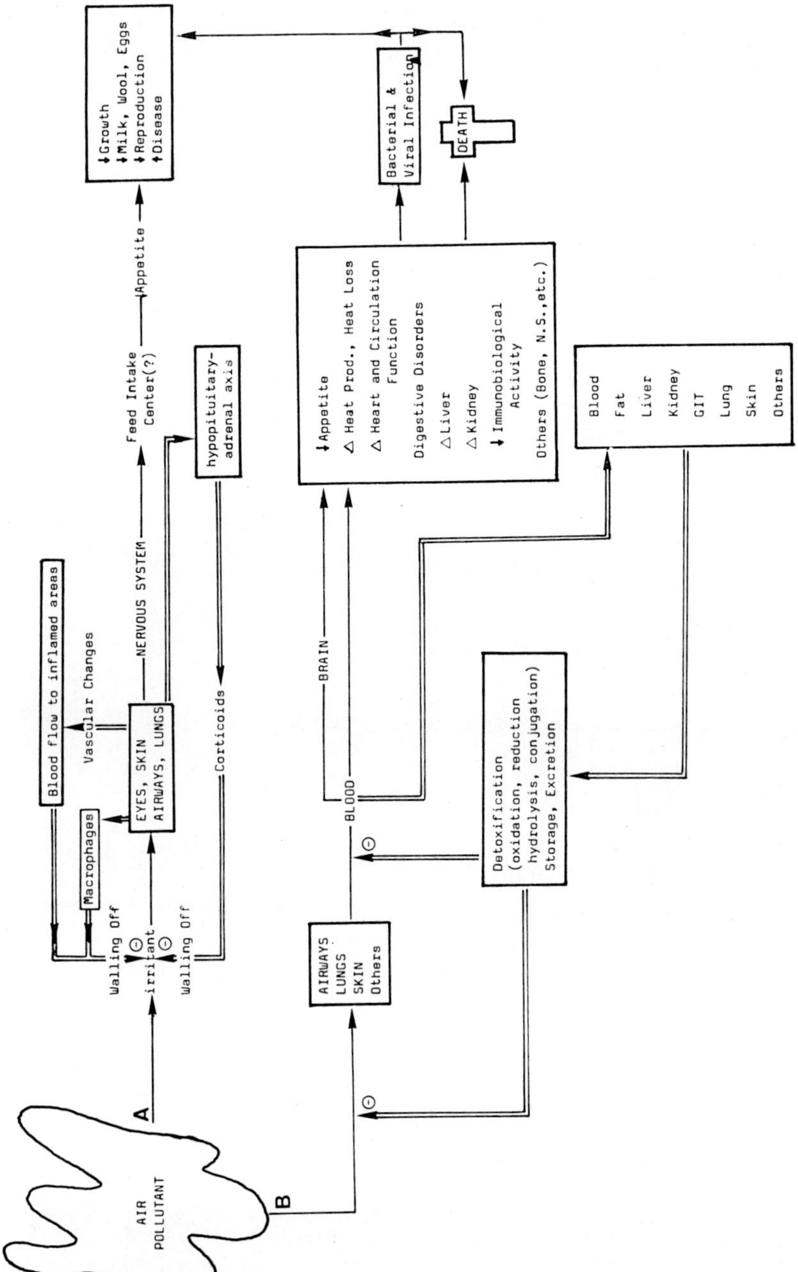

Fig. 3. A schematic summary of the general effect of air pollutants in animals with tolerance and adaptative responses. Single lines are the effects and sites of toxication while doubles lines suggest methods and sites of detoxification, tolerance and adaptation. Route A indicates short-term action and tolerance to air pollutants and Route B indicates long-term action, secondary effects and adaptative responses. Symbols include: $\downarrow$, a decreased response; $\uparrow$, an increased response; $\triangle$, a change in activity; and Θ indicates actions which decrease the activity of pollutants.

Table 3. Estimated tolerance levels of some pollutants for domestic animals

TOXICANT	CATTLE	SWINE	SHEEP & GOATS	POULTRY	HORSES	RABBITS	DOGS	MAN
F	† hay: 10 ppm safe † forage: 35 ppm safe † 2 mg/kg BW/day-toxic	high sol. salts † 8 ppm/BW/day or † 100/ppm/day-safe poorly sol. salts † 100-200 ppm/day-safe	high sol. salts † 1.5 ppm/BW/day or † 100 ppm/day-safe poorly sol. salts † 100-200 ppm/day-safe	high sol. salts Boilers: † 300-400 ppm/day Layers: † 100-250 ppm/day-safe		* HF 37 ppm, 6 hr/day for 6 days-no effec		* HF TLV, MAK 3 ppm
NH_3	* 100 ppm-accepted	* 100 ppm-irritant * 500 ppm-hazardous		* Layers: 20 ppm-↓ production				* TLV, MAK 50 ppm
CO				* Chicks: 600 ppm-distress in 30 min				* MAK 50 ppm
H_2S	* Dairy cows: 20 ppm OK	* 1000 ppm-deaths				* 1000 ppm-deaths in 2 hr		* TLV 10 ppm

Pb	* 0.15 mg/m^3-toxic	† 10-20 g/day-toxic	† 25 g/day-toxic		† 500-700 g/day-toxic	* 0.2 mg/m^3-toxic	* TLV 0.2 mg/m^3
Hg	* 1 mg/m^3-toxic		* 1 mg/m^3-toxic				* TLV, MAK 0.1 mg/m^3
Mo	* 1.1 mg/m^3 or † 230 ppm-toxic			† 300 ppm-toxic			* TLV 5 mg/m^3
O_3			* 15 ppm-toxic			* 1 ppm-irritant	* TLV 0.1 ppm
SO_2		* 40 ppm-hazardous					* TLV, MAK 5 ppm
CO_2	* 3500 ppm-accepted	* 250,000 ppm-deaths					* TLV, MAK 5000 ppm
CH_4						* 200,000 ppm slight effect	

Symbols: * amount in air
† amount in feed
ppm (Parts per million)
sol. (Soluble)

BW = body weight in kg
TLV = Threshold Limit Value (U.S.A.)
MAK = Maximum Work-site Concentration (Germ.)

more work must be done in this field in order to estimate realistic tolerance levels for air pollutants in respect to atmospheric levels and amounts in feed and water. When reviewing the literature it is important to realize the units describing the toxic load of the environment. Usually these are reported as ppb (parts per billion), ppm (parts per million), mg/m^3 all of which refer to amounts in the air and ppm in the feed, ppm in feed intake per day and ppm per body weight per day.

6. CONCLUSIONS

Due to the increased pressure of the confinement of agriculture and animal husbandry within urban areas more animals per unit area of land and air and more animals per person in urban areas create and are subjected to a greater amount of air pollution. These pollutants include smoke, internal engine exhausts, dusts, air-borne industrial wastes, organic pesticides, animal respiratory gases, gases from excreta and micro-organisms. Tolerance levels for these pollutants for animals have been estimated from field surveys and experimental work with either the animal of concern or with other species. In many cases much of this information is difficult to extrapolate for farm animals in the field.

The toxicity of air pollutants must be examined either as effects due to short-term exposures with high concentrations or to long-term exposures with low concentrations since in many cases the effects can be different. Long-term exposure increase the probability of introducing secondary effects of the pollutant such as decreased resistance to other stressors.

Many other environmental and biological factors can modify the effect of air pollutants. These in turn demand a greater amount of knowledge of the basic responses in the animal as well as their interactions with each other. Consequently, much more work must be done in the area of pollution toxicity and tolerance levels for farm animals.

REFERENCES: see p. 292.

Chapter 7, Section 2

Effects of Lead on the Growth of the Laboratory Rat

by

M. A. Lessler and G. L. Wright

INTRODUCTION

Lead was one of the first metals known and worked by man. It came into wide use during the Roman era, and as early as 370 B.C., physicians recognized plumbism in individuals who worked with lead. It has been suggested that lead intoxication, from food, water, and wine, may have been a contributing factor to the fall of Rome (Gilfillan, 1965). A considerable body of knowledge exists regarding the effects of exposure to lead in man and other animals, but harmful effects have been observed only at lead levels well above those that apply to the population in general. In fact, international monitoring of lead levels in air, water, and food have indicated that lead is far less important as a contaminant than other heavy metals.

POISONING WITH AIRBORNE LEAD

The natural concentration of lead in air has been estimated at approximately $5 \times 10^{-4}\ \mu g$ per cubic meter of air. In cities, airborne dust contains an average of 5 to 15 parts per million of lead depending on a number of factors such as automobile exhaust emission (Chow and Earl, 1970), industrial operations, abrasion of autos on roads, and the removal of paint from structural steel before repainting, etc. The burning of coal has been a major source of atmospheric lead, and probably adds more than 920 tons of lead per year to the atmosphere (Lead, 1970). Significant amounts of lead, also are released into the atmosphere from the smelting of lead ores (Murozumi, 1969). Estimates of airborne lead for 1968 indicate that approximately 985 tons of lead were emitted from smelting, and about 18,000 tons were released by the consumption of leaded fuels.

Aerosol lead is emitted into the atmosphere as dust, fumes, mist, or vapor depending on the source material and the method of generation of the aerosol (Emmerson, 1970). Studies have shown that the emission products from leaded gasoline are mainly lead bromochloride particles 2 to 10 μm in diameter, but lead sulfate and mixed lead halide oxides are also released in minor quantities (Habibi, 1970). If phosphorous compounds have been added to the gasoline, up to 20% of the emitted lead may be in the form of lead-phosphate-halide compounds. The amount of tetraethyl or tetramethyl lead that escapes into the atmosphere directly is extremely small, except at the site of manufacture and distribution of gasoline products, and does not significantly modify airborne lead concentrations.

Although the use of lead fuel additions has increased with the increase in the number of automobiles on the road, the available data suggest that airborne lead, even in cities, is increasing very slowly, if at all. A recent series of studies has indicated that previous estimates of airborne lead in cities have been misinterpreted and that air lead content has not been going up in many major cities in spite of greater use of leaded gasoline. This is probably due to the more efficient exhaust emission devices on automobiles, and the fact that lead is readily removed from air by aggregation and precipitation. Thus the lead content of rainfall is as variable as the lead aerosol content of the region in which the rainfall occurs. Lazrus and coworkers (1970) sampled atmospheric precipitation of lead at 32 U.S. stations and found an average lead concentration of 34 μg/liter of rain. Many stations reported lead concentrations of 10 μg/liter or less. The epidemiological studies of blood-lead by Goldsmith and Hexter (1967), which have been vigorously criticized, imply that longterm increases in atmospheric lead could result in higher blood lead levels in exposed populations. Recent reviews of their data, as well as new data, indicate that there is no correlation between airborne lead levels and blood lead levels.

The natural concentration of lead in soils is primarily due to their source of origin. There seems to be a geologic mechanism by which lead tends to move upward in the earth's crust, leading to higher concentrations of lead in the surface than in the subsurface layers. Many studies of lead content of soils along highways have shown that the lead content drops precipitously with distance from the highway (Lagerwerff and Specht, 1970). Thus animals grazing within 3 meters of a major road may be eating grass with about 5 times the lead concentration of grass 30 meters away. Relatively little of the lead in atmospheric aerosols is derived from the weathering of rocks and soil. In most cases, lead aerosol adds most of the lead to the topsoil, and also increases the lead content of the dust on plant materials by precipitation on them. It has been shown that soil lead is largely unavailable for uptake by plants. A number of investigators have indicated that the primary source of lead in plant materials probably comes from the roadside dust and rainfall and not from the soil (Lead, 1970).

Excessive lead intake is one of the causes of poisoning in domestic animals. It is usually recognized only when specific clinical signs of lead poisoning are apparent. Recent reports have shown that lead can cross the placenta into a developing embryo (Carpenter et al, 1973), and suggest that this may be instrumental in the decreased fertility noted in experimental animals exposed to relatively high doses of lead (McClain and Becker, 1970). The natural curiosity, grawing, and licking habits of many animals make any lead containing material a potential source of lead poisoning. For example, lead soldered cages and cages painted with lead based paint, have been implicated as possible sources of lead in lead poisoning of zoo animals and rodents (Bazell, 1971). Several outbreaks of lead poisoning in herds of domestic animals have been traced to the metal contamination of pasture and crops by industrial lead operations. Pastures can become contaminated by aerosols and dusts from an industrial plant that settle on the surrounding countryside. When the grasses are eaten by grazing animals, they ingest sufficient lead to cause lead poisoning. It has been reported that the presence of lead and arsenic dusts near the smelters in East Helena, Montana have made it impossible to raise horses there for several decades (Gordon, 1968). The minimum fatal dose of lead for cattle has been estimated by Hammond and Aronson (1964) at 6 to 7 mg/kg per day (a concentration of 300 ppm in the total diet). There is evidence that horses are more susceptible than cattle to the chronic ingestion of lead. Rats, on the other hand, are considerably more resistant to lead poisoning than the above mentioned species (Lessler and Cardona, 1970).

POISONING WITH LEAD-CONTAINING FEED

GROWTH OF LEAD-FED RATS

Recent awareness of problems concerned with increased lead in the environment have resulted in studies of this element's influence on a wide range of biological functions. It has long been known that lead intoxication interferes with the normal growth of mammals, but specific information on the influence of lead on mammalian development is sparse. In order to obtain quantitative information on the effect of chronic lead poisoning on young, rapidly growing rodents, 1% lead acetate was added to the diet of a selected group of rats, and their growth and hematology was studied over a period of 6 months. It should be emphasized that although the rats quickly became lead intoxicated, and were off their feed from time to time, none of the animals died during the 24 week experimental period. Based on their average daily food intake, each rat ingested approximately 135 mg of lead acetate per day or about 68 mg Pb/day. Kehoe's (1964) studies on the intake and excretion of lead by man indicated that only 5% to 10% of lead ingested in the diet is retained. If the excretion of dietary lead is the same for man and rat, our lead-fed animals accumulated between 3.4 mg and 6.8 mg Pb/day.

Healthy, young, male Wistar rats weighing 160 $\pm$ 10 g at the start of the experiment, were fed granulated Purina Lab Chow to which 1% lead acetate was added, and a group of paired controls were fed the same diet without added lead. Animals were housed 3 per cage in air conditioned rooms maintained at 23 $\pm$ 1°C, with a 12 hr light/12 hr dark photoperiod. Food intake of each cage was monitored daily, and all animals were weighed weekly. Blood samples were taken periodically from the tail vein of each rat with heparinized microhematocrit tubes for hematologic studies and blood lead determinations. The lead content of blood from control and lead-fed animals was determined on 10 μl samples by the nitric acid wet ash-Delves cup atomic absorption technique (Delves, 1970).

As expected, the control animals showed very rapid growth during the first 5 weeks and then a relatively slower growth rate (Fig. 1). The growth rate of the control rats from the 5th through the 23rd weeks showed an average increase of 10 g per week. The lead-fed rats grew only slightly slower than the controls during the first 4 weeks, but by the 5th week had significantly lower weights than the control animals. Calculation of the linear regression line for growth from the 5th to the 23rd weeks showed an average growth rate of only 7 g/week for the lead poisoned rats. Thus the lead intoxicated rats on the average grew only 70% as well as the controls and after 23 weeks had an average weight of 385 g as compared to 500 g for the rats that had no added lead in their diet.

Blood lead levels never rose above 25 μg/100 ml of blood in the controls, but in the first 2 days reached 140 μg to 150 μg% in the lead-fed animals. Although there was

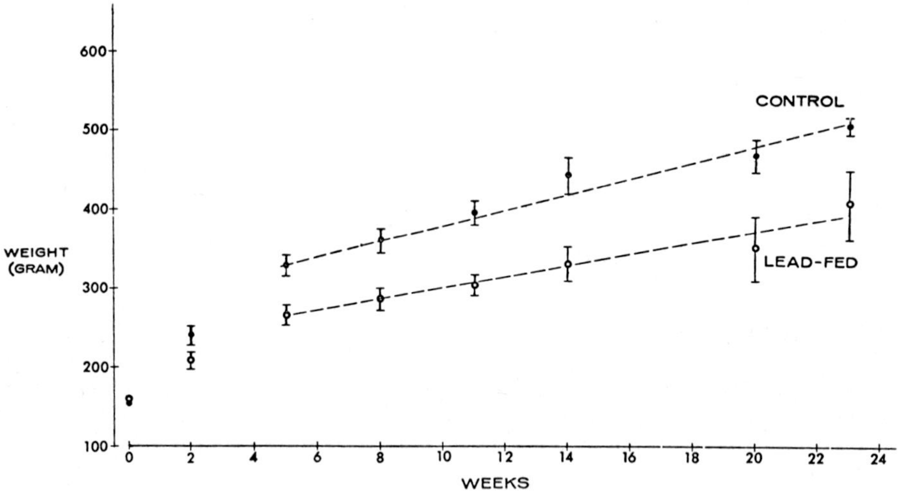

Fig. 1. Growth rate of young rats: 160 g at the start of the 24 week growth period. Controls were fed water and Purina Lab Chow ad libitum and the lead-fed rats the same diet with 1% lead acetate added. The dashed lines are linear regressions calculated from the weights obtained weekly for the 5th through the 23rd weeks. The bars through each mean indicate the standard deviation.

some variation from week to week the lead intoxicated animals averaged 150 μg of lead/100 ml blood during most of the experimental period. Most authorities consider blood lead levels above 100 μg/100 ml to indicate lead poisoning (Chisolm, 1971). In spite of the high blood lead levels in the lead-fed rats, hematologic signs of lead poisoning, such as decreased hematocrit and hemoglobin content, were not evident during the first 8 weeks, but there was a transient reticulocytosis during the 2nd and 8th weeks. The lead-fed rats showed definite signs of anemia by the 11th week, but appeared to adjust to the mild anemia during subsequent weeks. Thus the presence of 1% lead acetate in the diet of young, rapidly growing rats causes lead poisoning which retards their growth and development. This is probably due to several factors such as the lead itself, the anorexia and general lassitude induced by the lead poisoning, and the anemia. It is important to note that although the animals became sick, they ate enough to maintain reasonable growth, and somehow adjusted to the high lead content in their diet.

In order to study the effect of lead on the growth of mature rats, we repeated the procedures described above with a colony of mature rats that averaged 335 $\pm$ 10 g at the start of the experiment. The ingestion of Lab Chow with 1% lead acetate for one week caused an elevation in blood lead to 85 μg/100 ml, a level usually associated with borderline lead intoxication, but well below the 140 μg to 150 μg observed with the young rats. The mature rats maintained average blood lead levels of 75 μg for the first 5 weeks, and averaged 100 μg/100 ml during their last 3 weeks on the lead containing diet (Fig. 2). A reduction in food intake could have accounted for the lower blood

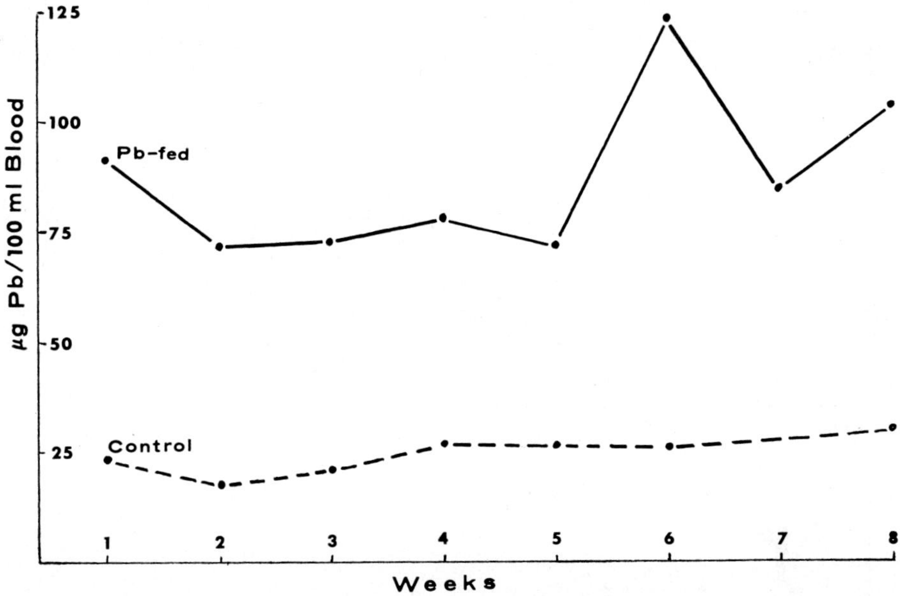

Fig. 2. Mean blood lead values for the mature rats who averaged 335 g at the start of the 8 week growth period. The dashed line indicates the control group, and the solid line the rats maintained on a diet containing 1% lead acetate.

lead levels of the mature rats, but the growth rate of both lead-fed and control animals was about the same for the first 3 weeks (Fig. 3) indicating that some mechanism other than anorexia was involved. During the 4th through the 8th weeks the growth rate of the mature lead-fed rats was significantly below that of the control rats, although both groups ate approximately the same amount of food.

These studies suggest that the mature rat is better able to handle a lead load than the young rat. They could do this by absorbing less lead from the gastrointestinal tract, by more efficiently excreting lead via the kidney, by sequestering more lead into an inactive form in bone, or by combinations of these mechanisms. Present data on the metabolism of lead do not allow us to distinguish among the three mechanisms which may act singly or in concert to play a protective role in the regulation of lead uptake and excretion in lead intoxication.

DIET AND LEAD POISONING

It has been suggested that the deleterious effects of lead on tissue metabolism may be exaggerated in phosphate deficiency. This is probably due to the protective effect of inorganic phosphate on mitochondrial damage by lead (Cardona et al., 1971). The absorption of lead from the gastrointestinal tract as well as the partitioning of lead into the various body compartments appears to be regulated, at least in part, by the same mechanisms that control calcium and phosphate metabolism. For example, Six and Goyer (1970) showed that rats placed on a calcium deficient diet with 200 µg lead/ml

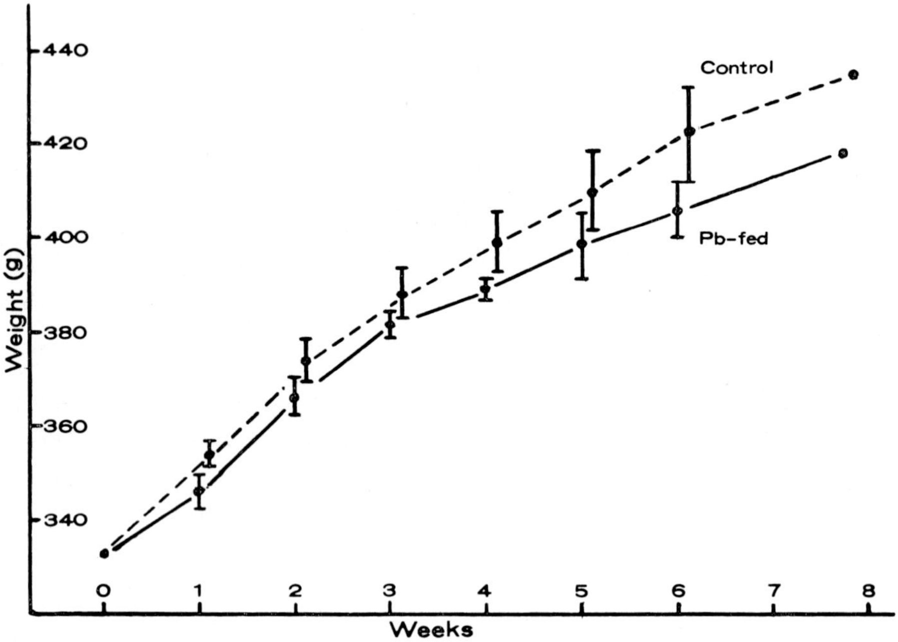

Fig. 3. Growth rate of mature rats: 335 g at the start of the 8 week period. The 1st week represents the mean weight of 30 animals for each group, and in subsequent weeks an average based on a decrease of 3 animals per group per week. The bars through each point show the standard error of the mean.

added to their drinking water had less fecal lead, but higher blood and urine lead than a comparable group of rats on a normal calcium diet. This amount of lead in the drinking water did not produce any measurable signs of lead toxicity with the normal calcium diet, but the rats showed inhibition of hemoglobin synthesis and severe anemia when on a low calcium diet. Indeed, Kostial and coworkers (1971) have shown that the addition of calcium and phosphate to milk fed to newborn rats actually reduced their body burden of lead. Thus it is evident that blood and urinary lead levels, as related to degree of lead poisoning are controlled by a multifactored mechanism.

SPECULATIONS ON THE RELATIVE RESISTANCE OF RATS TO LEAD POISONING

The presence of an increased body burden of lead and its effect on the growth and development of animals is much more complex than originally conceived. We now know that lead absorption, excretion, and storage varies from species to species, and within species based on dietary variations of the animals under study. Thus, it is dangerous to compare the relative susceptibility of animals to lead poisoning without definitive studies under controlled conditions. With this in mind, I would like to speculate on the relative resistance of rats to lead poisoning. It has been estimated that the minimal cumulative fatal dose of lead for cattle is 6 to 7 mg/kg/day, and about half this as the fatal dose for horses (Hammond and Aronson, 1964). Based on a mean weight of 250 g for the rat and average daily lead intake of 68 mg lead/day our rats had no fatalities due to lead poisoning for 24 weeks in the first experiment and 8 weeks in the second experiment. We calculated an average daily dose rate of 272 mg lead/kg/day or 40 to 45 times the chronic daily lethal dose for cows or horses. This indicates that the laboratory rat may be a species which has become adapted to elevated lead levels in the environment.

ACKNOWLEDGEMENT. The growth studies were supported, in part, by NIH General Research Support grant 5409.

REFERENCES: see pp. 292-293.

Effects of Ozone, Carbon Monoxide, Cadmium and Nitrogen Dioxide on the Cardiovascular System of Animals

by

Alice Y. S. P'an

INTRODUCTION

Chemical air pollutants are numerous; relatively few have been identified as toxicants. Usually chemicals present in the air do not attract attention until their toxic properties threaten personal survival; that is why pollutants causing death and illness to industrial workers were the first to be investigated while ambiant air pollutants causing a continuum of low grade insults with no acute biological response remained long ignored. Only after several acute air pollution episodes – caused by extraordinary meteorological conditions which confined pollutants to limited areas – had produced overt and wholesale illness among the population did toxicologic research become interested in air pollutants. Up to now the actions of some two dozen major air pollutants have been studied in animals; experiments usually focus on the respiratory tract and most of the information we have regarding cardiovascular effects were fortuitous findings.

When talking about atmospheric pollutants, it is important to keep in mind that their levels are very low: most of them are in the range of parts per hundred millions(pphm). Because of their low concentrations, their action on living organisms is insidious, slowly progressive. Their long-range effects on health depend mainly on their chemical characteristics, on their levels of penetration into the body, on the organ into which they accumulate and on their residence time there. Generally speaking, the cardiovascular system can come under attack under three circumstances: when the pollutant injures blood vessels (direct attack), when the pollutant, through injury of the renal tissue, induces high blood pressure (indirect attack), and when the pollutant, through deterioration of the lung tissue, causes pulmonary hypertension leading to right ventricular failure (indirect attack). These three mechanisms of action will be discussed under the headings of three air borne chemical pollutants: carbon monoxide, which damages arteries, cadmium, which injures the kidneys and nitrogen dioxide, which deteriorates the lung tissue. The examples presented in this chapter can be used as an approach to estimate the possible cardiovascular effects of any other chemical pollutant. Before these chemical pollutants are discussed we should like to mention the toxicity of the photochemical pollutant ozone.

OZONE

The upward trend of photochemical pollution in urban atmospheres has called attention to the toxic potentiality of ozone, the major component of this modern type of air pollution. The acute toxicity of high concentrations of ozone has been inown for years in occupational medicine but the insidious effects of long-term exposure to low ambient levels of this strong oxidant are still under study.

People living in areas polluted by photochemical smog complain of eye irritation; epidemiological studies have shown that their lung function is also affected (Moltey et al, 1959; Remners and Balchum, 1965). Studies using laboratory animals indicated that ozone can produce a series of pulmonary disorders among which bronchiolitis, emphysema and septal fibrosis are the most common (Stokinger, 1965; Wethamer et al., 1970). Early investigators had centered their toxicological studies on the lungs and the respiratory tract, which was a logical way to proceed since these constitute the primary target of inhaled ozone, but since Brinkman and Lamberts (1958) expressed the possibility of ozone's having radiomimetic effects, research was extended to extrapulmonary fields. Ozone was then found to accelerate the process of cell aging (Brinkman et al., 1964), to influence enzymatic activity (Chow and Tappel, 1973), to weaken pulmonary defense mechanisms (Dowell et al., 1970; Goldstein et al., 1972) and to intervene in various ways with physiologic functions and pathologic phenomena (Stokinger and Coffin, 1967). Our knowledge of ozone induced biological changes is far from exhaustive. Up to now, very little is known about the effects of this oxidant pollutant upon the cardiovascular system.

Several studies, aimed at the effect of ozone upon the respiratory system, have shown that this gas can cause vascular damage. In high concentrations, ozone is a determinant of pulmonary edema and hemorrhage both in man and animals (Stokinger and Coffin, 1967). Scheel et al. (1959) found no edema in the lungs of mice exposed to 1 ppm ozone for 4 hours, however, they observed that the blood vessels were moderately engorged and the capillaries filled with leucocytes. Bils (1970), using electron microscopy, detected that ozone affects specifically the endotelial cells of the capillaries. He noticed a proliferation of endothelial cells in capillary cross sections and protrusions of cytoplasmic processes into the lumina.

Brinkman et al. (1964) studied the effect of 0.2 ppm ozone upon the circulating erythrocytes of man and animals and upon the myocardial fibres of rabbits and mice. The results of the former investigation indicated that ozone, just like X irradiation, greatly speeds up the sphering of erythrocytes. The young erythrocyte enters blood circulation as a very flat disc and becomes more and more spherical as it approaches the final stage of its life cycle, therefore acceleration of the sphering of erythrocytes is equivalent to acceleration of its ageing process.

Brinkman et al. (1964) also made a histological study of the myocardial tissue of rabbits exposed to 0.2 ppm ozone for several weeks and killed one week after the discontinuance of exposures. Many myocardial fibres had suffered nuclear damage: destruction of the envelope and extrusion of the nuclear content. This type of nuclear damage, often encountered during X ray treatment, was not found in animals sacrificed 4 weeks after ozone exposure had been stopped.

P'an and Jegier (1972) demonstrated that the small pulmonary arteries of rabbits intermittently exposed to 0.4 ppm ozone for 180 days had thickened walls and narrowed lumina. Using a micrometer eyepiece over a 10X objective, which gave a magnification of 100X, small pulmonary arteries cut at right angles to their course and having exterior diameters ranging between 105 μ and 1,050 μ, were measured according to the technique described by Morlock (1939). The wall of each vessel was measured in four opposite places (Fig. 1: ab, cd, ef and gh) and the lumen in two perpendicular diameters (Fig. 1: bc and fg). The average thickness of the wall and the average diameter of the lumen were calculated; the ratio of the thickness of the wall to the diameter of the lumen gave an accurate notion of the degree of thickness of each vessel. Table 1 shows how the ratios of the exposed animals contrasted with those of the controls.

Table 1. Mean ratio of thickness of wall to diameter of lumen of small pulmonary arteries.

Rabbits	Mean Ratio of Wall Thickness to Lumen Diameter of 18 Vessels	Mean Ratio for Whole Groups	SD	Coefficient of Security
Exposed				
1	1:1,9			
2	1:1,4			
3	1:1,7	1:1,7	0,21	
4	1:1,4			
5	1:1,4			
6	1:2,7			
Control				P < .005
1	1:5,1			
2	1:4,9			
3	1:5,2	1:4,9	0,08	
4	1:4,8			
5	1:4,7			
6	1:4,9			

The walls of the small pulmonary arteries of the ozone exposed rabbits were very much thicker than those of the small pulmonary arteries of the control rabbits. The thickening was due to an increase in the thickness of the tunica media vasorum (see Figs 2, 3, 4 and 5). In some vessels the middle layer appeared to be edematous; in others, there was cellular hyperplasia. This type of vascular damage is frequently seen after radiotherapy; it is another argument in favour of the radiomimetic properties of ozone (Brinkman et al., 1964).

The radiomimetic aspect of ozone is still being investigated and nothing definite can be said about the direct action of ozone upon living cells. In the case of the thickening of small arteries, it can only be conjectured that ozone attacked first the endothelial lining

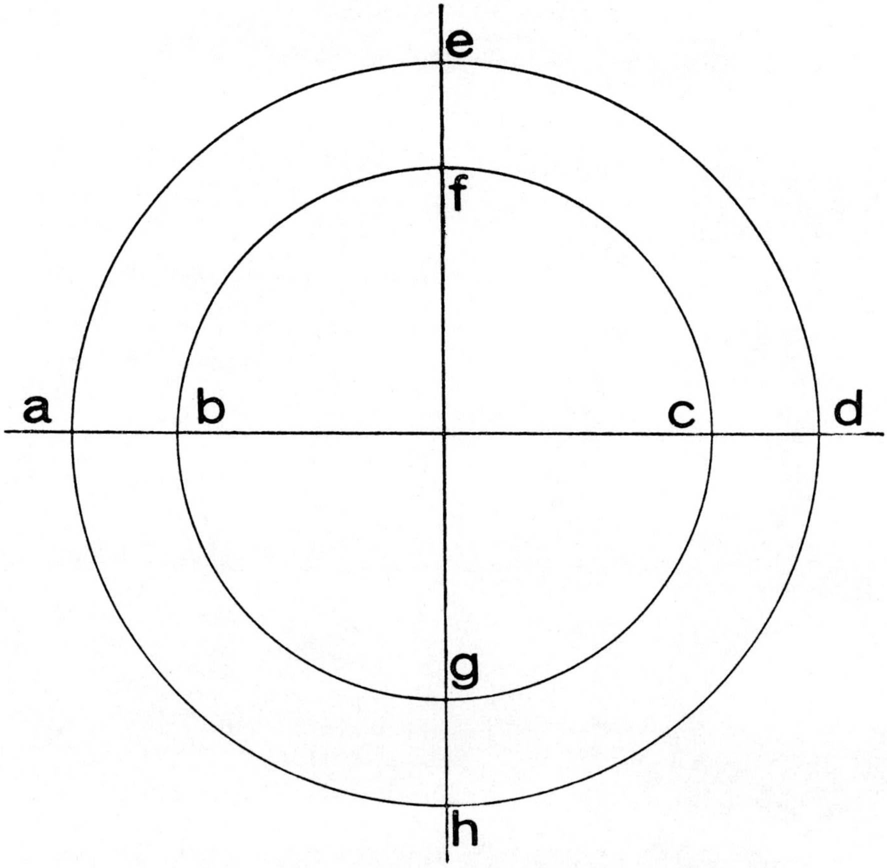

Fig. 1. Cross section of a small artery illustrating the Morlock method of measurement of blood vessels. Thickness of the wall is abtained by averaging ab, cd, ef and gh. Diameter of the lumen is obtained by averaring bc and fg.

(Bils, 1970); through this primary lesion, subintimal edema developed, splitting up the fibrillar texture. This should be an early reversible type of arterial thickening that could disappear upon cessation of administration of ozone. As exposure to ozone was continued, the edemagenic response probably acted as a noxious stimulus inducing cellular proliferation and collagen formation. The thickening of the muscular layer narrowed the lumen and blood flow was diminished. As the inner half of the tunica media vasorum of small arteries is nourished by direct permeation from the lumen, such morphological changes would soon lead to hypoxia and by this time, the lesion might have become irreversible.

Reduced lumina in the small pulmonary arteries can progressively lead to pulmonary hypertension which in turn will stress the right ventricle. When the pulmonary tissue is

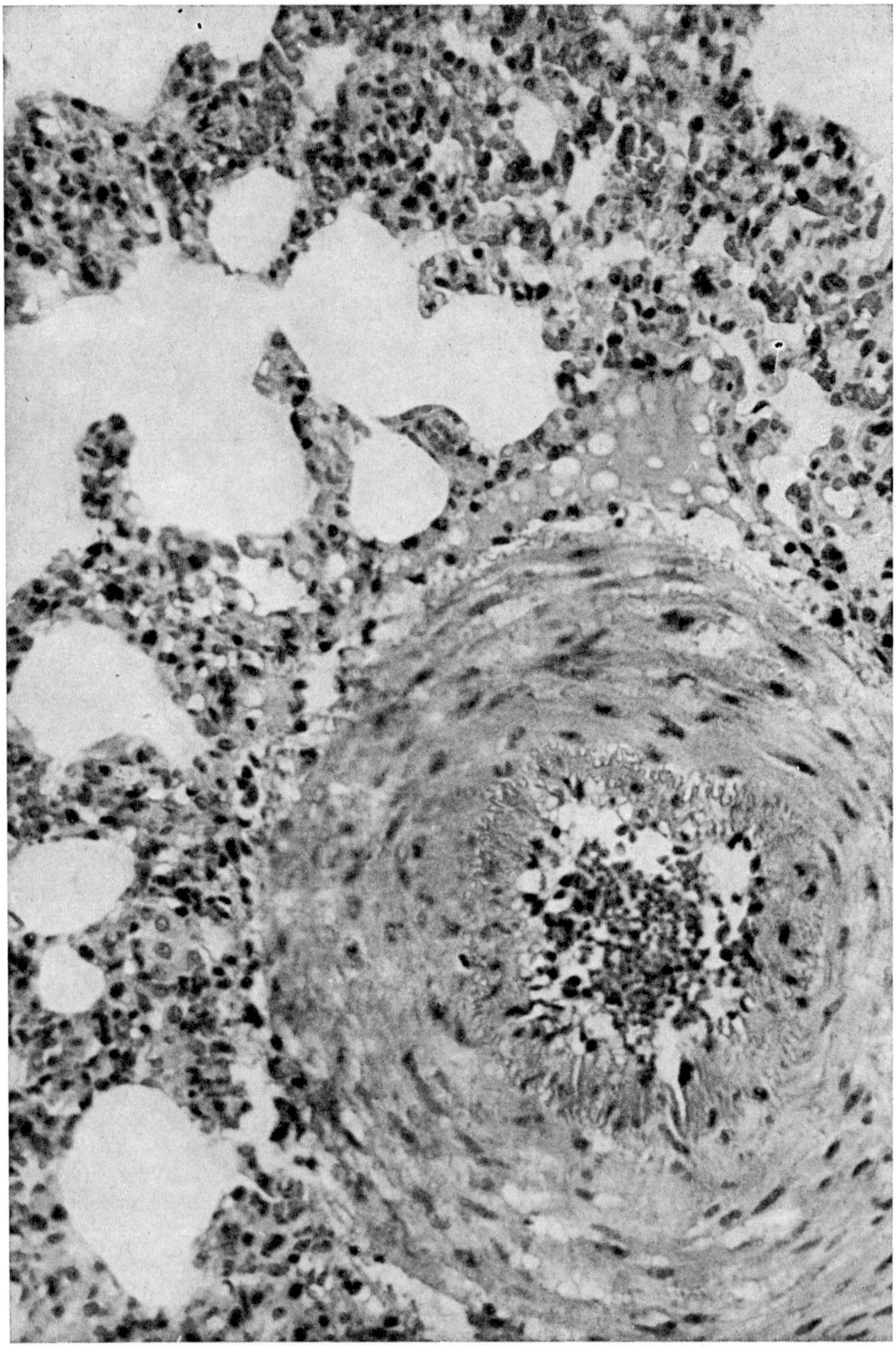

Fig. 2

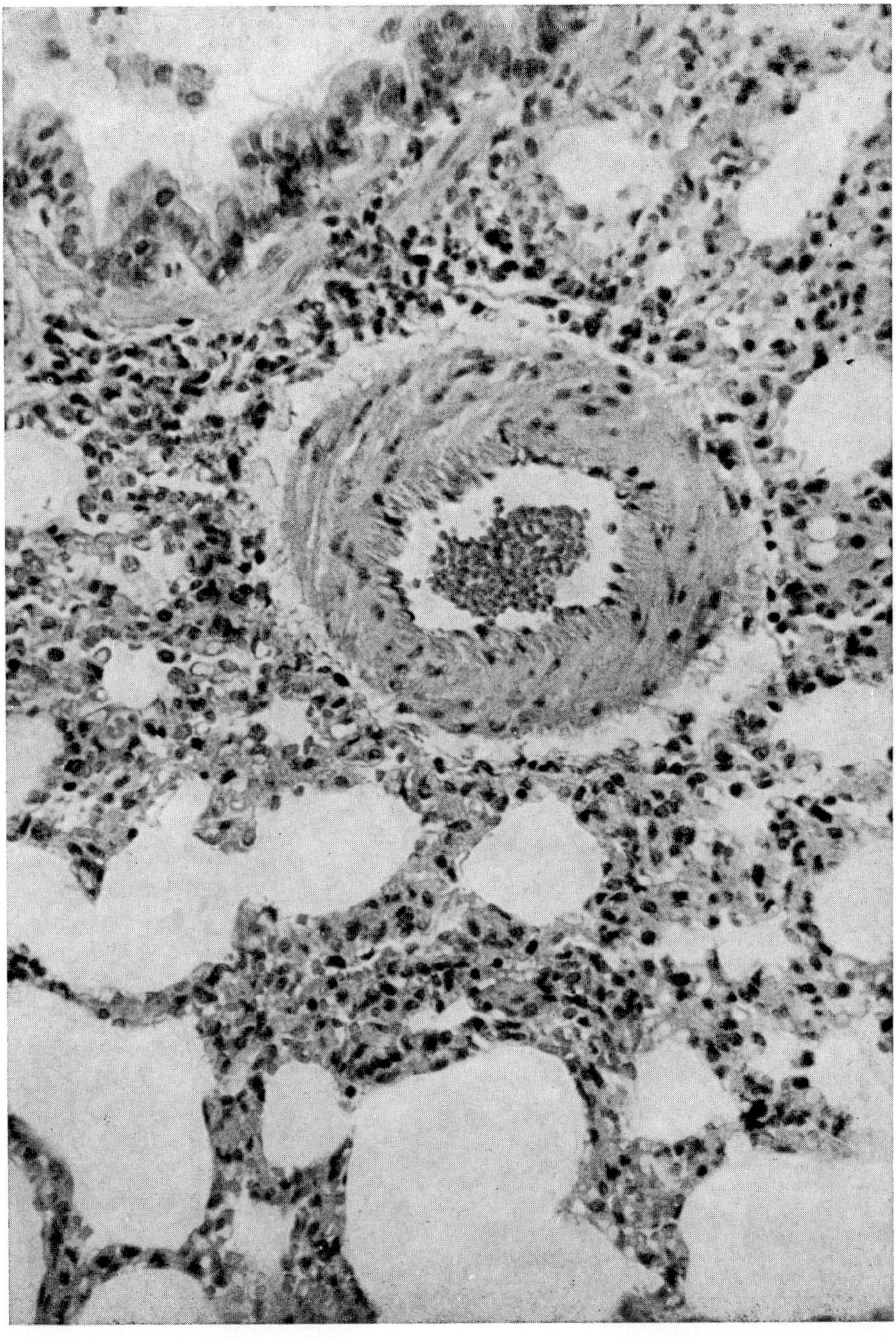

Fig. 3

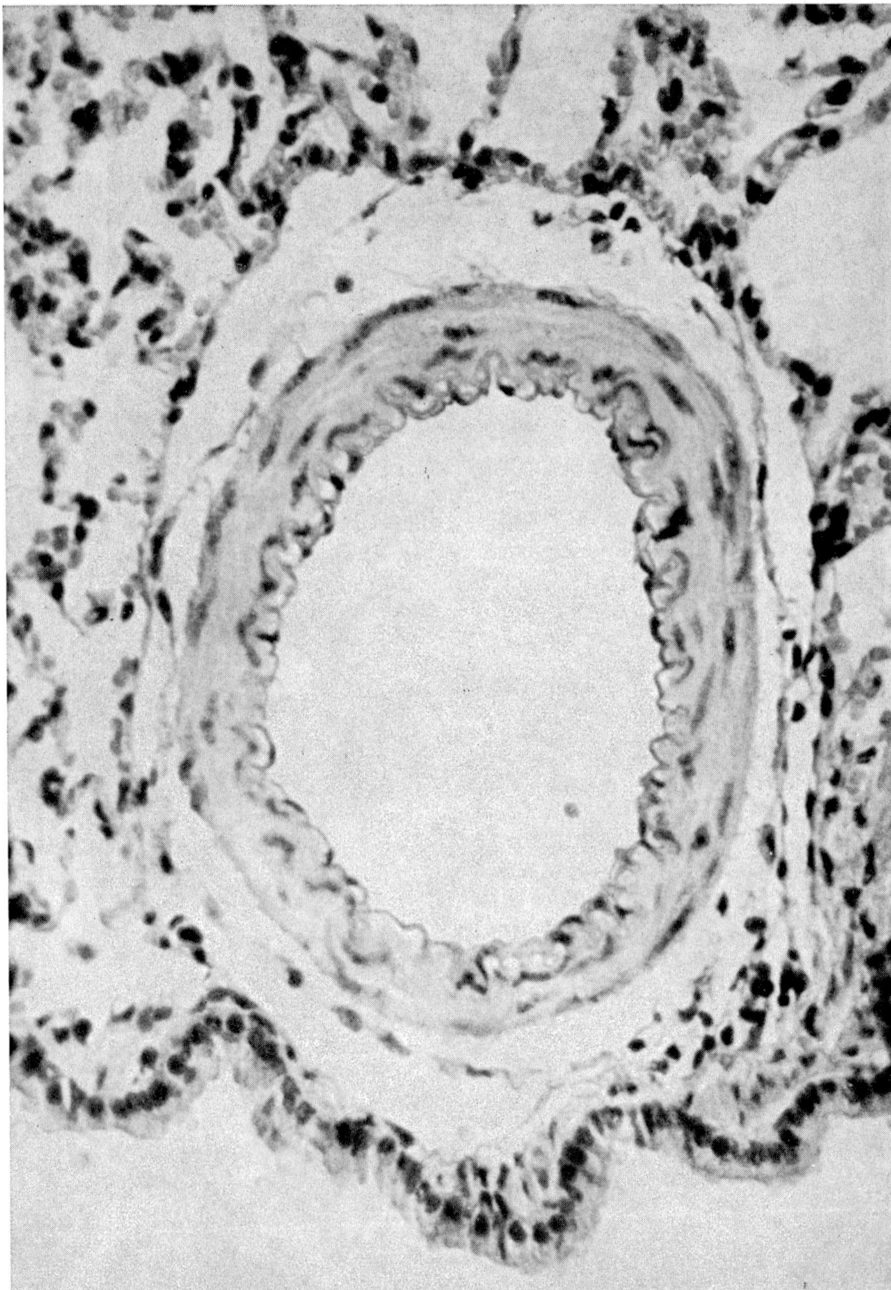

Fig. 4

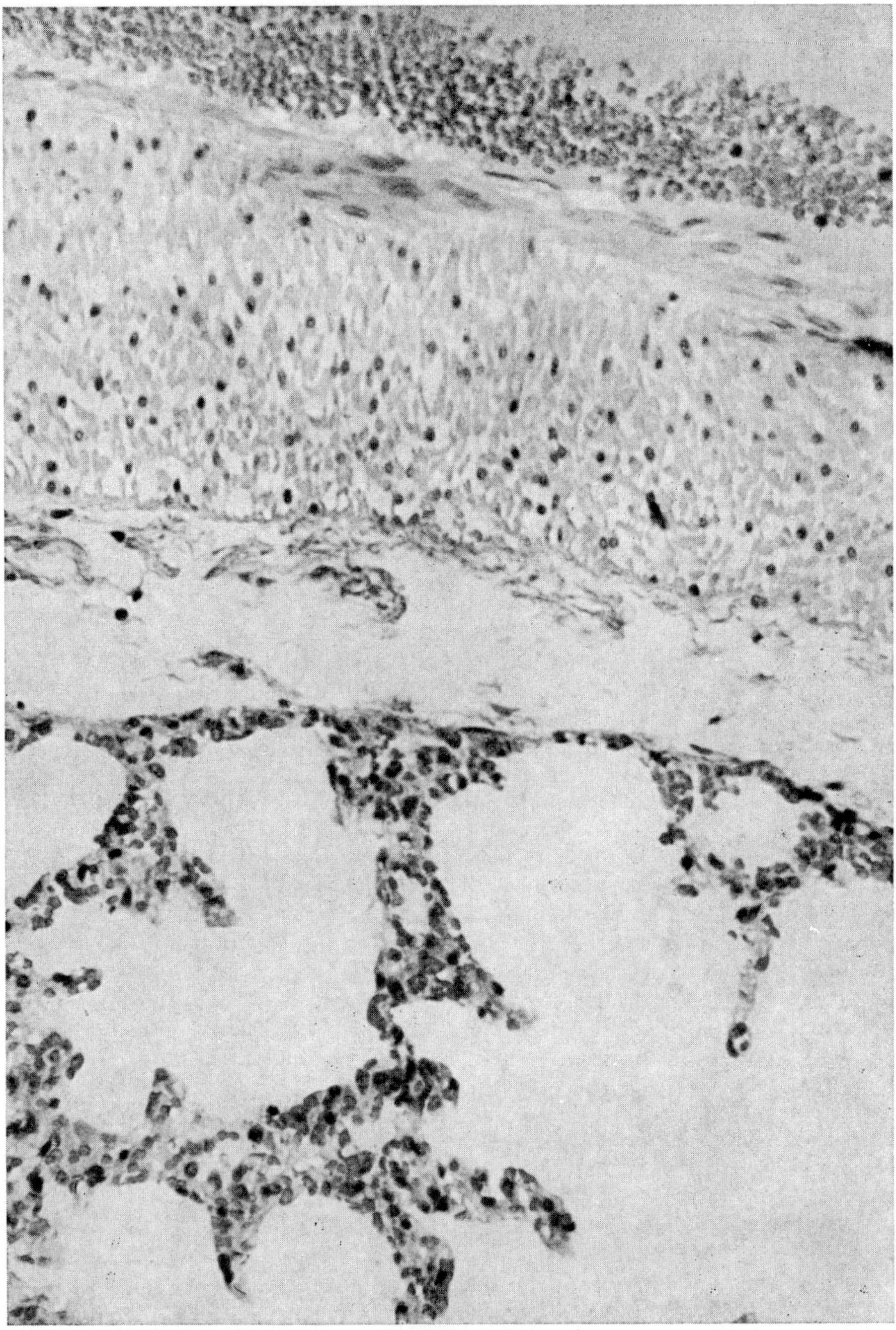

Fig. 5

also damaged – in P'an and Jegier's experiment (1972), the lungs of the exposed rabbits showed emphysematous lesions characterized by the apparent loss of alveolar walls and the clubbing of their ruptured extremities – it may end up with cor pulmonale.

Ozone, which exists in airplane cabins at altitudes flying above 30,000 ft, in urban atmospheres polluted by photochemical oxidants, and in homes where "fresh air" is provided by commercial ozonisers, is a health hazard that should not be ignored. More investigations should be conducted on the insidious effects of long-term exposure to ambiant concentrations of ozone.

CARBON MONOXIDE

Carbon monoxide – CO – is a major gaseous pollutant; as its combining power for hemoglobin (Hb) is 210 times greater than that of oxygen (O_2), it reduces rapidly the O_2-carrying capacity of the blood and tissue respiration is impaired because of hypoxia. Acute CO poisoning, often resulting in collapse and death, has been recognized for ages but chronic CO intoxication is a rather recent concept. It was believed that since the carboxyhemoglobin complex (HbCO) dissociates rapidly by the respiration process after exposure to CO ceases, this hazardous gas could not have cumulative effects. Kamatsu was the first, in 1955, to consider repeated inhalation of CO in low concentrations as a possible cause of heart disease. More recently, the studies of Astrup (1967) and Wanstrup (1969) showed that CO increases deposition of lipids in the inner layers of large blood vessels. Before proceeding any further it may be useful to remind the reader that the toxicity of pollutants is rarely if ever limited to a single "target" organ. According to present knowledge, CO affects the O_2-carrying function of the hemoglobin, the central nervous system, psychomotor abilities, time-interval estimation and the integrity of arterial walls. Only the latter effect falls within the scope of this Section.

The Astrup group noticed that atherosclerotic smokers had higher HbCO levels than smokers without atherosclerosis; this led them to suspect that continuous inhalation of small doses of CO could play a role in the origin of the disease. Atherosclerosis is a chronic arterial disease which affects principally the aorta, the coronary, cerebral and peripheral arteries of the lower extremities; it becomes increasingly common as age advances and is often associated with hypertension. The essential lesion is the plaque or atheroma: a local accumulation of cholesterol and other lipids in the intimal tissue. As the disease progresses, the artery loses its elasticity, its lumen becomes narrower and narrower and serious impairment of the blood supply to important structures may result. Conorary atherosclerosis, for instance, is responsible for angina pectoris. Ulceration of the plaque may lead to thrombosis and arterial occlusion. Cardiac infarction, cerebral stroke and senile gangrene are examples of this type of complication.

Astrup and his collaborator (1967) tested their hypothesis by exposing cholesterol-fed rabbits to CO; the animals developped atheromas. In order to study the formation process of such arterial lesions, Astrup et al. exposed another batch of rabbits, fed on a normal diet, to 0.009% of CO for three months (Wanstrup et al., 1969). The results of this experiment indicated that there was a higher degree of intimal-subintimal changes in the aorta arch of the exposed animals (edema, cellular prolifcration and breakdown, splitting of fibrils and collagen formation) than in the aorta arch of the control rabbits. The hearts of the exposed rabbits showed hyperaema and stasis in the subendocardial and subpericardial regions; the peritoneal, pleural and pericardial cavities were filled with serous fluid.

Carbon monoxide is known to increase endothelial permeability for albumin (Siggaard-Andersen et al., 1968), therefore the observed vascular changes may have originated in localized edematous areas in the aorta. Since the inner coats of arteries receive oxygen and nutrients by direct permeation from the lumen, local swelling and splitting up of the intima and subintima would increase the diffusion distance for oxygen, thus creating local hypoxia. Hypoxic tissues incorporate more lipids than well oxygenated tissues do and because lipid incorporation increases oxygen consumption, local hypoxia becomes more severe. The aortic arch is the segment of the vessel having the highest oxygen demand; it is thereby most susceptible to hypoxia and hypoxia-induced plaque formation. The hypoxia-atherosclerosis "vicious cycle" has been summarized by Lazzarini-Robertson (1968) in the following diagram:

Local changes
↓
Local hypoxia
↑ ↓
Increased oxygen consumption ← Increased lipid incorporation (atherogenesis)

At an early stage, edema caused vascular wall hypoxia may be reversible but once plaque formation arises on the damaged parts of the intima, atheromatosis sets in. Hypoxia at the blood-tissue interface is the best explanation one has for the development of atheromas; it appears to be the vascular mode of action of carbon monoxide.

CADMIUM

Over a century ago, cadmium (Cd) was labelled toxic in industry because exposure to fumes or dust of Cd caused acute pulmonary edema among the workers. Now Cd is in steadily increasing consumption and it has become a pollutant of urban atmospheres as well as a contaminant of food and water. Cigarette smoking adds to the cadmium body burden of the smoker.

Study of the long term effects of repeated exposures to low concentrations of Cd started in the 1950's (Friberg, 1959); existing literature reports that Cd is implicated in testicular damage (Niemi, 1968), in emphysema (Friberg, 1959), in renal lesions (Axelsson and Piscator, 1966), in hypertension (Schroeder, 1965) and in a variety of other diseases. The only effect to be discussed here is cadmium-hypertension.

Animal studies have shown that Cd is absorbed mostly from the respiratory tract; absorption from the gastrointestinal tract is poor. When administered subcutaneously or intraveinously, it is rapidly removed from circulating blood and deposited primarily in the kidneys and the liver (Lucis & Lucis, 1969). In the early stage of Cd exposure, the metal deposited in the kidney is found in the proximal tubular epithelium; in later stages, deposits are also seen in the distal tubular epithelium (Axelsson et al., 1968). In the cell Cd is bound to a cytoplasmic macromolecule named metallothionein (Shaikh and Lucis, 1971). As clinical hypertension is commonly associated with renal involvement, the accumulation of Cd particles in the kidneys was readily believed to have some kind of relation with cadmium-induced hypertension.

Hypertension is a chronic disease characterized by an elevated diastolic pressure associated with generalized arteriolar vasoconstriction. Vasoconstriction might be due to neurogenic stimuli (increased sympathetic tone, for instance) or to humoral stimuli (a circulating pressor substance acting directly on the smooth muscles of the arterioles). Vasoconstriction is always more severe in the renal vessels than elsewhere. Hypertension has been produced experimentally in animals by partial occlusion of one renal artery. Renal ischaemia hypertension is independent of nerve supply; it is due to the liberation into the circulation of a pressor substance formed in the damaged kidney: renin, Renin is an enzyme which enters the blood stream by the renal vein and reacts with a globulin precursor in the plasma to produce an active pressor agent named angiotonin. Reduced renal blood flow is also associated with human hypertension but it is not clear whether the kidney is primarily or secondarily involved. Since vasoconstriction increases peripheral resistance, the heart has to increase its force of contraction to maintain a normal blood supply to the tissues; in due course, cardiac hypertrophy develops. Later arteriolar walls thicken and lose their elasticity; this arterioslerotic process can also develop in larger vessels. Coronary artery disease and cerebrovascular accidents are frequent complications of hypertension.

The investigations of Schroeder et al. (1970) have shed some light on the mechanism of cadmium hypertension. These authors found that rats receiving cadmium by mouth or by injection had higher blood pressures than age control rats, however when the two groups of animals were administered norepinephrine or angiotensin (two pressor agents), the mean rise of blood pressure was less in the treated animal group than in the control group. Injection of the chelate of zinc Na_2Zn CDTA (cyclohexane-1,2-diamino NNN′N′-tetraacetate) resulted in lower pressures in control and cadmium-treated animals and vascular response to norepinephrine was restored in the latter group. These results suggest two interesting conclusions:

1. Cadmium decreases the sensitivity of blood pressure to normally present pressor agents, therefore the mechanism of cadmium hypertension must be different from that of renal ischaemia hypertension.
2. Zinc as contained in the chelate Na_2Zn CDTA, antagonizes the vascular effects of cadmium, thus cadmium toxicity may depend not only on its concentration in the organism but also on the concurrent level of zinc.

The fact that workers exposed to cadmium are at the same time exposed to zinc (cadmium occurs mainly in zinc sulfides sphalerite and wurtzite) may be the reason why so few cases of hypertension have been found among them. The renal Cd/Zn ratio is believed to be important (Schroeder, 1962, 1965).

NITROGEN DIOXIDE

Nitrogen dioxide (NO_2) is an important end product of the burning of fuel in engines and furnaces; it is an ubiquitous air pollutant.
In high concentrations, NO_2 produces acute pulmonary edema; those who survive the attack end up with diffuse alveolitis and bronchiolitis obliterans (Lowry and Schuman, 1956). Animal studies have shown that NO_2, in low concentrations, destroys the lung tissue and stimulates the growth of the bronchiolar epithelium (Buckley and Loosli, 1969). In due course, emphysema and partial occlusion of the terminal bronchioles lead to heart failure.

The basic plan of the lung is that of a tube with many branches ending in approximately 300 millions of tiny expansible sacs called alveoli. Each microscopic alveolus is surrounded by a tight fitting network of capillaries. Capillary walls and alveolar cells lie in direct contact side by side with air moving on one side and blood flowing past on the other. Through this thin double membrane O_2 enters the blood stream and carbon dioxide (CO_2) makes its way out of the body. The expansion of the alveoli provides the necessary surface for the vital gaseous exchange: some 70 square meters.

In emphysema, the walls between the alveoli break down, the capillary network is destroyed, the lungs lose their elasticity; the small clusters of alveoli are replaced by large air sacs with less surface for gas exchange and a diminished blood flow. Moreover attendant obstruction of the bronchioles allows very little O_2 to get into the expanded alveolar sac while CO_2 accumulates in it. Not only is the exchange of CO_2 for O_2 slowed down but since CO_2 is 20 times more soluble than O_2, it diffuses more rapidly, causing chronic hypoxia. Hypoxia leads to reflex constriction of the pulmonary capillaries, constriction of the pulmonary capillaries leads to pulmonary hypertension and pulmonary hypertension leads to the dilatation of the hard working right heart. In the long run, the pumping ability of the enlarged heart fails; respiratory infection, or stress from other types of disease condition, will precipitate acute heart failure, often resulting in cardiac shock and death. During episodes of high atmospheric pollution such as those which have occurred in the Meuse Valley (Firket, 1931), Donora, Pa (Schrenk et al. 1949), London (Logan, 1953) and New York (Greenburg et al., 1962), the increase in mortality has been highest among patients suffering from chronic respiratory diseases and cardiac diseases. There is a probability that the air pollutants which caused their death were also at the origin of their unduly susceptible condition.

CONCLUSIONS

The effect of ozone upon the cardiovascular system has not been systematically investigated, however certain experiments have demonstrated that this potent oxidant can accelerate the sphering of erythrocytes, cause thickening of the walls of the small pulmonary arteries and rupture the nuclear membrane of the myocardial fibres.

Carbon monoxide, cadmium and nitrogen dioxide can attack the cardiovascular system in three ways: by injuring the internal layer of the blood vessels, by provoking a rise in

the blood pressure or by inducing pulmonary hypertension.
Carbon monoxide damages the arterial walls; cadmium causes hypertension and nitrogen dioxide produces emphysema and pulmonary hypertension.

REFERENCES: see pp. 293-294.

Effects of Sulphur Dioxide and Related Compounds on Monkeys

by

Y. Alarie

INTRODUCTION

Monkeys have been used to investigate the possible deleterious effects of airborne chemical irritants such as sulfur dioxide (Alarie et al., 1972), sulfuric acid mist (Alarie et al., 1973b), fly ash (MacFarland et al., 1971), mixtures of fly ash and sulfur dioxide (Alarie et al., 1973c) and nitrogen dioxide (Alarie et al., 1971a and unpubl.). The various groups of animals and exposure concentrations for these pollutants are listed in Table 1. All these studies were conducted in controlled environmental chambers of a similar type (Alarie et al., 1972) with continuous exposure for periods of 18 to 24 months. For evaluation of the effects of these pollutants pulmonary function tests were performed in a similar fashion for all groups. Monkeys have also been used to evaluate the effect of carbon monoxide (Eckard et al., 1972 and DeBias et al., 1973) and cigaretter smoking (Binns and Clark, 1973).

INSTRUMENTATION DEVELOPED

Various measurements were made in monkeys in much the same manner as made in humans. These respiratory function studies in Cynomolgus monkeys prior to and during exposure to various pollutants, consisted of the measurements of: respiratory rate, tidal volume, pulmonary resistance, pulmonary compliance, work of breathing, diffusing capacity of the lung, distribution of pulmonary ventilation, arterial blood gas tensions and acidity (Alarie et al., 1972; Macfarland et al., 1971). Experimental arrangements were simply scaled down and adapted for monkeys (Alarie et al., 1970, 1971b, 1971c) but data was processed on-line with a digital computer system in order to handle the large volume of tests to be performed and to obtain immediate results as the animals were being tested at frequent intervals during exposure to the pollutants.

EXPERIMENTAL FINDINGS

Of the pollutants or mixture of pollutants listed in Table 1, sulfaric acid mist and nitrogen dioxide induced deleterious effects on pulmonary function and induced changes in pulmonary tissues. It is interesting to note that both pollutants which are considered as deep lung irritants as opposed to sulfur dioxide which is primarily an upper respiratory tract irritant (Alarie, 1973a) induced similar type of effects. The distribution of pulmonary ventilation was impaired, the respiratory rate was increased and decreases in arterial blood oxygen tension were observed. Modifications of the

Table 1. Summary of main findings in cynomolgus monkeys exposed to various pollutants

Pollutants	Duration	Concentration		Particle	Pulmonary Microscopic Alterations*						Pulmonary Function		
	months	ppm	mg/m³	Size MMD μ	A	B	C	D	E	F	Distribution of Ventilation	Respiratory Rate	Arterial Oxygen Tension
Control	18				No Significant Alteration Present								
Sulfur Dioxide	18	0.11			No Significant Alteration Present						No Significant Alteration		
Sulfur Dioxide	18	0.64			No Significant Alteration Present						No Significant Alteration		
Sulfur Dioxide	18	1.28			No Significant Alteration Present						No Significant Alteration		
Sulfur Dioxide	7	4.69									No Significant Alteration		
Sulfur Dioxide**		4.69			9/9 (++)	9/9 (++)	9/9 (++)	8/9 (++)			Deterioration (+++)	Increase	Decrease
Fly Ash	18		0.16	2.73	Presence of Fly Ash Detected No Significant Reaction Detected						No Significant Alteration		
Fly Ash	18		0.46	2.63	Presence of Fly Ash Detected No Significant Reaction Detected						No Significant Alteration		
Nitrogen Dioxide	21	0.5			No Significant Alteration Present						No Significant Alteration		
Nitrogen Dioxide	24	6.8			7/9 (++)	7/9 (++)			6/9 (+)	2/9 (+)	Deterioration (++)	Increase	Decrease at Beginning and Recovery

Sulfuric Acid Mist	18		0.38	2.15	5/8 (+)	3/8 (+)		No Change	Increase	No Change
Sulfuric Acid Mist	18		2.43	3.60	8/8 (++)	8/8 (++)	8/8 (++)	Deteriora- tion (++)	Increase	Decrease
Sulfuric Acid Mist	18		0.48	0.54	No Significant Alteration Present			Deteriora- tion (+) at Some Time Intervals	No Change	No Change
Sulfuric Acid Mist	18		4.79	0.73	8/8 (++)	8/8 (+++)		Deteriora- tion (++)	Increase	No Change
Sulfur Dioxide & Fly Ash	18	0.11	0.62	2.62	Presence of Fly Ash Detected No Significant Reaction Detected			No Significant Alteration		
Sulfur Dioxide & Fly Ash	18	1.02	0.52	3.38	Presence of Fly Ash Detected No Significant Reaction Detected			No Significant Alteration		
Sulfur Dioxide & Fly Ash	18	5.11	0.54	2.76	Presence of Fly Ash Detected No Significant Reaction Detected			No Significant Alteration		

* A: Bronchiolar epithelial hyperplasia
B: Thickening of walls of respiratory bronchioles
C: Thickening of alveolar walls
D: Bronchiectasis, bronchiolectasis
E: Bronchitis
F: Focal emphysema
Number under each column indicate animals affected in each group (+) slight, (++) moderate, (+++) moderate to severe.

** This group was exposed to 4.69 ppm SO_2 for 7 months prior to an overexposure accident (1 hour $\simeq$ 200-1,00 ppm). Animals were then kept under control conditions for eleven months prior to sacrifice and histopathological examination. Results on pulmonary function are those obtained during these eleven months.

distribution of ventilation indicates an effect primarily on the peripheral conducting airways (Macklem, 1971). The increase in respiratory rate may be attributed to stimulation of pulmonary receptors although other factors may have contributed (Alarie, 1973a). With sulfur dioxide at low concentrations or with sulfur dioxide and fly ash mixtures no deleterious effects could be observed. However, with an accidental exposure to a very high concentration the effects observed were similar to those obtained with nitrogen dioxide and sulfuric mist thus indicating that such an exposure resulted in irritation of the peripheral pulmonary airways.

What is interesting in this case is that the abnormal pulmonary function detected immediately after the accidental exposure persisted for a period of 48 weeks during which the animals were maintained in filtered air and no apparent recovery occurred during this period. Histopathological examination at the end of this 48 week period revealed pulmonary lesions almost similar in nature as that obtained following acute or subacute exposure to very high concentration of this gas, apparently no healing process had taken place.

In all three instances where deleterious effects were detected, involvement of the peripheral airways can be compared with the action of other pulmonary irritants such as phosgene and ozone (Alarie, 1973a, Alarie et al., 1973b).

DISCUSSION

The use of monkeys seems to be appropriate for evaluation of the effects of various respiratory irritants commonly found in either urban atmospheres or industrial situations. Following a period of adaptation to the restraining chair and face mask pulmonary function tests can be done in these animals in the same manner as in humans. Monkeys either in a sitting or upright position will remain quiet and breathe normally during the various tests without need for anesthetics. This is an obvious advantage since anesthesia has a profound effect on respiratory function and will introduce confusing artifacts in studies on the effects of low concentrations of pulmonary irritants. However, to obtain arterial blood samples for determination of oxygen and carbon dioxide tranquilization should be used (Banerjee et al., 1968).

Cynomolgus monkeys have been used in these studies instead of rhesus monkeys because they are less susceptible to tuberculosis. Also infestations with lung mites is much lower or non-existent. This is advantageous for histopathological evaluation. Whether or not monkeys have some advantages over other species in terms of predicting the effects to be observed in man from exposure to the various pollutants listed in Table 1 remains unanswerable at this time. Upper respiratory tract retention and alveolar deposition of aerosols seems similar in monkey and man (Palm et al., 1956) and their tidal volume and respiratory frequency as well as other mechanical properties of the lung are very close to those found in young infants. However, other aspects are less well known.

As with other animal species pulmonary function tests such as determination of maximum ventilatory capacity which requires cooperation of human subjects cannot be made in monkeys. A serious drawback in the use of monkeys is their increasing scarcity and cost. Also a period of at least three months should be observed after a quarantine period for tuberculosis detection so that the animals will be well adapted to their new laboratory environment and be in good health. Monkeys are fragile animals and diarrhea and dehydration must be prevented or controlled promptly.

The design of chronic studies to evaluate the effect of pulmonary irritants at concentrations which can be observed in either urban atmosphere or industrial situations is still in its infancy. What kind of tests should be used and how frequently they should be performed still remains to be adequately defined. Certainly pulmonary function tests currently applied in humans should be more rapidly introduced in animals and hopefully correlation between epidemiological finding and laboratory findings could be obtained. However, several of the tests used in epidemiological studies are of the type demanding collaboration of the subject and impossible of exact duplication in animals.

In long-term studies statistical analysis of the data as rapidly as possible following each evaluation interval is mandatory and the use of sequential analysis techniques should be expanded. For examples, with nitrogen dioxide and sulfuric acid mist it was rapidly found that deleterious effects had occurred with the first six months of exposure. Further testing, up to 18 or 24 months, revealed that after initial insult there was no further deterioration in pulmonary function. However, in these studies no attempt was made to study any possible recovery following exposure.

Although there is presently a great need to perform chronic studies on a variety of airborne chemicals and mixtures of some of them in order to establish certain criteria for human exposure this must be balanced by the cost of undertaking such studies. Indeed well designed acute and subacute experiments should be conducted to delineate what kinds of chemicals or mixtures are likely to be of importance so that we can reserve for chronic experiments only those where shorter term studies are truly inadequate.

REFERENCES: see pp. 294-295.

Effects of Pesticides on Animals

by

S. C. Bhatia

INTRODUCTION

The discovery of DDT in 1939 opened the era of modern synthetic pesticides and thus provided man with a powerful tool to control his environment. Using pesticides to control diseases carried by rodents and insects has improved the quality of life in our modern world by eliminating many devastating disorders. They have helped in eradicating malaria almost completely and have contributed to the control of an impressive list of diseases including viral encephalitis, louse-bornetyphus, bacillary and amoebic dysentery, cholera, Rocky Mountain spotted fever, etc. With growing emphasis on increasing agricultural outputs, the use of these compounds has become widespread and rather indiscriminate. The addition of persistent chemicals to the environment continues by millions of pounds each year. This may provide immediate benefits but at the same time may create short and long term problems of deep consequences. Hence the benefits of pesticides have recently been questioned in the light of their effects on the environment, man and animals.

Pesticides or economic poisons are defined as "any substance or mixture of substances intended for preventing, destroying, repelling or mitigating any insects, rodents, nematodes, fungi or weeds or any other forms of life declared to be pests; any substance or mixture of substances intended for use as a plant regulator, defoliant or desiccant".

Among the vast majority of pesticidal chemicals, the synthetic organic group of insecticides have commanded much attention. This group comprises of the organochlorine, the organophosphorus and the carbamate compounds. Of these, the chlorinated hydrocarbons such as dieldrin, DDT, aldrin, endrin etc., are most popular. They owe their insecticidal success to (1) their ability to kill a wide variety of insects, (2) they work as contact poisons and do not have to be ingested like the inorganic insecticides, (3) they are powerful and low doses can kill insects and so cost less to use, (4) they are chemically stable and retain their poisonous properties for several months or even decades and (5) they are cheap to produce, particularly DDT, because it is not patented.

Besides their persistence on food crops, residues of pesticides are now known to be present almost everywhere – in the atmosphere, soil, fish and in the body-fat of man and other animal species. Being fat-soluble compounds, the concentration and transmission of organochlorine insecticides takes place in the fatty phases of living things. Daphnia, a lower link in the food chain, can absorb and concentrate up to 20,000 times the residue of DDT in the water surrounding it. Fish then feed upon daphnia and again concentrate residues. A bird eating some of these fish again delivers an increasing concentration of chlorinated hydrocarbon insecticides and so the transmission of residues proceeds. Since organochlorine insecticides are not readily biodegradeablc their residues can be found in various biological systems for generations. Levels of residues in man and animals are shown to be dependent on factors such as age, sex, race, location, diet, exposure, species and heredity. Effects of pesticides in animals include disturbances in the endocrine, hepatic, nervous and reproductive systems as well as subtle effects on mutagenicity, teratogenicity and carcinogenicity.

The aim of this Section is to present the known effects of dieldrin, DDT and related insecticides on the various physiological systems of animals. An awareness of the toxicity of these chemicals being of recent origin we are handicapped by the limitation of quantitative data on several animal species. Because of the easy availability, most of the experimental studies employ rats as their model.

ADRENOCORTICAL FUNCTION

The presence of an insecticide in the body, particularly after an acute exposure, constitutes a condition of stress and the common physiologic response which follows is an increased stimulation of the pituitary and adrenal glands. Stimulation of the adrenals is indicated by several parameters including adrenal hypertrophy, a decrease in the ascorbic acid and cholesterol contents of the gland (Sayers, 1950; Kitabchi, 1967) and an increase in the concentration of serum corticosteroids. In our laboratory we studied the metabolic effects of a single subacute dose of dieldrin (30 mg/kg body weight) in rats and observed a significant decrease in the adrenal cholesterol and vitamin C contents (Bhatia et al., 1971a; Table 1) and an elevated plasma corticosterone level (Bhatia et al., 1972b; Table 3). It has been postulated that stress leads to increased pituitary adrenocortical activity by increasing the requirement of peripheral tissue cells for cortical hormones and subsequently leads to hypertrophy of the adrenal glands. Hypertrophy has not been observed in our investigation nor has it been reported for rats ingesting DDT (Wassermann et al., 1969). It maybe accepted, however, that organochlorine compounds affect the chain of metabolic processes involved in the equilibrium of steroid hormones at more than one stage.

Table 1. Effect of feeding dieldrin* on cholesterol and ascorbic acid levels in adrenals (From Bhatia et al., 1971a)

Estimation	Control (6)	Dieldrin (6)	p
Adrenal wt, mg	19.83 ± 0.7	21.47 ± 0.8	—
Cholesterol, mg/g tissue	11.86 ± 0.45	6.91 ± 0.36	<0.01
Ascorbic acid, mg/g tissue	6.06 ± 0.17	4.22 ± 0.13	<0.01

* Dieldrin was administered in a single po dose of 30 mg/kg to Wistar strain male albino rats weighing 100-120 g. Controls received 0.2-0.3 ml of carrier groundnut oil only. Rats were killed 24 hrs after treatment. Values given in the table are mean ±SE. Figures in parentheses indicate the number of animals used. A p value $\leqslant$ 0.05 is considered statistically significant.

Ingestion of dieldrin enhances the ascorbic acid levels of rat plasma and various tissues viz, brain, liver, kidney, heart and spleen (Bhatia et al., 1971a). This is in keeping with the view that animals exposed to harmful substances show an increased rate of biosynthesis of this vitamin, and hence this test can be used as an index of the response and adaptation of the organism to toxic compounds.

HEPATIC FUNCTION

Liver being the chief site of metabolic activity, is the organ largely affected by organochlorine insecticides. The blood presents a good index of the metabolic changes occurring in the liver. Tables 2 and 3 show the effect of dieldrin on major chemical constituents of rat liver and blood respectively. Dieldrin ingestion causes a significant increase in liver weight and consequently in liver: body weight ratio. Many investigators have reported similar enlargement of liver and an increase in liver: body weight ratio in rats and other animal species treated with dieldrin (Hodge et al., 1967; Kimbrough et al., 1971). Jefferies and French (1969) observed that avian hepatic tissue is also markedly enlarged after treatment with DDT. Since no change has been observed in the moisture and protein contents, the increase in liver weight induced by dieldrin and other organochlorine insecticides is apparently the result of a stimulation of growth or an increase in the hepatic cell mass. The latter possibility finds support from the observation that dieldrin treatment causes marked proliferation of hepatic smooth endoplasmic reticulum (Hutterer et al., 1968; Kimbrough et al., 1971; Lucier and McDaniel, 1972). The increase in hepatic weight has been interpreted to represent an adaptive mechanism of the animal and may not be a toxic manifestation of the insecticide.

One of the striking changes in the liver occurs in the carbohydrate metabolism. Dieldrin treated rats have 300% more fasting glycogen content than the corresponding controls (Table 2). The accumulation of glycogen is brought about by the stimulation of the synthetic pathway coupled with a decreased breakdown of the polysaccharide (Bhatia et al., 1973). Hepatic gluconeogenesis is another process which is enhanced by organochlorine insecticides. Dieldrin (Bhatia et al., 1971b; 1972a; 1973), DDT and chlordane (Kacew and Singhal, 1973a, b) have been shown to increase the activities of important gluconeogenic enzymes in rat liver. However, while the gluconeogenic effect of dieldrin is thought to be mediated by the cortical hormones, that of DDT has been shown to be independent of the adrenal hormones. Organochlorine insecticides, when administered in subacute doses, create a hyperglycemic condition in the animals. Dieldrin, over and above, has also been shown to decrease the glucose tolerance of the animal during an oral glucose loading test which indicates an impairment in the uptake and utilization of glucose by the tissues (Bhatia et al., 1972b). Such an effect of the insecticides may be secondary to the stimulation of the adrenocortical activity, but nevertheless manifests a condition comparable to diabetes.

Another gross pathological finding observed during organochlorine insecticide toxicity is the fatty infiltration of the liver of mammals. This has been reported for dogs fed a mixture of aldrin and DDT (Deichmann et al., 1969) and for rats given a single subacute dose of dieldrin (Bhatia and Venkitasubramanian, 1972). The mechanism of fat

Table 2. Effect of dieldrin on liver weight and cellular constituents (From Bhatia et al., 1973)

Constituents	Control	Dieldrin	p
Moisture (per cent)	70.9 ± 0.3 (9)	69.3 ± 0.4 (8)	
Glycogen:			
mg/g wet tissue	1.6 ± 0.11 (24)	5.1 ± 0.37 (24)	0.01
mg/100 g body weight	5.2 ± 0.34 (12)	20.5 ± 1.54 (12)	0.01
Protein:			
mg/g wet tissue	148 ± 3.2 (5)	153 ± 2.7 (5)	0.05*
mg/100 g body weight	478 ± 13 (5)	611 ± 20 (5)	0.01
Total lipid:			
mg/g wet tissue	57.9 ± 1.2 (6)	63.0 ± 1.7 (6)	0.05
mg/100 g body weight	189 ± 8.53 (6)	249 ± 9.74 (6)	0.01
RNA:			
mg/g wet tissue	8.5 ± 0.18 (6)	9.2 ± 0.17 (6)	0.02
mg/100 body weight	27.60 ± 1.37 (6)	36.2 ± 0.89 (6)	0.01
DNA:			
mg/g wet tissue	1.3 ± 0.09 (6)	1.3 ± 0.09 (6)	0.05*
mg/100 body weight	4.2 ± 0.27 (6)	4.9 ± 0.33 (6)	0.05*
Lactic acid:			
μg/g wet tissue	99 ± 9 (6)	208 ± 20 (6)	0.01
μg/100 g body weight	338 ± 25 (6)	839 ± 72 (6)	0.01

* Not significant.

Table 3. Blood constituents in control and dieldrin-fed rats (From Bhatia et al., 1972b)

Constituents	Control	Dieldrin	p
Corticosterone (μg/100 ml plasma)	22.3 ± 1.87 (4)	29.1 ± 0.06 (4)	0.05
Glucose (mg/100 ml blood)	43.0 ± 0.94 (5)	65.0 ± 2.24 (5)	0.01
NEFA (μEq/litre plasma)	572 ± 33 (4)	834 ± 32 (4)	0.01
Lactic acid (mg/100 ml blood)	8.9 ± 0.46 (6)	14.6 ± 0.58 (6)	0.01
Pyruvic acid (mg/100 ml blood)	2.1 ± 0.08 (8)	3.70 ± 0.18 (8)	0.01

For explanation of Tables 2 and 3 see legend of Table 1

accumulation in the liver proposed by Bhatia and Venkitasubramanian (1972) is that dieldrin first brings about the mobilization of free fatty acids (FFA) from the adipose tissue. This effect may be either direct since the insecticide is fat-soluble and is stored in the adipose tissue, or mediated by adrenocortical hormones. Nevertheless, mobilization of FFA has been observed in DDT treated rats also (Schwabe, 1964) and appears to be a common finding during organochlorine insecticide toxicity. Thus the increased availability of FFA together with the increased synthesis of α-glycerophosphate through the gluconeogenic pathway, create conditions favourable for triglyceride synthesis in the liver. The other factors like enhanced lipogenesis and impairment of hepatic triglyceride secretory mechanism, which may also cause fat accumulation, are not affected by dieldrin.

Evidence is now becoming available incriminating the insecticides with damage to the subcellular structure. Kimbrough et al. (1971) have observed atypical mitochondria in hepatic cells of rats fed dieldrin (2-6 mg/kg/day for 8 weeks). The morphological changes brought about by the insecticide include swelling, increase in the number of cristae, a double outer membrane or loss of a portion of the outer membrane and confluence of two mitochondria. Dieldrin administered in an acute dose brings about a certain degree of liver injury. It has a marked effect on the parenchymal cells of the liver as evidenced by the leaching out of the enzyme alkaline phosphatase into the serum (Onikienko, 1963; Bhatia et al., 1972a). In long-term studies on dogs fed mixture of aldrin and DDT, cloudy swelling, hypertrophy and fatty degeneration of hepatic cells were observed (Deichmann et al., 1969).

CENTRAL NERVOUS SYSTEM

Insecticidal chemicals acting on the central nervous system create a stress condition which may have secondary effects important to the ultimate survival or death of the poisoned animal. Both dieldrin and DDT are known to produce convulsions in man and laboratory animals. Dieldrin-induced seizures have been compared to the "grand mal" seizures of epilepsy. During these convulsions a mixture of betaine coenzyme-A esters is released through damage to mitochondria of the brain cells. The oxidation of succinate is also impaired as a result of this damage (Hosein and Proulx, 1960). In a 2-day study of the effect of a lethal dose of dieldrin or DDT on rat brain Ghazal (1963) observed a slight increase in the level of glutamic acid in the dieldrin treated rats and of glutamine in the DDT group of animals. The increase in glutamine was correlated with a protective action against the convulsions. Witter and Farrior (1964) in similar studies report an increase in the concentration of α-alanine in the rat brain and ascribe it to the inhibition of oxidative mechanisms in the brain, possibly at the α-ketoglutarate stage. According to Peters et al. (1972) serotonin is the secondary substance released during DDT – induced hyperpyrexia.

The hyperactivity of the rat brain during dieldrin toxicity has also been associated with the increased concentration of lactic and pyruvic acids in the organ (Hathway et al., 1965). Further, it has been observed that dieldrin liberates ammonia in the brain before the ontset of the convulsions and throughout their course. The insecticide apparently inhibits glutamine synthesis in the brain since the ammonia binding mechanism is inadequate during the seizures.

Several recent papers suggest that organochlorine insecticides may bring about poisoning of various species by inhibiting the brain microsomal adenosine triphosphate phosphohydrolases (ATPases) (Koch et al., 1969; 1971; Akera et al., 1971; Cutkomp et al., 1971; Jackson and Gardner, 1973). However it is observed that the efficacy of the various insecticides as inhibitors of in vitro ATPase activity do not correspond well with their in vivo toxicities. Some of the nontoxic metabolites of DDT also inhibit brain ATPase whereas dieldrin, known to be toxic in vivo, does not inhibit the brain enzyme in vitro.

REPRODUCTION

In studies on birds Woodwell (1967) and Wurster and Wingate (1968) have determined DDT residues in a number of species and have implicated contamination by organochlorine insecticides as a probable major cause of the decline in reproduction in several species. Other groups of workers (Ratcliffe, 1967; Bitman et al., 1968; Jefferies and French, 1969) while incriminating insecticides in reproduction have inferred that DDT mimics the in vivo action of estrogen and hence causes a disturbance in the estrogen-parathormone regulation of calcium metabolism in certain birds. Another point of view which may be considered is that by stimulating the activity of drug and steroid metabolizing enzymes, the insecticides increase the metabolism of sex hormones thereby resulting in a decreased action of the hormone in its target organ. In the light of this view it is interesting to note that in the male species, insecticide administration is reported to decrease testicular growth, inhibit the development of secondary sexual characteristics and the process of spermatogenesis (Burlington and Lindeman, 1950; McFarland and Lacy, 1969).

In studies on the cellular metabolism of reproductive organs DDT was found to enhance uterine water inbibition and wet weights as well as the incorporation of glucose-^{14}C into uterine protein, lipid and RNA (Bitman et al., 1968; Welch et al., 1969). In addition Singhal et al. (1970) have demonstrated that DDT produces a marked stimulation of uterine glycogen formation as well as of several enzymatic responses in estrogen-deprived animals.

BODY DEFENSE SYSTEMS

It is now well established that organochlorine insecticides influence to a considerable extent the defense mechanisms of the organism. Whatley and Glick (1966) in their study on chicks treated with DDT have observed a decease in the level of serum antibody titer to a foreign protein. Wassermann et al. (1969) report a decreased serum antibody titer and γ-globulin fraction in ovalbumin-immunized rats fed DDT (200 ppm) in their drinking water. A similar immunologic response has also been observed in rabbits ingesting DDT (Wassermann et al., 1971) or dieldrin (Wassermann et al., 1972). The authors have attributed the decrease in serum antibody to atrophic changes of lymphoid tissue. A direct effect of the organochlorine insecticides on the antibody-forming system may also be inferred from its inhibitory action on lymphoplasma cells in dogs, rabbits and rats following parenteral administration of DDT for extended periods of time (Solarino and Tripodo, 1954). However, parenteral administration of DDT to guinea pigs (Gabliks et al., 1973; Askari and Gabliks, 1973) prior to and following their immunization with diphtheria toxoid did not affect the protective immunity nor the serum antitoxin levels, but did reduce the severity of anaphylactic shock when the immunized animals were challenged with the toxoid. The moderation of anaphylactic shock has been ascribed to a release of histamine from mast cells by DDT resulting in a lowered histamine content in lungs.

Again a marked reduction in the spleen weight has been observed in rats chronically exposed to DDT (Wassermann et al., 1969) and to a subacute dose of dieldrin (Bhatia, 1972) but not in DDT administered guinea pigs (Gabliks et al., 1973). The involvement of the spleen in the immunologic processes is well documented. The activity of the reticuloendothelial system is thought to be moderated by the feedback control of antibody synthesis. This control may operate by affecting either the cellular transformation or the synthetic activity of the antibody-forming cells (Dubiski and Fradette, 1966). Thus, the decrease in spleen weight manifested by organochlorine insecticides argues in favour of a quantitative diminution of its lymphoid tissue. The capacity of these chemicals to influence certain enzymatic processes, as has been discussed earlier under hepatic function, may play a role in the above effect on the immunologic systems.

Toxicologic research has of late become sophisticated through an appreciation of cellular biochemistry, immunology and cytogenetics. Though research on pesticides has never been undertaken as extensively from an epidemiologic and toxicologic view point as it has been in the last few years, it may be said that it is still in its infancy. A large number of variables exist which control pesticides residues in various organisms. Certain immunologic responses reported for rats and rabbits have not been observed for guinea pigs. Discrepancies may be attributed to species difference and to the route of insecticide administration. Short-term studies employing large doses of the insecticides are necessary to gain an insight into their mechanism of toxicities, however, caution must be exercised while comparing high test dose levels in experimental studies to the relatively low doses encountered in the environment to which man and animals are actually exposed. One must also bear in mind that the cumulative adverse effects of pesticide residues may not necessarily be evident on the generation directly exposed to the poisons but may be passed on to the next generation of insects, mammals and other forms of life. It is of paramount importance to elucidate the mechanism of action of pesticidal chemicals both in target and nontarget organisms before any meaningful judgment can be passed on them. DDT, dieldrin and some of the commonly used herbicides have created certain survival problems in some forms of life, enough to warrant national legislations. Although now banned in a few countries these chemicals are still used in areas where the need for increased food output is greater than an awareness of their toxicity.

REFERENCES: see pp. 295-297.

REFERENCES

CHAPTER 2

EFFECTS OF LIGHT ON ANIMALS

Chapter 2, Section 1b (1): Circadian Rhythms of Protein Metabolism in the Rat Brain (M. B. TER HAAR)

BARRACLOUGH, C. A. (1968): Alterations in reproductive function following prenatal and early postnatal exposure to hormones. In: Advances in Reproductive Physiology, 3. A. McLaren (ed.). Logos Press, London: 81-112.

BROWN-GRANT, K., DAVIDSON, J. M. and GREIG, F. (1973): Induced ovulation in albino rats exposed to constant light. J. Endocr., 57: 7-22.

BURNET, F. R., TER HAAR, M. B. and MacKINNON, P. C. B. (1974): The effects of exogenous oestradiol on protein synthetic activity in discrete areas of the rat brain. Proc. roy. Soc. Med., 67: 39-40.

BURNET, F. R. and MacKINNON, P. C. B. (1975): Restoration by oestradiol benzoate of a neural and hormonal rhythm in the ovariectomized rat. J. Endocr., 64: 27-35.

CALIGARIS, L., ASTRADA, J. J. and TALEISNIK, S. (1971): Release of luteinizing hormone induced by estrogen injection into ovariectomized rats. Endocrinology, 88: 810-815.

CRITCHLOW, V. and BAR-SELA, M. E. (1967): Control of the onset of puberty. In: Neuroendocrinology, 2. L. Martini and W. F. Ganong (eds). Academic Press, New York: 101-162.

DONOVAN, B. T. and VAN DER WERFF TEN BOSCH, J. J. (1965): In: Physiology of puberty. Monogr. physiol. Soc., 15. Edward Arnold, London.

EVERETT, J. W. and SAWYER, C. H. (1949): Neural timing factor in the mechanism by which progesterone advances ovulation in the cyclic rat. Endocrinology, 45: 581-595.

EVERETT, J. W., SAWYER, C. H. and MARKEE, J. E. (1949). A neurogenic timing factor in control of the ovulatory discharge of luteinizing hormone in the cyclic rat. Endocrinology, 44: 234-250.

EVERETT, J. W. and SAWYER, C. H. (1950): A 24-hour periodicity in the "LH-release apparatus" of female rats, disclosed by barbiturate sedation. Endocrinology, 47: 198-218.

GORSKI, M. E. and LAWTON, I. W. (1973): Adrenal involvement in determining the time of onset of puberty in the rat. Endocrinology, 93: 1232-1234.

HALASZ, B. (1969): The endocrine effects of isolation of the hypothalamus from the rest of the brain. In: Frontiers in Neuroendocrinology. W. F. Ganong and L. Martin (eds). Oxford University Press, New York: 307-342.

HARRIS, G. W. and CAMPBELL, H. C. (1966): The regulation of the secretion of luteinizing hormone and ovulation. In: The pituitary gland, 2. G. W. Harris and B. T. Donovan (eds). Butterworths, London: 99-165.

HARRIS, G. W. (1970): Hormonal differentiation of the developing central nervous system with respect to patterns of endocrine function. Phil. Trans. B., 259: 165-177.

HEMMINGSEN, A. M. and KRARUP, N. B. (1937): Rhythmic diurnal variation in the oestrous phenomena of the rat and their susceptibility to light and dark. Kgl. Danske Vidensk. Selsk. Biol. Medd., 13: 1-61.

HIROSHIGE, T., ABE, K., WADA, S. and KANEKA, M. (1973): Sex differences in circadian periodicity of CRF activity in the rat hypothalamus. Neuroendocrinology, 11: 306-320.

PFEIFFER, C. A. (1936): Sexual differences of the hypophyses and their determination by the gonads. Amer. J. Anat., 58: 195-225.

RICHARDSON, K. and ROSE, P. R. (1971): A diurnal rhythmicity in incorporation of lysine into rat brain regions. Nature (Lond.), 233: 182-184.

TER HAAR, M. B. and MacKINNON, P. C. B. (1973): Changes in gonadotrophin concentrations, and in protein concentration and in vivo incorporation of [^{35}S] methionine into protein of discrete brain areas and the anterior pituitary of the rat during the oestrous cycle. J. Endocr., 58: 563-576.

TER HAAR, M. B., MacKINNON, P. C. B. and BULMER, M. G. (1974): Sexual differentiation in the phase of the circadian rhythm of [^{35}S] methionine incorporation into cerebral proteins and of serum gonadotrophin levels. J. Endocr., 62: 257-265.

THOMAS, P. J. (1973): Steroid hormones and their receptors. J. Endocr., 57: 333-359.

WEISZ, J. and GUNSALUS, P. (1973): Estrogen levels in immature female rats: true or spurious – ovarian or adrenal? Endocrinology, 93: 1057-1065.

Chapter 2, Section 1b (2): Circadian Rhythms in Food Intake and Control Processes Thereof in the Rat (M. B. TER HAAR)

ALBRECHT, P., VISSCHER, M. B., BITTNER, J. J. and HALBERG, F. (1956): Daily changes in 5-hydroxytryptamine concentration in mouse brain. Proc. Soc. exp. Biol. (N.Y.), 92: 703-706.

ANAND, B. K., CHHINA, G. S., SHARMA, K. N., DUA, S. and SINGH, B. (1964): Activity of single neurons in the hypothalamic feeding centers: effect of glucose. Amer. J. Physiol., 207: 1146-1154.

BAILE, C. A., SIMPSON, C. W., BEAN, S. M., McLAUGHLIN, C. L. and JACOBS, H. L. (1973): Prostaglandins and food in rats: A component of energy balance regulation? Physiol. a. Behav., 10: 1077-1085.

BALAGURA, S. and DEVENPORT, L. D. (1970): Feeding patterns of normal and ventromedial hypothalamic lesioned male and female rats. J. comp. physiol. Psychol., 71: 357-364.

BELL, D. D. and ZUCKER, I. (1971): Sex differences in body weight and eating; organization and activation by gonadal hormones in the rat. Physiol. a. Behav., 7: 27-34.

BROBECK, J. R. (1948): Food intake as a mechanism of temperature regulation. Yale J. Biol. Med., 20: 545-552.

BROWN-GRANT, K., EXLEY, D. and NAFTOLIN, F. (1970): Peripheral plasma oestradiol and luteinizing hormone concentrations during the oestrous cycle in the rat. J. Endocr., 48: 295-296.

COLIN-JONES, D. G. and HIMSWORTH, R. L. (1970): The location of the chemoreceptor controlling gastric acid secretion during hypoglycaemia. J. Physiol. (Lond.), 206: 397-409.

FRIEDMAN, A. H. and WALKER, C. A. (1968): Circadian rhythms: rat midbrain and caudate nucleus biogenic amine levels. J. Physiol. (Lond.), 197: 77-86.

GROSSMAN, S. P. (1968): Hypothalamic and limbic influences on food intake. Fed. Proc., 27: 1349-1360.

HANIN, I., MASSARELLI, R. and COSTA, E. (1970): Acetylcholine concentrations in rat brain: diurnal oscillation. Science, 170: 341-342.

HIMSWORTH, R. L. (1970): Hypothalamic control of adrenaline secretion in response to insufficient glucose. J. Physiol. (Lond.), 206: 411-417.

HIROSHIGE, T., ABE, K., WADA, S. and KANEKO, M. (1973): Sex differences in circadian periodicity of CRF activity in the rat hypothalamus. Neuroendocrinology, 11: 306-320.

KAKOLEWSKI, J. W., COX, V. C. and VALENSTEIN, E. S. (1968): Sex differences in body-weight change following gonadectomy of rats. Psychol. Rep., 22: 547-554.

KENNEDY, G. C. (1953): The role of depot fat in the hypothalamic control of food intake in the rat. Proc. roy. Soc., 140: 578-592.

KINSON, G. A. and CHUNG-CHING, L. (1973): Diurnal variation in plasma testosterone of the male laboratory rat. Horm. metab. Res., 5: 233-234.

LE MAGNEN, J., DEVOS, M., GANDILLIÈRE, J., LOUIS-SYLVESTRE, J. and TALLON S. (1973): J. comp. physiol. Psychol., 84: 1-23.

LEUNG, P. M. and ROGERS, Q. R. (1969): Food intake: regulation by plasma amino acid pattern. Life Sci., 8: 1-9.

MAYER, J. (1955): Regulation of energy intake and body weight: the glucostat theory and the lipostat hypothesis. Ann. N.Y. Acad. Sci., 63: 15-43.

MELLINKOFF, S. M., FRANKLAND, M., BOYLE, D. and GREIPEL, M. (1956): Relationship between serum amino acid concentration and fluctuations in appetite. J. appl. Physiol., 8: 535-538.

MENAKER, M. (1971): Rhythms, reproduction and photoreception. Biol. Reprod., 4: 295-308.

MERRITT, J. H. and SULKOWSKI, T. S. (1970): Rhythmicity of RNA polymerase activity and RNA levels in nuclei of rat cerebral cortex. J. Neurochem., 17: 1327-1328.
MOORE, R. Y., HELLER, A., BHATNAGER, R. K., WURTMAN, R. J. and AXELROD, J. (1968): Central control of the pineal gland visual pathways. Arch. Neurol., 18: 208-218.
QUAY, W. B. (1965): Regional and circadian differences in cerebral cortical serotonin concentration. Life Sci., 4: 379.
REIS, D. J. and WURTMAN, R. J. (1968): Diurnal changes in brain noradrenalin. Life Sci., 7: 135-145.
RICHTER, C. P. (1927): Animal behaviour and internal drives. Quart. Rev. Biol., 2: 307-343.
RICHTER, C. P. (1965): Biological Clocks in Medicine and Psychiatry. C. C. Thomas, Springfield, Ill.
RICHTER, C. P. (1968): Inherent twenty-four hour and lunar clocks of a primate – the squirrel monkey. Comm. Behav. Biol., 1: 305-332.
RICHTER, C. P. (1971): Inborn nature of the rat's 24-hour clock. Physiol. Psychol., 75: 1-14.
ROLLS, E. T. and ROLLS, B. J. (1973): Altered food preferences after lesions in the basolateral region of the amygdala in the rat. J. comp. physiol. Psychol., 83: 248-259.
SIEGEL, P. S. and STUCKEY, H. L. (1947): The diurnal course of water and food intake in the normal mature rat. J. comp. physiol. Psychol., 40: 365-370.
STELLAR, E. (1954): The physiology of motivation. Physiol. Rev., 61: 5-22.
STROMINGER, J. L. and BROBECK, J. R. (1953): A mechanism of regulation of food intake. Yale J. Biol. Med., 25: 383-390.
TARTELLIN, M. F. and GORSKI, R. A. (1971): Influence of ovarian steroids on the regulation of food intake in the rat. Fed. Proc., 30: 353.
TER HAAR, M. B. (1972): Circadian and oestral rhythms in food intake in the rat. Horm. Behav., 3: 213-219.
TER HAAR, M. B., MacKINNON, P. C. B. and BULMER, M. G. (1974): Sexual differentiation in the phase of the circadian rhythm of [^{35}S]methionine incorporation into cerebral proteins and of serum gonadotrophin levels. J. Endocr., 62: 257-265.
WADE, G. N. and ZUCKER, I. (1967): Taste preferences of female rats: modification by neonatal hormones, food deprivation and prior experience. Physiol. a. Behav., 4: 935-943.
WADE, G. N. and ZUCKER, I. (1970): Modulation of food intake and locomotor activity in female rats by diencephalic hormone implants. J. comp. physiol. Psychol., 72: 328-336.
YOSHINAGA, K., HAWKINS, R. A. and STOCKER, J. F. (1969): Estrogen secretion by the rat ovary in vivo during the estrous cycle and pregnancy. Endocrinology, 85: 103-112.
ZUCKER, I. (1968): Hormonal determinants of sex differences in saccharin preference, food intake and body weight. Physiol. a. Behav., 4: 595-602.

See also references in P.H.B., Part IA, Chapter 6, Section 4b, pp. 261-272.

Chapter 2, Section 1c: Photoperiodic Effects on Neuro-Endocrine Functions (G. C. JACKSON)

ALLEVA, J. J., WALESKI, M. V. and ALLEVA, F. R. (1970): The Zeitgeber for ovulation in rats: non-participation of the pineal gland. Life Sci., 9, 1: 241-246.
BARFUSS, D. W. and ELLIS, L. C. (1971): Seasonal cycles in melatonin synthesis by the pineal gland as related to testicular function in the house sparrow (*Passer domesticus*). Gen. comp. Endocr., 17: 183-193.
DONOVAN, B. T. and VAN DER WERFF TEN BOSCH, J. J. (1959): The relationship of the hypothalamus to estrous in the ferret. *J. Physiol.*, 147: 93-108.
EVERETT, J. W. (1969): Neuroendocrine aspects of mammalian reproduction. Ann. Rev. Physiol., 31: 383-416.
FOLLETT, B. K. and SHARP, P. J. (1969): Circadian rhythmicity in photoperiodically induced gonadotrophin release and gonadal growth in the quail. Nature (Lond.), 223: 968-970.
FRAPS, R. M. (1961): Ovulation in the domestic fowl. In: Control of Ovulation, C. A. Villee (ed.), Pergamon Press, Oxford/New York: 133-162.

GREELY, F. and MEYER, R. K. (1953): Seasonal variation in testis-stimulating activity of male pheasant pituitary glands. Auk, 70: 350-358.
JACKSON, G. L., ROCHE, J. F., FOSTER, D. L. and DZIUK, P. J. (1971): Luteinizing hormone releasing activity in the hypothalamus of anestrous and cyclic ewes. Biol. Reprod., 5: 5-12.
JACKSON, G. L. (1974): Absence of a critical period in estrogen-induced release of LH in the anestrous ewe. Endocrinology, 94, in press.
JONAS, H. A., SALAMONSEN, L. A., BURGER, H. G., CHAMLEY, W. A., CUMMING, I. A., FINDLAY, J. K. and GODING, J. R. (1973): Release of FSH after administration of gonadotropin releasing hormone or estradiol to the anestrous ewe. Endocrinology, 92: 862-865.
KRIEGER, D. and RIZZO, F. (1969): Serotonin mediation of circadian periodicity of plasma 17-hydroxycorticosteroids. Amer. J. Physiol., 217: 1703-1707.
MANSHARDT, J. and WURTMAN, R. J. (1968): Daily rhythm in the noradrenaline content of rat hypothalamus. Nature (Lond.), 217: 574-575.
MENAKER, M. (1971): Rhythms, reproduction, and photoreception. Biol. Reprod., 4: 295-308.
MOORE, R. Y. and EICHLER, V. B. (1972): Loss of a circadian adrenal corticosterone rhythm following suprachiasmatic lesions in the rat. Brain Res., 42: 201-206.
MOORE, R. Y. and LENN, N. J. (1972): A retinohypothalamic projection in the rat. J. comp. Neurol., 146: 1-14.
QUAY, W. B. (1968): Differences in circadian rhythms in 5-hydroxytryptamine according to brain region. Amer. J. Physiol., 215: 1448-1453.
REIS, D. J., WEINBURN, M. and CORVELLI, A. (1968): A circadian rhythm of norepinephrine regionally in the cat brain: its relationship to environmental lighting and to regional diurnal variations in brain serotonin. J. exp. Pharmacol. exp. Therap., 164: 135-145.
REITER, R. J. (1973): Comparative physiology: pineal gland. Ann. Rev. Physiol., 35: 305-328.
SCAPAGNINI, U., MOBERG, G. P., VANLOON, G. R., DE GROOT, J. and GANONG, W. F. (1971): Relation of brain 5-hydroxytryptamine content to diurnal variation in plasma corticosterone in the rat. Neuroendocrinology, 7: 90-96.
SCHWARTZ, N. B. and MCCORMACK, C. E. (1972): Reproduction: gonadal function and its regulation. Ann. Rev. Physiol., 34: 425-472.
STETSON, M. H. and ERICKSON, J. E. (1970): A daily periodicity in pituitary gonadotropin in white-crowned sparrows. Z. vergl. Physiol., 68: 263-267.
VAN TIENHOVEN, A. (1968): Reproductive Physiology of Vertebrates. W. B. Saunders Co., Philadelphia: pp. 388-425.
YAMAMOTO, M., DIEBEL, N. D. and BOGDANOVE, E. M. (1970): Radioimmunoassay of serum and pituitary LH and FSH levels in intact male rats and of serum and pituitary LH in castratrated rats of both sexes – apparent absence of diurnal rhythms. Endocrinology, 87: 798-806.
WURTMAN, R. J., AXELROD, J. and KELLEY, D. E. (1968): The Pineal. Academic Press, New York: 107-144.

Chapter 2, Section 1d: Photoperiodic Effects on Mammalian Reproduction (R. J. REITER)

BISSONNETTE, T. H. (1932): Modification of mammalian sexual cycles: reactions of ferrets (*Putorius vulgaris*) of both sexes to electric light added after dark in November and December. Proc. roy. Soc. B, 110: 322-336.
CRITCHLOW, V. (1963): The role of light in the neuroendocrine system. In: Advances in Neuroendocrinology. A. V. Nalbandov (ed.). University of Illinois Press, Urbana: 377-402.
EVERETT, J. W. and SAWYER, C. H. (1950): A 24-hour periodicity in the "LH-release apparatus" of female rats disclosed by barbiturate sedation. Endocrinology, 47: 198-218.
HAFEZ, E. S. E. (1964): Environment and reproduction in domesticated species. Int. Rev. gen. exp. Zool., 1: 113-164.

HERBERT, J. (1971): The role of the pineal gland in the control by light of the reproductive cycle of the ferret. In: The Pineal Gland. G. E. W. Wolstenholme and J. Knight (eds). Churchill Livingstone, London: 303-327.
HOFFMAN, R. A. and REITER, R. J. (1965): Pineal gland: influence on gonads of male hamsters. Science, 148: 1609-1611.
HOFFMANN, J. C. (1967): Effects of light deprivation on the rat estrous cycle. Neuroendocrinology, 2: 1-10.
HOFFMANN, K. (1973): The influence of photoperiod and melatonin on testes size, body weight, and pelage colour in the djungarian hamster (*Phodopus sungorus*). J. comp. Physiol., 85: 267-282.
JÖCHLE, W. (1967): The circadian rhythm in female reproduction. In: Reproduction in the Female Mammal. G. E. Lamming and E. C. Amoroso (eds). Plenum Press, New York: 267-279.
REITER, R. J. and FRASCHINI, F. (1969): Endocrine aspects of the mammalian pineal gland: a review. Neuroendocrinology, 5: 219-255.
REITER, R. J. (1972): The role of the pineal in reproduction. In: Reproductive Biology. H. Balin and S. Glasser (eds). Excerpta Medica, Amsterdam: pp. 71-114.
REITER, R. J. (1973a): Comparative physiology: pineal gland. Ann. Rev. Physiol., 35: 305-328.
REITER, R. J. (1973b): Pineal control of a seasonal reproductive rhythm in male golden hamsters exposed to natural daylight and temperature. Endocrinology, 92: 423-430.
SORRENTINO, S., Jr. and REITER, R. J. (1970): Pineal-induced alterations of estrous cycles in blinded hamsters. Gen. comp. Endocr., 15: 39-42.

Chapter 2, Section 1e: Seasonal Rhythms in Animals (J. L. CLOUDSLEY-THOMPSON)

See references in P.H.B., Part IB, Chapter 6, Sect. 22b (1), pp. 578-580.

Chapter 2, Section 1f: Seasonal Molting Cycles in Wild Mammals (ELIZABETH JOHNSON)

BAKER, B. L. and WHITAKER, W. L. (1949): Relationship of the adrenal cortex to inhibition of growth of hair by oestrogen. Amer. J. Physiol., 159: 118-123.
BASSETT, C. F., PEARSON, O. P. and WILKIE, F. (1944): The effect of artificially increased length of day on molt, growth and priming of silver fox pelts. J. exp. Zool., 96: 77-83.
BASSETT, C. F. and LLEWELLYN, L. M. (1948): The molting and fur growth pattern in the adult silver fox. Amer. Midl. Nat., 39: 597-601.
BASSETT, C. F. and LLEWELLYN, L. M. (1949): The molting and fur growth pattern in adult mink. Amer. Midl. Nat., 42: 751.
BISSONNETTE, T. H. (1935): Relations of hair cycles in ferrets to changes in the anterior hypophysis and to light cycles. Anat. Rec., 63: 159-168.
BONSMA, J. C. and PRETORIUS, A. J. (1943): Farming in South Africa., 18: 101.
BOROWSKI, S. (1958): Variations in the density of coat during the life cycle of *Sorex araneus*. Acta theriol., 2: 286-289.
BOSSE, K. (1965): Growth and replacement of hair in the guinea-pig. In: Comparative Physiology and Pathology of the Skin. A. J. Rook and G. S. Walton (eds). Blackwell, Oxford: 151-160.
BURKHARDT, J. (1947): Transition from anoestrus in the mare and the effects of artificial lighting. J. agric. Sci. (Camb.), 37: 64-68.
BUTCHER, E. O. (1940): The effect of irritants and thyroxin on hair growth in albino rats. Amer. J. Physiol., 129: 553.
COOP, I. E. and CLARK, V. R. (1958): The effect of thyroxine on wool growth and lamb production. N.Z. J. Agric. Res., 1: 365-381.
COTT, H. B. (1940): Adaptive Coloration in Animals. Methuen, London.

DIEKE, S. (1948): Effect of removing various endocrine glands on the hair cycles of black rats. Endocrinology, 42: 315-319.
DRY, F. W. (1926): The coat of the mouse (*Mus musculus*). J. Genet., 16: 287-340.
EBLING, F. J. and JOHNSON, E. (1961): Systemic influence on activity of hair follicles in skin homografts. J. Embryol. exp. Morph., 9: 285-293.
EBLING, F. J. and JOHNSON, E. (1964): The action of hormones on spontaneous hair growth cycles in the rat. J. Endocr., 29: 193-201.
ECKE, D. and KINNEY, A. (1956): Ageing meadow mice (*Microtus californicus*) by observation of molt progression. J. Mammal., 37: 249-254.
FLEROV, K. K. (1952): Fauna of U.S.S.R. Mammals. 1, Musk Deer and Deer. Nat. Sci. Found. (Wash.) (Translation, 1960).
FLUX, J. E. (1970): Colour change of mountain hares (*Lepus timidus scoticus*) in north-east Scotland. J. Zool. (Lond.), 162: 345-358.
FRENCH, C. E., McEWEN, L. C., MAGRUDER, N. D., RADER, T., LONG, T. A. and SWIFT, R. W. (1960): Responses of white-tailed bucks to added artificial light. J. Mammal., 41: 23-29.
HAITLINGER, R. (1968): Seasonal variation of pelage in representatives of *Apodemus* found in Poland. Zool. Pol., 18: 330-345.
HAMMOND, J. (1952): Control of reproductive and pelt changes in ferrets: some experiments with animals kept entirely on artificial light. J. agric. Sci. (Camb.), 42: 293-303.
HART, J. S. (1964): Geography and season: mammals and birds. In Handbook of Physiology. Sect. 4, Adaptation to the Environment. Amer. physiol. Soc. (Wash.).
HARVEY, N. E. and MACFARLANE, W. V. (1958): The effects of day length on the coat-shedding cycles, body weight, and reproduction of the ferret. Aust. J. biol. Sci., 11: 187-199.
HEROUX, O. (1960): Adjustments of the adrenal cortex and thyroid during cold acclimation. Fed. Proc., 19: 82.
HEWSON, R. (1958): Moults and winter whitening in the mountain hare (*Lepus timidus scoticus*). Proc. zool. Soc. (Lond.), 131: 99-108.
HOOKER, C. W. and PFEIFFER, C. A. (1943): Effects of sex hormones upon body growth, skin, hair and sebaceous glands in the rat. Endocrinology, 32: 69-76.
HOUSSAY, A. B. and HIGGINS, G. M. (1949): The effect of gonadectomy on the growth of hair in mice. Proc. Mayo Clin., 24: 269.
IRVING, L., KROG, H. and MONSON, M. (1955): The metabolism of some Alaskan animals in winter and summer. Physiol. Zool., 28: 173-185.
JACKES, A. D. and WATSON, A. (1975): Winter whitening of Scottish mountain hares (*Lepus timidus scoticus*) in relation to daylength, temperature and snow lie. J. Zool. Lond. 176: 403-409.
JOHNSON, E. (1958a): Quantitative studies of hair growth in the albino rat. II. The effect of sex hormones. J. Endocr., 16: 351-359.
JOHNSON, E. (1958b): Quantitative studies of hair growth in the albino rat. III. The role of the adrenal glands. J. Endocr., 16: 360-368.
JOHNSON, E. and HORNBY, J. (1975): Seasonal changes of pelage in the roe deer *Capreolus capreolus* and its role in thermoregulation. J. nat. Hist. (in press), (previously unpublished observations).
JOHNSON, E. (1972): Moulting cycles. Mammal Rev., 1: 198-208.
KHATEEB, A. A. and JOHNSON, E. (1971a): Seasonal changes of pelage in the vole (*Microtus agrestis*). 1. Correlation with changes in the endocrine glands. Gen. comp. Endor., 16: 217-228.
KHATEEB, A. A. and JOHNSON, E. (1971b): Seasonal changes of pelage in the vole (*Microtus agrestis*). 2. The effect of daylength. Gen. comp. Endocr., 16: 229-235.
KHATEEB, A. A. and JOHNSON, E. (1971c): Seasonal changes of pelage in the vole (*Microtus agrestis*). 3. The role of the endocrine system. Gen. comp. Endocr., 16: 236-240.
KIRTON, A. M., CRESWELL, E., COCKREM, F. R. and BUTLER, G. W. (1959): Some effects of thyroxine on live weight, metabolism and wool growth of Romney ewes. N.Z. J. Agric. Res., 2: 1143-1158.
KLIGMAN, A. M. (1959): The human hair cycle. J. invest. Derm., 33: 307-316.
KOPONEN, T. (1964): The sequence of pelages in the Norwegian lemming (*Lemmus lemmus*). Arch. Soc. Zool. Bot. Fenn. (Vanamo.), 18: 260-277.

LABBAN, F. M. (1957): The effects of L-thyroxine on sheep and wool production. J. agric. Sci. (Camb.), 49: 26-50.
LING, J. K. (1970): Pelage and molting in wild mammals with special reference to aquatic forms. Quart. Rev. Biol., 45: 16-54.
LYMAN, C. P. (1943): Control of coat color in the varying hare, *Lepus americanus*. Bull. Mus. comp. Zool. (Harv.), 93: 393-461.
LYNE, A. J. (1966): The development of hair follicles. Aust. J. Sci., 28: 370-377.
MAZAK, V. (1962): Spring moult in *Equus hemionus kiang* and a contribution to the phylogenesis of moulting in the subfamily Equinae. Zool. Anz. 168: 164-170.
MEIJERE, J. C. H. DE (1894): Über die Haare der Säugetiere, besonders über ihre Anordnung. Morph. Jb, 21: 312-424.
MOHN, M. P. (1958): The effects of different hormonal states on the growth of hair in rats. In: The Biology of Hair Growth W. Montagna and R. A. Ellis (eds). Academic Press, New York, 336 pp.
PAVLOVIC, M. (1965): Investigations of roe-buck skin: 8th Congr. Union International des Biologistes du Gibier: 119-123.
PINTER, A. J. (1968): Hair growth responses to nutrition and photoperiod in the vole, *Microtus montanus*. Amer. J. Physiol., 215: 828-831.
RAND, C. G., RIGGS, D. S. and TALBOT, N. B. (1952): The influence of environmental temperature on the metabolism of the thyroid hormone in the rat. Endocrinology, 51: 562-569.
REINEKE, E. P., TRAVIS, H. F. and DOLNICK, E. M. (1962): The effects of thyroid gland destruction and replacement therapy on fur growth in mink (*Mustela vison*) given a thyroxine free diet. Amer. J. vet. Res., 23: 121-127.
ROTHSCHILD, M. (1942): Change of pelage in the stoat. Nature (Lond.), 149: 78.
RUST, C. C. (1965): Hormonal control of pelage cycles in the short-tailed weasel. Gen. comp. Endocr., 5: 222-231.
RUST, C. C. and MEYER, R. K. (1969): Hair color, molt and testis size in male short-taled weasels treated with melatonin. Science, 165: 921-922.
RYDER, M. L. (1970): Structure and seasonal change of coat in Scottish wild goats. J. Zool. (Lond.), 161: 355-362.
SCHOLANDER, P. F., WALTERS, V., HOCK, R. and IRVING, L. (1950): Body insulation of some arctic and tropical mammals and birds. Biol. Bull. 99: 259-271.
SEALANDER, J. A. (1951): Survival of *Peromyscus* in relation to environmental temperature and acclimation at high and low temperatures. Amer. Midl. Nat., 46: 257-309.
SEELE, E. (1968): Haut und Haar der Chinchillidae. Zool. Anz., 181: 60-75.
SEVERAID, J. H. S. (1945): Pelage changes in the snowshoe hare, *Lepus americanus*. J. Mammal., 26: 41-63.
SHANKS, C. E. (1948): The pelt-primeness method of aging muskrats. Amer. Midl. Nat., 39: 179-187.
SPEED, J. G. (1960): The importance of the coat in Exmoor and other mountain and moorland ponies living out of doors. Brit. vet. J., 116: 91-98.
STEVENS, C. E., D'ANGELO, S. A., PASCHKIS, K. E., CANTAROW, A. and SUNDERMAN, F. W. (1955): The response of the pituitary-thyroid system of the guinea-pig to low environmental temperature. Endocrinology, 56: 143-155.
STEVENSON, J. M. J. (1962): Moult in mink and silver fox. Mamm. Soc. Bull., 17: 13-14.
WATSON, A. (1963): The effect of climate on the colour change of mountain hares in Scotland. Proc. zool. Soc. (Lond.), 141: 823-825.
WRIGHT, P. L. (1942): A correlation between spring molt and spring changes in the sexual cycle of the weasel. J. exp. Zool., 91: 103-110.

Chapter 2, Section 1g: Photoperiodic Effects on Hair and Wool Growth of Domestic Animals (J. C. D. HUTCHINSON)

ALLEN, T. E., BENNETT, J. W., DONEGAN, S. M. and HUTCHINSON, J. C. D. (1970): Moisture, its accumulation and site of evaporation in the coats of sweating cattle. J. agric. Sci. (Camb.), 74: 247-258.

BASSETT, C. F., PEARSON, O. P. and WILKE, F. (1944): The effect of artificially increased length of day on moult, growth and priming of silver fox pelts. J. exp. Zool., 96: 77-83.
BELJAEV, D. K., UTKIN, L. G. and KULIČKOV, B. A. (1964): Effect of the light regime on development of the fur in mink. Izv. sib. Otdel. Akad. Nauk SSSR, no. 4: 91-100. From Anim. Breed. Abstr. (1968), 36, 637.
BENNETT, J. W., HUTCHINSON, J. C. D. and WODZICKA-TOMASZEWSKA, M. (1962): Annual rhythm of wool growth. Nature (Lond.), 194: 651-652.
BENOIT, J. (1964): The role of the eye and of the hypothalamus in the photostimulation of gonads in the duck. Ann. N.Y. Acad. Sci., 117: 204-216.
BISONNETTE, T. H. and WILSON, E. (1939): Shortening daylight periods between May 15 and September 12 and the pelt cycle of the mink. Science, 89: 418-419.
BROWN, G. D. and WILLIAMS, O. B. (1970): Geographical distribution of the productivity of sheep in Australia. J. Aust. Inst. agric. Sci., Sept.: 182-198.
BÜNNING, E. (1960): Circadian rhythms and the time measurement in photoperiodism. Cold Spr. Harb. Symp. quant. Biol., 25: 249-256.
BURKHARDT, J. (1947): Transition from anoestrus in the mare and the effects of artificial lighting. J. agric. Sci. (Camb.), 37: 64-68.
CHAPMAN, S. (1951): The Earth's Magnetism, 2nd Ed. Methuen and Co. Ltd., London.
DONEY, J. M. (1964): The fleece of the Scottish Blackface sheep IV. The effects of pregnancy, lactation and nutrition on seasonal wool production. J. agric. Sci., (Camb.), 62: 59-66.
DONEY, J. M. (1966): Breed difference in response of wool growth to annual nutritional and climatic cycles. J. agric. Sci. (Camb.), 67: 25-30.
DOWLING, D. F. and NAY, T. (1960): Cyclic changes in the follicles and hair coat in cattle. Aust. J. agric. Res., 11: 1064-1071.
DOWNES, A. M. and HUTCHINSON, J. C. D. (1969): Effect of low skin surface temperature on wool growth. J. agric. Sci., (Camb.), 72: 155-158.
FARNER, D. S. (1970): Predictive functions in the control of annual cycles. Environm. Res., 3: 119-131.
FERGUSON, K. A., CARTER, H. B. and HARDY, M. H. (1949): Studies of comparative fleece growth in sheep. I. The quantitative nature of inherent differences in wool growth rate. Aust. J. sci. Res., B2: 42-81.
FERGUSON, K. A., WALLACE, A. L. C. and LINDNER, H. R. (1965): Hormonal regulation of wool growth. In: Biology of the Skin and Hair Growth. A. G. Lyne and B. F. Short (eds). Angus and Robertson Ltd, Sydney, Chapt. 41.
FOLLET, B. K. and SHARP, P. J. (1969): Circadian rhythmicity in photoperiodically induced gonadotrophin release and gonadal growth in the quail. Nature (Lond.), 223: 968-971.
GANONG, W. F., SHEPHERD, M. D., WALL, J. R., VAN BRUNT, E. E. and CLEGG, M. T. (1963): Penetration of light into the brain of mammals. Endocrinology, 72: 962-963.
GOSS, R. J. (1969a): Photoperiodic control of antler cycles in deer. I. Phase shift and frequency changes. J. exp. Zool., 170: 311-324.
GOSS, R. J. (1969b): Photoperiodic control of antler cycles in deer. II. Alterations in amplitude. J. exp. Zool., 171: 223-234.
GREENWOOD, A. W. (1962): An experiment with a constant environment for the domestic fowl. Anim. Prod., 4: 80-90.
HAGENBECK, C. (1911): Dyr og Mennesker. Transl. Dreyer, W. Gyldendalske Boghandel Nordisk Forlag: Chapt. 12.
HAMMOND, J. Jr (1951): Control by light of reproduction in ferrets and mink. Nature (Lond.), 167: 150-151.
HAMMOND, J. Jr (1952): Control of reproductive and pelt changes in ferrets: some experiments with animals kept entirely upon artificial light. J. agric. Sci., (Camb.), 42: 293-303.
HAMMOND, J. Jr (1953): Effects of Artificial Lighting on the Reproductive and Pelt Cycles of Mink. Heffer and Sons, Cambridge.
HART, D. S. (1955): Photoperiodic and hormone response of wool growth in sheep. Proc. N.Z. Soc. Anim. Prod., 15: 57-65.
HART, D. S. (1961): The effect of light-dark sequences on wool growth. J. agric. Sci (Camb.), 56: 235-242.
HART, D. S., BENNETT, J. W., HUTCHINSON, J. C. D. and WODZICKA-TOMAS-

ZEWSKA, M. (1963): Reversed photoperiodic seasons and wool growth. Nature (Lond.), 198: 310-311.
HART, J. S. (1956): Seasonal changes in insulation of the fur. Canad. J. Zool., 34: 53-57.
HARVEY, N. E. and MACFARLANE, W. V. (1958): The effects of daylength on the coat-shedding cycles, body weight and reproduction of the ferret. Aust. J. biol. Sci., 11: 187-199.
HAYMAN, R. H. and NAY, T. (1961): Observations on hair growth and shedding in cattle. Aust. J. agric. Res., 12: 513-527.
HAYMAN, R. H. (1965): Hair growth in cattle. In: "Biology of the Skin and Hair Growth. A. G. Lyne and B. F. Short (eds). Angus and Robertson Ltd, Sydney: Chapt. 35.
HINES, C. O., PAGHIS, I., HARTZ, T. R. and FEJER, J. A. (eds) (1966): The Physics of the Earth's Upper Atmosphere. Prentice Hall, New Jersey.
HUTCHINSON, J. C. D. and WODZICKA-TOMASZEWSKA, M. (1961): Climate physiology of sheep. Anim. Breed. Abstr., 29: 1-14.
HUTCHINSON, J. C. D. (1962): The annual rhythm of egg production in fowls. In: Proc. 12th Wld Poultry Congr., 2: 124-129.
HUTCHINSON, J. C. D. (1965): Photoperiodic control of the annual rhythm of wool growth. In: Biology of the Skin and Hair Growth. A. G. Lyne and B. F. Short (eds). Angus and Robertson Ltd, Sydney: Chapt. 34.
IGNATOVICH, A. D. (1970): The effect of decreased illumination on growth and on pelt maturation and quality in arctic foxes. Trudy nauchno-issled. Inst. sel'. Khoz. Kraĭn. Sev., 18: 188-201. From Anim. Breed. Abstr. (1972), 40, No. 830.
LOFTS, B. (1970): Animal Photoperiodism. Edward Arnold Ltd, London.
MARSHALL, A. J. (1960): Annual periodicity in the migration and reproduction of birds. Cold Spr. Harb. Symp. quant. Biol., 25: 499-505.
MORRIS, L. R. (1961): Photoperiodicity of seasonal rhythm of wool growth in sheep. Nature (Lond.), 190: 102-103.
MORRIS, T. R. (1962): The effect of changing daylengths on the reproductive responses of the pullet. Proc. 12th Wld Poultry Congr., 2: 115-124.
NAGARCENKAR, R. and BHATTACHARYA, P. (1964a): Effect of environmental factors on wool growth. I. Effect of normal climatic factors. Indian J. vet. Sci., 34: 89-99.
NAGARCENKAR, R. and BHATTACHARYA, P. (1964b): Effect of environmental factors on wool growth. II. Effect of photoperiodicity. Ind. J. vet. Sci., 34: 100-107.
PITTENDRIGH, C. S. (1960): Circadian rhythms and the circadian organization of living systems. Cold Spr. Harb. Symp. quant. Biol., 25: 159-184.
PITTENDRIGH, C. S. and MINIS, D. H. (1964): The entrainment of circadian oscillations by light and their role as photoperiodic clocks. Amer. Nat., 98: 261-294.
RATCLIFFE, J. A. (ed.) (1960): Physics of the Upper Atmosphere. Academic Press Inc. Ltd, London.
ROUGEOT, J. (1957): Action de la variation saisonnière de la durée quotidienne d'éclairement sur la mue de certaines fibres de la toison de la race ovine Limousine. C. R. Soc. Biol. (Paris), 151: 834-837.
ROUGEOT, J. (1961): Actions comparées des variations périodiques, annuelles et semestrielles, de la durée quotidienne de l'éclairement sur les cycles des follicules des jarres courts de la toison des brebis Limousines. Relations avec leur cycle de reproduction. Ann. Biol. Anim. Biochem. Biophys., 1: 385-402.
RYDER, M. L. and STEPHENSON, S. K. (1968): Wool Growth. Academic Press Inc. Ltd, London.
SLEE, J. (1965): Seasonal patterns of moulting in Wiltshire Horn sheep. In: Biology of the Skin and Hair Growth. A. G. Lyne and B. F. Short (eds). Angus and Robertson Ltd, Sydney: Chapt. 33.
STOUT, F. M. and ADAIR, J. (1972): New results on accelerated fur growth. Dtsch. Pelztierzüchter, 46: 84-87. From Anim. Breed. Abstr. (1972), 40, No. 4915.
SYMINGTON, R. B. (1959): Light regulation of coat shedding in a tropical breed of hair sheep. Nature (Lond.), 184: 1076.
THIBAULT, C., COUROT, M., MARTINET, L., MAULEON, P., DU MESNIL DU BUISSON, F., ORTAVANT, R., PELLETIER, J. and SIGNORET, J. P. (1966): Regulation of breeding season and estrous cycles by light and external stimuli in some mammals. J. Anim. Sci. suppl., 25: 119-142.

TURNER, H. G. and SCHLEGER, A. V. (1970): An analysis of growth processes in cattle coats and their relations to coat type and body weight gain. Aust. J. biol. Sci., 23: 201-218.

WEBSTER, A. J. F., HICKS, A. M. and HAYS, F. L. (1969): Cold climate and cold temperature induced changes in the heat production and thermal insulation of sheep. Canad. J. Physiol., 47: 553-562.

WEBSTER, A. J. F., CHLUMECKY, J. and YOUNG, B. A. (1970): Effects of cold environments on the energy exchanges of young beef cattle. Canad. J. Anim. Sci., 50: 89-100.

WILDMAN, A. B. (1957): Photoperiodicity and wool growth in Romney rams and wethers. Nature (Lond.), 180: 296-297.

WODZICKA, M. (1960): Seasonal variations in wool growth and heat tolerance of sheep. I. Wool growth. Aust. J. agric. Res., 11: 75-84.

WURTMAN, R. J., AXELROD, J. and KELLY, D. E. (1968): The Pineal. Academic Press, Inc. Ltd, London.

YEATES, N. T. M. (1955): Photoperiodicity in cattle. I. Seasonal changes in coat character and their importance in heat regulation. Aust. J. agric. Res., 6: 891-902.

YEATES, N. T. M. (1957): Photoperiodicity in cattle. II. The equatorial light environment and its effect on the coat of European cattle. Aust. J. agric. Res., 8: 733-739.

YEATES, N. T. M. (1958): Observations on the role of nutrition in coat shedding in cattle. J. agric. Sci (Camb.), 50: 110-112.

YEATES, N. T. M. (1965): Modern Aspects of Animal Production. Butterworth Scientific Publications, London: Chapt. 9 and 10.

Chapter 2, Section 2a: Effects of Solar Radiation on the Sensible Heat Exchange of Mammals (J. C. D. HUTCHINSON, G. D. BROWN and T. E. ALLEN)

ABBIE, A. A. (1969): The Original Australian. A. H. and A. W. Reed, Pty Ltd, Wellington, Chapt. 4.

ALEXANDER, G. and WILLIAMS, D. (1971): Heat stress and development of the conceptus in domestic sheep. J. agric. Sci. (Camb.), 76: 53-72.

ALLEN, T. E., BENNETT, J. W., DONEGAN, S. M. and HUTCHINSON, J. C. D. (1970): Moisture, its accumulation and site of evaporation in the coats of sweating cattle. J. agric. Sci. (Camb.), 74: 247-258.

BENNETT, J. W. (1964): Thermal insulation of cattle coats. Proc. Aust. Soc. Anim. Prod., 5: 160-166.

BENNETT, J. W. and HUTCHINSON, J. C. D. (1964): Thermal insulation of short lengths of Merino fleece. Aust. J. agric. Res., 15: 427-445.

BERMAN, A. and VOLCANI, R. (1961): Seasonal and regional variations in coat characteristics of dairy cattle. Aust. J. agric. Res., 12: 528-538.

BIRKEBAK, R. C. (1966): Heat transfer in biological systems. Int. Rev. gen. exp. Zool., 2: 269-344.

BLAXTER, K. L. and WAINMAN, F. W. (1964): The effect of increased air movement on the heat production and emission of steers. J. agric. Sci. (Camb.), 62: 207-214.

BLIGH, J. and HARTHOORN, A. M. (1965): Continuous radiotelemetric records of the deep body temperature of some unrestrained African mammals under near-natural conditions. J. Physiol. (Lond.), 176: 145-162.

BLIGH, J. and LAMPKIN, G. H. (1965): A comparison of the deep body temperature of Hereford and Zebu cows recorded continuously by radiotelemetry under similar field conditions. J. agric. Sci. (Camb.), 64: 221-227.

BONSMA, J. C. (1943): Influence of colour and coat vocer on adaptability of cattle. Farming in S. Africa, Febr.: 101-120.

BROWN, G. D. (1971a): The responses of sheep and kangaroos to climate in a natural environment using multichannel telemetry. Proc. Symp. Biotelemetry, South African C.S.I.R., Pretoria: 285-302.

BROWN, G. D. (1971b): Thermal status of sheep at pasture in western New South Wales. Aust. J. agric. Res., 22: 797-808.

BROWN, G. D. (1973): The biology of marsupials of the Australian arid zone. Aust. Mammal., 1: 2.

BURTON, A. C. and EDHOLM, O. G. (1955): Man in a Cold Environment. Edward Arnold Ltd, London.
CENA, K. (1966): Badania nadpochłanianiem promieniowania słonecznego u krów i koni o różnej masci. Acta agra. silvest., 6: 93-138.
CENA, K. (1974): Radiative loss from animals and man. In: Heat Loss from Animals and Man. J. L. Monteith and L. E. Mount (eds.). Butterworths sci. Publ., London, in press.
CENA, K. and MONTEITH, J. L. (1975): Transfer processes in animal coats 1. Radiative Transfer. Proc. roy. Soc. B, 188: 377-393.
CLAPPERTON, J. L., JOYCE, J. P. and BLAXTER, K. L. (1965): Estimates of the contribution of solar radiotion to the thermal exchanges of sheep at a latitude of 55° north. J. agric. Sci. (Camb.), 64: 37-49.
CLARK, J. A., CENA, K. and MONTEITH, J. L. (1973): Measurements of the local heat balance of animal coats and human clothing. J. appl. Physiol., 35: 751-754.
DAWSON, T. J. and BROWN, G. D. (1970): A comparison of the insulative and reflective properties of the fur of desert kangaroos. Comp. Biochem. Physiol., 37: 23-38.
DAWSON, T. J. and DENNY, M. J. S. (1969): A bioclimatological comparison of the summer day microenvironments of two species of arid zone kangaroo. Ecology, 50: 328-332.
DAWSON, T. J. (1972): Likely effects of standing and lying on the radiation heat load experienced by a resting kangaroo on a summer day. Aust. J. Zool., 20: 17-22.
DOWLING, D. F. (1959): The medullation characteristic of the hair coat as a factor in heat tolerance of cattle. Aust. J. agric. Res., 10: 736-743.
DRUMMOND, A. J., FRITZ, S., MÖLLER, F., MÖRIKOFER, W., ROBINSON, G. D. and SCHÜEPP, W. (1958): Radiation instruments and measurement. Ann. int. geophys. Yr, 5: 365-466.
ELAM, C. J. (1971): Problems related to intensive indoor and outdoor beef productive systems. J. Anim. Sci., 32: 554-559.
FINCH, V. A. (1972a): Thermoregulation and heat balance of the East African eland and hartebeest. Amer. J. Physiol., 222: 1374-1379.
FINCH, V. A. (1972b): Energy exchanges with the environment of two East African antelopes, the eland and the hartebeest. Symp. Zool. Soc. (Lond.), 31: 315-326.
FINCH, V. A. (1973): Thermoregulation and heat balance in some East African herbivores. Ph.D. Thesis, Univ. Nairobi, Kenya.
FOX, D. L. (1953): Animal Biochromes and Structural Colours. Cambridge University Press, Cambridge.
FOX, H. M. and VEVERS, G. (1960): The Nature of Animal Colours. Sidgwick and Jackson Ltd, London.
HAMMEL, H. T. (1956): Infrared emissivities of some Arctic fauna. J. Mammal., 37: 375-378.
HUTCHINSON, J. C. D. and BROWN, G. D. (1969): Penetrance of cattle coats by radiation. J. appl. Physiol., 26: 454-464.
HUTCHINSON, J. C. D., BROWN, G. D. and ALLEN, T. E. (1973): Effect of the coat on the heat dissipation of cattle in a radiant environment. 1st Aust. Conf. Heat and Mass Transfer, Sect. 3: 15-24.
JOYCE, J. P., BLAXTER, K. L. and PARK, C. (1966): The effect of natural outdoor environments on the energy requirements of sheep. Res. vet. Sci., 7: 342-259.
KOVARIK, M. (1964): Flow of heat in an irradiated protective cover. Nature (Lond.), 201: 1085-1087.
LAND, M. F. (1972): The physics and biology of animal reflectors. Progr. Biophys., 24: 77-106.
MACFARLANE, W. V., MORRIS, R. J. H. and HOWARD, B. (1958): Heat and water in tropical Merino sheep. Aust. J. agric. Res., 9: 217-228.
MACFARLANE, W. V., HOWARD, B. and MORRIS, R. J. H. (1966): Water metabolism of Merino sheep shorn during summer. Aust. J. agric. Res., 17: 219-225.
MOUNT, L. E. (1968): Climatic physiology of the Pig. Edward Arnold Ltd, London, Chapt. 7.
MUKHERJEE, D. P., DUTTA, S. and BHATTACHARYA, P. (1961): Studies on the draught capacity of Hariana bullocks. Ind. J. vet. Sci. Anim. Husb., 31: 39-50.
NISHI, Y. and GAGGE, A. P. (1970): Direct evaluation of convective heat transfer coefficient by naphthalene sublimation. J. appl. Physiol., 29: 830-838.
OJO, O. (1971): Bovine energy balance climatology and livestock potential in Nigeria. Agric. Meteor., 8: 352-359.

ØRITSLAND, N. A. (1971): Wavelength-dependent solar heating of harp seals (*Pagophilus groenlandicus*). Comp. Biochem. Physiol., 40A: 359-361.
ØRITSLAND, N. A.: A windchill and solar radiation index for homeotherms. J. theoret. Biol., in press.
PORTER, W. P. and GATES, D. M. (1969): Thermodynamic equilibria of animals with environment. Ecol. Monogr., 39: 227-245.
PRIESTLEY, C. H. B. (1957): The heat balance of sheep standing in the sun. Aust. J. agric. Res., 8: 271-280.
RIEMERSCHMID, G. (1943): Some aspects of solar radiation in its relation to cattle in South Africa and Europe. Onderstepoort J. vet. Res., 18: 327-353.
RIEMERSCHMID, G. and ELDER, J. S. (1945): The absorptivity for solar radiation of different coloured hairy coats of cattle. Onderstepoort J. vet. Res., 20: 223-234.
ROBERTSHAW, D. and TAYLOR, C. R. (1969): A comparison of sweat gland activity in eight species of East African bovids. J. Physiol., (Lond.), 203: 135-143.
SCHLEGER, A. V. and TURNER, H. G. (1960): Analysis of coat characters of cattle. Aust. J. agric. Res., 11: 875-885.
SCHLEGER, A. V. and BEAN, K. G. (1971): Factors determining sweating competence of cattle skin. Aust. J. biol. Sci., 24: 1291-1300.
SCHMIDT, P. J. (1969): Behaviour of cattle in a hot dry climate. M. Sc. Thesis, Univ. New England, N.S.W.
SCHMIDT-NIELSEN, K. (1964): Desert Animals: Physiological Problems of Heat and Water. Oxford University Press, London.
SCHOLANDER, P. F., WALTERS, V., HOCK, R. and IRVING, L. (1950): Body insulation of some arctic and tropical mammals and birds. Biol. Bull., 99: 225-236.
SHELTON, M. and HUSTON, J. E. (1968): Effects of high temperature stress during gestation on certain aspects of reproduction in the ewe. J. Anim. Sci., 27: 153-158.
SMITH, I. D. and ALEXANDER, G. (1966): Perinatal mortality in Shorthorn cattle in northern Queensland. Proc. Aust. Soc. Anim. Prod., 6: 63-65.
STEWART, R. E. (1953): Absorption of solar radiation by the hair of cattle. Agric. Eng. (St. Joseph), 34: 235-238.
TURNER, H. G. and SCHLEGER, A. V. (1960): The significance of coat type in cattle. Aust. J. agric. Res., 11: 645-663.
TURNER, H. G. and SCHLEGER, A. V. (1970): An analysis of growth processes in cattle coats and their relations to coat type and body weight gain. Aust. J. biol. Sci., 23: 201-218.
WEINER, J. S. (1971): Man's Natural History. Weidenfeld and Nicolson, London.
YEATES, N. T. M. (1965): Modern Aspects of Animal Production. Butterworths sci. Publ., London: 151.
YEATES, N. T. M. and SCHMIDT, P. J. (1974): Beef Cattle Production. Butterworths sci. Publ., London.

Chapter 2, Section 2b (1): Effects of Light Intensity on Reproduction of Mammals (J. C. HOFFMANN)

ASCHOFF, J. (1961): Exogenous and endogenous components in circadian rhythms. In: Biological Clocks, Cold Spr. Harb. Symp. quant. Biol. Vol. 25. Biol. Lab., Cold Spr. Harb., N.Y., 11-28.
BARTHOLOMEW, G. A. (1949): The effect of light intensity and day length on reproduction in the English sparrow. Bull. Mus. comp. Zool. Harvard Coll., 101: 433-476.
BENOIT, J. (1961): Opto-sexual reflex in the duck: physiological and histological aspects. Yale J. Biol. Med., 34: 97-116.
BROWMAN, L. G. (1937): Light in its relation to activity and estrous rhythms in the albino rat. J. exp. Zool., 75: 375-388.
DIAMOND, M., DIAMOND, A. L. and MAST, M. (1972): Visual sensitivity and sexual arousal levels during the menstrual cycle. J. nerv. ment. Dis., 155: 170-176.
DOBIE, J. B. (1946): Poultry lighting for egg production. Wash. State Coll. Agric. exp. Stat. Bull., 471.
EVERETT, J. W. (1961): The mammalian female reproductive cycle and its controlling

mechanisms. In: Sex and Internal Secretions, 3rd Ed., 1, W. C. Young (ed.). The Williams and Wilkins Co., Baltimore: 497-555.

FARNER, D. S. (1961): Comparative physiology: photoperiodicity. Ann. Rev. Physiol., 23: 71-96.

HERTIG, B. A. (1968): Measurement of the physical environment. In: Adaptation of Domestic Animals, E. S. E. Hafez (ed.) Lea & Febiger, Philadelphia: 324-337.

HOFFMANN, J. C. (1968): Effect of photoperiod on estrous cycle length. Endocrinology, 83: 1355-1357.

HOFFMANN, J. C. (1970): Light and reproduction in the rat: effects of photoperiod length on albino rats from two different breeders. Biol. Reprod., 2: 255-261.

HOFFMANN, J. C. (1973): Light and reproduction in the rat: effects of early lighting on responses measured in adult females. Biol. Reprod., 8: 473-480.

LOGAN, E. R. (1954): Influence of environmental lighting conditions on the estrous cycle of the rat. Ph. D. Thesis, Northwestern University: 73-80.

MAGEE, K., BASINSKA, J., QUARRINGTON, B. and STANCER, H. C. (1970): Blindness and menarche. Life Sci., 9: 7-12.

MARSHALL, F. H. A. (1940): The experimental modification of the oestrous cycle in the ferret by different intensities of light irradiation and other methods. J. exp. Biol., 17: 139-146.

McGUIRE, R. A., RAND, W. M. and WURTMAN, R. J. (1973): Entrainment of the body temperature rhythm in rats: effect of color and intensity of environmental light. Science: 181: 956-957.

OISHI, T. and LAUBER, J. K. (1973): Photoreception in the photosexual response of quail II. Effects of intensity and wavelength. Amer. J. Physiol., 225: 880-886.

PIACSEK, B. E. and HAUTZINGER, G. M. (1974): Effects of duration, intensity and spectrum of light exposure on sexual maturation time of female rats. Biol. Reprod., 10: 380-387.

ROWAN, W. (1938): Light and seasonal reproduction in animals. Biol. Rev. Camb. phil. Soc., 13: 374-402.

WURTMAN, R. J. (1967): Effects of light and visual stimuli on endocrine function. In: Neuroendocrinology, II. L. Martini and W. F. Ganong (eds). Academic Press, N.Y.: 19-59.

ZACHARIAS, L. and WURTMAN, R. J. (1964): Blindness: its relation to age of menarche. Science, 144: 1154-1155.

Chapter 2, Section 2b (2): Effects of Light Spectra on Reproduction of Mammals and Birds (D. P. CARDINALI)

BARBANTI-SILVA, E. (1932): Influenza di alcune loci colorate sulle funzioni della reproduzione e del accrescimento. Monitore Ostet.-ginec., 4: 145-155.

BENOIT, J. and OTT, L. (1944): External and internal factors in sexual activity. Effect of irradiation and different wave-lengths on the mechanisms of photostimulation of the hypophysis and on testicular growth in the immature duck. Yale J. Biol. Med., 17: 27-46.

BENOIT, J. (1970): Etude de l'action des radiations visibles sur la gonadostimulation et de leur pénétration intra-cranïenne chez les oiseaux et les mamifères. In: La Photoregulation de la Réproduction chez les Oiseaux et les Mamifères. J. Benoit and I. Assenmacher (eds). C.N.R.S. (Paris): 121-150.

BISSONETTE, T. H. (1932): Studies on the sexual cycle in birds. VI. Effects of white, green and red lights of equal luminous intensity on the testes activity of the European starling (*Sturnus vulgarus*). Physiol. Zoöl., 5: 92-98.

BISSONETTE, T. H. (1934): Light and sexual cycles in starling and ferrets. Quart. Rev. Biol., 8: 201-208.

CARDINALI, D. P. (1974): Melatonin and the endocrine role of the pineal organ. In: Current Topics in Experimental Endocrinology 2. V. H. T. James and L. Martini (eds.). Academic Press, New York: 107-128.

CARDINALI, D. P. and WURTMAN, R. J. (1975a): The effects of light on man. In: Physiological Anthropology. A. Doman (ed.). Oxford University Press, New York: 13-38.

CARDINALI, D. P. and WURTMAN, R. J. (1975b): Control of melatonin synthesis in the

mammalian pineal organ. In: Frontiers of Pineal Physiology. M. D. Altschule (ed.). The MIT Press Cambridge (U.S.A.): 12-41.
CARDINALI, D. P., LARIN, F. and WURTMAN, R. J. (1972a): Control of rat pineal gland by light spectra. Proc. nat. Acad. Sci. (Wash.), 69: 2003-2005.
CARDINALI, D. P., LARIN, F. and WURTMAN, R. J. (1972b): Action spectra for effects of light on hydroxyindole-O-methyl transferases in rat pineal, retina and harderian gland. Endocrinology, 91: 877-886.
HARRISON, P. C., McGINNIS, J., SCHUMAIER, G., and LAUBER, J. (1969): Sexual maturity and subsequent reproductive performance of White Leghorn chickens subjected to different parts of the light spectrum. Poultry Sci., 47: 878-887.
HARRISON, P. C., LATSHAW, J. D., CASEY, J. M. and McGINNIS, J. (1970): Influence of decrease length of different spectral photoperiods on testis development of domestic fowl. J. Reprod. Fertil. 22: 269-275.
MARSHALL, F. H. A. and BOWDEN, F. P. (1936): The further effects of irradiation on the oestrus cycle of the ferret. J. exp. Biol., 13: 383-386.
MENAKER, M. (1972): Non-visual light reception. Sci. Amer. 226: 22-29.
MOORE, R. Y. (1969): Visual pathways controlling neuroendocrine functions. In: Progress in Endocrinology. C. Gual (ed.) Excerpta med. (Amst.), pp. 490-494.
MOORE, R. Y. (1973): Retinohypothalamic projections in mammals: a comparative study. Brain Res., 49: 403-409.
REITER, R. J. and KLEIN, D. C. (1971): Observations on the pineal gland, the harderian glands, the retina, and the reproductive organs of adult female rats exposed to continuous light. J. Endocr., 51: 117-125.
RINGOEN, A. R. (1942): Effects of continuous green and red light illumination on gonadal response in the English sparrow (*Passer domesticus* Linn.). Amer. J. Anat., 71: 99-106.
SCOTT, H. M. and PAYNE, L. F. (1937): Light in relation to the experimental modification of the breeding season of turkeys. Poultry Sci., 16: 90-96.
VRIEND, J. and LAUBER, J. K. (1973): Effects of light intensity, wavelength and quanta on gonads and spleen of the deer mouse. Nature (Lond.), 244: 37-38.
WOLSTENHOLME, G. E. W. and KNIGHT, J. (eds) (1971): The Pineal Gland. A Ciba Found. Symp., Churchill Livingstone, London.
WOODARD, A. E., MOORE, J. A. and WILSON, W. O. (1969): Effect of wave length of light on growth and reproduction in Japanese quail (*Coturnix coturnix*). Poultry Sci., 48: 118-123.
WURTMAN, R. J. (1967): Effects of light and visual stimuli on endocrine function. In: Neuroendocrinology, 2. L. Martini and W. F. Ganong (eds). Academic Press, New York: 19-59.
WURTMAN, R. J., AXELROD, J. and KELLY, D. E. (1968): The Pineal. Academic Press, New York.
WURTMAN, R. J. and WEISEL, J. (1969): Environmental lighting and neuroendrocrine function Relationship between spectrum of light source and gonadal growth. Endocrinology, 85: 1218-1220.
WURTMAN, R. J. (1970): Effects of light on metabolic processes. Birth Defects: Original Art. Ser., 6: 60-62.

Chapter 2, Section 2c: Phototoxicity in Animals (T. W. DUKES)

ALLISON, A. C., MAGNUS, I. A. and YOUNG, M. R. (1966): Role of lysosomes and cell membranes in photosensitization. Nature (Lond.), 209: 874-878.
BAY, W. W., GLEISER, C. A., DUKES, T. W. and BROWN, R. S. (1970): The experimental production and evaluation of drug induced phototoxicity in swine. Toxicol. appl. Pharm., 17: 538-547.
CORNELIUS, C. E., ARIAS, I. M. and OSBURN, B. I. (1965): Hepatic Pigmentation with photosensitivity. A syndrome in Corriedale Sheep resembling Dubin-Johnston Syndrome in man. J. Amer. vet. med. Ass., 146: 709-713.
CORNELIUS, C. E. and GRONWELL, R. R. (1968): Congenital photosensitivity and hyperbilirubinemia in Southdown sheep in the United States Amer. J. vet. Res. 29: 291-295.

CLARE, N. T. (1955): Photosensitization in Animals. Advanc. vet. Sci., 2: 182-211.
GLENN, B. L., PANCIERA, R. J. and MONLUX, A. W. (1965): A hepatogenous photosensitivity disease of cattle II Histopathology and pathogenesis of the hepatic lesions. Path. vet., 2: 49-57.
HAKIM, R. E., FREEMAN, R. G., GRIFFIN, A., CLARK, J. M. and KNOX, J. M. (1961): Experimental toxicotogic studies on 8 – methoxypsoralen in animals exposed to the long ultraviolet. J. Pharmacol exp. Ther., 131: 394-399.
KANEKO, J. J. (1970): Porphyrin, heme, and erythrocyte metabolism. In: The porphyrias in Clinical Biochemistry of Domestic Animals, 2nd Ed. 1. Cornelius and Kaneko ed. Academic Press, New York.
LA VIA, M. and HILL, R. (eds.) (1971): Principles of Pathobiology. Oxford Univ. Press, New York: 62 and 231.
PEREZ-TAMAYO, R. (1961): Mechanisms of Disease. W. B. Saunders Co., Philadelphia: 171 and 188.
QUIN, J. I. (1933): The effect of surgical obstruction of the normal bile flow. Onderstespoort J. vet. Sci Anim. Ind., 1: 505-526.
SAMS, M. W. Jr (1966): The experimental production of drug phototoxicity in Guinea pigs. II Using artificial light sources. Arch. Derm., 94: 773-777.
SCHMID, R. (1966): The prophyrias. In: The Metabolic Basis of Inherited Disease. 2nd Ed. McGraw-Hill Book Co.
SLATER, T. F. and RILEY, P. A. (1966): Photosensitization and lysosomal damage. Nature (Lond.), 209: 151-154.
TOBIAS, G. (1964): Congenital porphyria in a cat. J. Amer. vet. med. Ass., 145: 462-463.

Chapter 2, Section 3: Effects of Light and Temperature on Certain Physiological Functions of Shrews (M. GĘBCZYŃSKI)

BOROWSKI, S. (1964): Moult of shrews (*Sorex* L.) under laboratory conditions. Acta theriol., 8: 125-135.
BUCKNER, C. H. (1964): Metabolism, food capacity, and feeding behavior in four species of shrews. Canad. J. Zool., 42: 259-279.
CROWCROFT, P. (1964): Note on the sexual maturation of shrews (*Sorex araneus* Linnaeus, 1758) in captivity. Acta theriol., 8: 89-93.
DEHNEL, A. (1949): Studies on the genus *Sorex* L. Ann. Univ. M. Curie-Skłodowska, Chapt. 4: 17-102 (In Polish with English abstr.).
DRYDEN, G. L., GÉBCZYÑSKI, M. and DOUGLAS, E. L. (1974): Oxygen consumption by Nursling and adult musk shrews. Acta theriol., 19: 453-461.
GĘBCZYŃSKI, M. (1965): Seasonal and age changes in the metabolism and activity of *Sorex araneus* Linnaeus, 1758. Acta theriol., 10: 303-331.
GĘBCZYŃSKI, M. (1971): The rate of metabolism of the lesser shrew. Acta theriol., 16: 329-339.
HYVÄRINEN, H. (1968): On the seasonal variation of the activity of alkaline phosphatase in the kidney of the bank vole (*Clethrionomys glareolus* Schr.) and the common shrew (*Sorex araneus* L.). Aquilo, 6: 7-11.
HYVÄRINEN, H. (1969): Seasonal variation and distribution of alkaline phosphatase and glucose-6-phosphatase activity in the liver of the common shrew (*Sorex araneus* L.) and of the bank vole (*Clethrionomys glareolus* Schr.). Aquilo, 9: 44-49.
HYVÄRINEN, H. and PASANEN, S. (1972): Seasonal changes in the cytochrome c content of some tissues in three small mammals active in winter. J. Zool. (Lond.), 170: 63-67.
JUDIN, B. S. (1971): Insectivorous Mammals of Siberia. Publ. House "Nauka", Novosibirsk, 170 pp.
MALZAHN, E. (1974): Tissue metabolism in the Common shrew and the bank vole. Acta theriol., 19: 301-314.
MARTINSEN, D. L. (1969): Energetics and activity patterns of short-tailed shrews (*Blarina*) on restricted diets. Ecology, 50: 505-510.
MEZHZHERIN, V. A. (1964): Dehnel's phenomenon and its possible explanation. Acta theriol., 8: 95-114 (In Russian with English and Polish abstr.).

MORRISON, P., RYSER, F. A. and DAWE, A. R. (1959): Studies on the physiology of the masked shrew *Sorex cinereus*. Physiol. Zoöl., 32: 256-271.

NEAL, C. M. and LUSTICK, S. I. (1973): Energetics and evaporative water loss in the short-tailed shrew *Blarina brevicauda*. Physiol. Zoöl., 46: 180-185.

PASANEN, S. (1969): On the seasonal variation of the weight and of the alkaline phosphatase activity of the brown fat in the common shrew (*Sorex araneus* L.). Aquilo, 8: 36-43.

PEARSON, O. P. (1948): Metabolism of small mammals, with remarks on the lower limit of mammalian size. Science, 108: 44.

PUCEK, Z. (1964): Morphological changes in shrews kept in captivity. Acta theriol., 8: 137-166.

PUCEK, Z. (1965): Seasonal and age changes in the weight of internal organs of shrews. Acta theriol., 10: 369-438.

PUCEK, Z. (1970): Seasonal and age change in shrews as an adaptive process. Symp. zool. Soc. (Lond.), 26: 189-207.

WOŁK, E. (1974): Variations of the haematological values of the shrews. Acta theriol., 19: 315-346.

CHAPTER 3

EFFECTS OF HIGH ALTITUDE ON ANIMALS

Chapter 3, Section 1: Domestic Animals at High Altitude (R. F. GROVER)

ALEXANDER, A. F. and JENSEN, R. (1963): Pulmonary vascular pathology of high altitude-induced pulmonary hypertension in cattle. Amer. J. vet. Res., 24: 1112-1121.

ALEXANDER, J. K., HARTLEY, L. H., MODELSKI, M. and GROVER, R. F. (1967): Reduction in stroke volume during exercise in man following ascent to 3,100 m altitude. J. appl. Physiol. 23: 849-858.

BANCHERO, N., SIME, F., PENALOZA, D., CRUZ, J., GAMBOA, R. and MARTICORENA, E. (1966): Pulmonary pressure, cardiac output, and arterial oxygen saturation during exercise at high altitude and at sea level. Circulation, 33: 249-262.

BANCHERO, N., GROVER, R. F. and WILL, J. A. (1971): Oxygen transport in the llama. Respir. Physiol., 13: 102-115.

BANCHERO, N. and GROVER, R. F. (1972): Effect of different levels of simulated altitude on O_2 transport in llama and sheep. Amer. J. Physiol., 222: 1239-1245, 1972.

BISGARD, G. E., RUIZ, A. V., GROVER, R. F. and WILL, J. A. (1973): Ventilatory control in the Hereford calf. J. appl. Physiol., 35: 220-226.

BISGARD, G. E., RUIZ, A. V., GROVER, R. F. and WILL, J. A. (1974): Ventilatory acclimatization to 3400 m altitude in the Hereford calf. Respir. Physiol., 21: 271-296.

BOUVEROT, P., CANADAS, V. and LIBERT, J. P. (1973): Role of the arterial chemoreceptors in ventilatory adaptation to hypoxia of awake dogs and rabbits. Respir. Physiol., 17: 209-219.

BROOKS, J. G., III, and TENNEY, S. M. (1968): Ventilatory response of llama to hypoxia at sea level and high altitude. Respir. Physiol. 5: 269-278.

BYRNE-QUINN, E., SODAL, I. E. and WEIL, J. V. (1972): Hypoxic and hypercapnic ventilatory drives in children native to high altitude. J. appl. Physiol., 32: 44-46.

CUBA, A. (1956): Estudio comparativa del mal de montaña. Anal. Fac. Med. (Lima), 39: 1104-1127.

DAWSON, T. J. and EVANS, J. V. (1962): Haemoglobin and erythrocyte potassium types in sheep and their influence on oxygen dissociation and haemoglobin denaturation. Aust. J. biol. Sci., 15: 371-378.

DAWSON, A. and GROVER, R. F. (1974): Regional lung function in natives and longterm residents at 3,100 m altitude. J. appl. Physiol., 36: 294-298.

DILL, D. B. and EVANS, D. S. (1970): Report barometric pressure. J. appl. Physiol., 29: 914-916.

GILFILLAN, R. S., CUTHBERTSON, E. M., HANSEN, J. T. and PACE, N. (1967): Surgical excision of the canine carotid bidoes and denervation of the aortic bodies. J. surg. Res., 7: 457-463.

GROVER, R. F. and REEVES, J. J. (1963): Pathophysiologic correlations in experimental pulmonary hypertension. Mem. IV Congr. Mundial de Cardiologia, Mexico, Galve, I-B: 165-169.

GROVER, R. F., REEVES, J. T., WILL, D. H. and BLOUNT, S. G., Jr. (1963). Pulmonary vasoconstriction in steers at high altitude. J. appl. Physiol., 18: 567-574.

GROVER, R. F. (1968): The high altitude resident of North America. Scientia (Milano), 103: 1-17.

GROVER, R. F., LUFSCHANOWSKI, R. and ALEXANDER, J. K. (1970): Decreased coronary blood flow in man following ascent to high altitude. Advanc. Cardiol., 5: 72-79.

GUENTER, C. A., JOERN, A. T., SHURLEY, J. T. and PIERCE, C. M. (1970): Cardiorespiratory and metabolic effects in men on the South Polar plateau. Arch. int. Méd. exp., 125: 630-637.

HANSEN, J. E., HARTLEY, L. H. and HOGAN, R. P., III (1972): Arterial oxygen increase by high-carbohydrate diet at altitude. J. appl. physiol., 33: 441-445.

HORNBEIN, T. F. and SORENSEN, S. C. (1969): Ventilatory response to hypoxia and hypercapnia in cats living at high altitude. J. appl. Physiol., 27: 834-836.
KELLOGG, R. H. (1963): The role of CO_2 in altitude acclimatization. In: The Regulation of Human Respiration. J. C. Cunningham and B. B. Lloyd (eds). Davis, Philadelphia: 379-395.
KLAUSEN, K. (1966): Cardiac output in man at rest and work during and after acclimatization to 3,800 m. J. appl. Physiol. 21: 609-616.
McMURTRY, I. F., FRITH, C. H. and WILL, D. H. (1973): Cardiopulmonary responses of male and female swine to simulated high altitude. J. appl. Physiol. 35: 459-462.
MINES, A. H. and SORENSEN, S. C. (1970): Ventilatory responses of awake normal goats during acute and chronic hypoxia. J. appl. Physiol., 28: 826-831.
MONGE-M., C. (1948): Acclimatization in the Andes. Johns Hopkins Press: Baltimore, 36-38.
PENALOZA, D. and SIME, F. (1971): Chronic cor pulmonale due to loss of altitude acclimatization (chronic mountain sickness). Amer. J. Med., 50: 728-743.
REEVES, J. T., GROVER, E. B. and GROVER, R. F. (1963a): Pulmonary circulation and oxygen transport in lambs at high altitude. J. appl. Physiol., 18: 560-566.
REEVES, J. T., GROVER, E. B. and GROVER, R. F. (1963b): Circulatory responses to high altitude in the cat and rabbit. J. appl. Physiol., 18: 575-579.
ROY, S. B. and SINGH, I. (1963): Acute mountain sickness. In: Biomedicine Problems of High Terrestrial Elevations. A. H. Hegnauer (ed.). U.S. Army Res. Inst. environm. Med. Natick, Mass.: 32.
RUIZ, A. V., BISGARD, G. E., TYSON, I. B., GROVER, R. F. and WILL, J. A. (1974): Regional lung function in calves during acute and chronic pulmonary hypertension. J. appl. Physiol., 37: 384-391.
SINGH, I. and ROY, S. B. (1969): High altitude pulmonary edema. In: Biomedicine Problems of High Terrestrial Elevations. A. H. Hegnauer (ed.). U.S. Army Res. Inst. Environm. Med. Natick, Mass.: 108-120.
SMITH, T. C. (1969): Toward better measurement. Anesthesioly, 30: 125-126.
SORENSEN, S. C. and SEVERINGHAUS, J. W. (1968a): Respiratory sensitivity to acute hypoxia in man born at sea level living at high altitude. J. appl. Physiol., 25: 211-216.
SORENSEN, S. C. and SEVERINGHAUS, J. W. (1968b): Irreversible respiratory insensitivity to acute hypoxia in man born at high altitude. J. appl. Physiol., 25: 217-220.
SORENSEN, S. C. and MINES, A. H. (1970): Ventilatory responses to acute and chronic hypoxia in goats after sinus nerve section. J. appl. Physiol., 28: 832-835.
SPÖRRI, H. and SCHLATTER, C. (1959): Blutdruckhöhungen im Lungenkreislauf. Schweiz. Arch. Tierheilk., 101: 525-541.
TUCKER, A., McMURTRY, I. F., REEVES, J. T., ALEXANDER, A. F., WILL, D. H. and GROVER, R. F. (1975): Lung vascular smooth muscle, a determinant of pulmonary hypertension at high altitude. Amer. J. Physiol., 228: 761-762.
VALDIVIA, E. (1958): Total capillary bed in striated muscle of guinea pigs native to the Peruvian mountains. Amer. J. Physiol., 194: 585-589.
VOGEL, J. A. and HANNON, J. P. (1966): Cardiovascular and metabolic responses of dogs to exercise at high altitude. J. appl. Physiol., 21: 1595-1601.
WEIDMAN, W. H., TITUS, J. L. and SHEPHERD, J. T. (1965): Effect of hypoxia on the pulmonary circulation of cats. Proc. Soc. exp. Biol., 118: 1158-1164.
WEIL, J. V., BYRNE-QUINN, E., SODAL, I. E., FRIESEN, W. O., UNDERHILL, B., FILLEY, G. F. and GROVER, R. F. (1970): Hypoxic ventilatory drive in normal man. J. clin. Invest., 49: 1061-1072.
WEIL, J. V., BYRNE-QUINN, E., SODAL, I. E., FILLEY, G. F. and GROVER, R. F. (1971): Acquired attenuation of chemoreceptor function in chronically hypoxic man at high altitude. J. clin. Invest., 50: 186-195.
WEIR, E. K., TUCKER, A., REEVES, J. T., WILL, D. H. and GROVER, R. F. (1974): The genetic factor influencing pulmonary hypertension in cattle at high altitude. Cardio vasc. Res., 8: 745-749.
WEST, J. B., LAHIRI, S., GILL, M. B., MILLEDGE, J. S., PUGH, L. G. C. D. and WARD, M. P. (1962): Arterial oxygen saturation during exercise at high altitude. J. appl. Physiol. 17: 617-621.

WILL, D. H., ALEXANDER, A. F., REEVES, J. T. and GROVER, R. F. (1962): High altitude-induced pulmonary hypertension in normal cattle. Circulat. Res., 10: 172-177

Chapter 3, Section 2: Altitude and Exercise Effects on Animals (P. D. ALTLAND)

ALTLAND, P. D. (1961): Altitude tolerance of chickens and pigeons. J. appl. Physiol., 16: 141-143.

ALTLAND, P. D. and HIGHMAN, B. (1963): Effects of age and exercise tolerance in rats. Physiological Effects of High Altitude. Pergamon Press, Oxford/New York, 301-309.

ALTLAND, P. D., HIGHMAN, B. and GARBUS, J. (1964): Exercise Training and Altitude Tolerance in Rats: blood, tissue, enzyme and isoenzyme changes. Aerospace Med., 35: 1034-1039.

ALTLAND, P. D., BRUBACH, H. F., PARKER, M. C. and HIGHMAN, B. (1967): Blood gases and acid-base values of unanesthetized rats exposed to hypoxia. Amer. J. Physiol., 212: 142-148.

ALTLAND, P. D., HIGHMAN, B. and DIETER, M. P. (1969): Reduced exercise performance of rats at sea level after altitude acclimatization: Changes in serum enzymes, glucose, corticosterone and tissue structure. Int. J. Biometeor., 13: 173-181.

ALTLAND, P. D. and HIGHMAN, B. (1971): Effects of polycythemia and altitude hypoxia on rat heart and exercise tolerance. Amer. J. Physiol., 221: 388-393.

BALKE, B. (1960): Work capacity at altitude. Science and Medicine in Sports. W. R. Johnson (ed.). Harper and Brothers Publ., New York: 339-347.

BALKE, B., NAGLE, F. J. and DANIELS, J. (1965): Altitude and maximum performance in work and sports activity. J. Amer. med. Ass., 194: 176-179.

BALKE, B., FAULKNER, J. A. and DANIELS, J. T. (1966): Maximum performance capacity at sea level and at moderate altitude before and after training at altitude. Schweiz. Z. Sportsmed., 14: 106-116.

BERT, P. (1943): Barometric Pressure. Columbus Book Co., Columbus, 57, 159, 686-688, 976-978. (Transl.).

DENZER, H. W. (1950): Comparative Altitude Physiology of Animals. German Aviation Medicine, 1, U.S. Gvt. Printing Off, (Wash). Supp. Doc.: 331, 337-338.

DILL, D. B. (1964): Handbook of Physiology. Sect. 4, Adaptation to the Environment. Amer. physiol. Soc. (Wash.): 1-1015.

FRAPS, R. M. and OLSEN, M. W. (1950): Survival of day-old chicks under conditions of simulated high altitude transport. Poultry Science, 29: 829-832.

FREGLY, M. J. (1970): Comments on Cross Adaptation. In: Physiology, Environment and Man. D. H. K. Lee and D. Minard (eds.). Academic Press, New York: 170.

GARBUS, J., HIGHMAN, B. and ALTLAND, P. D. (1964): Serum enzymes and lactic dehydrogenase isoenzymes after exercise and training in rats. Amer. J. Physiol., 207: 467-472.

HALE, H. B. (1970): Cross-Adaptation. In: Physiology, Environment and Man. D. H. K. Lee and D. Minard (eds.). Academic Press, New York, N.Y.: 158.

HALL, F. G., DILL, D. B. and BARRON, E. S. G. (1936): Comparative Physiology in High Altitudes. J. cell comp. Physiol., 8: 301-313.

HECHT, H. H., KUIDA, H., LANGE, R. L., THORNE and BROWN, A. M. (1962): Brisket Disease. II Clinical features and hemodynamic observations in altitude-dependent right heart failure of cattle. Amer. J. Med., 32: 171-183.

HIGHMAN, B. and ALTLAND, P. D. (1963): Effect of exercise and training on serum enzymes and tissue change in rats. Amer. J. Physiol., 205: 162-166.

HOCH, R. J. (1964): Animals in high altitudes: reptiles and amphibians. In: Handbook of Physiology. Sect. 4, Adaptation to the Environment. D. B. Dill (ed.). Amer. physiol. Soc. (Wash.): 841-842.

KLAUSEN, K., ROBINSON, S., MICAHEL, E. D. and MYRHE, L. G. (1966): Effect of high altitude on maximal working capacity. J. appl. Physiol., 21: 1191-1194.

MONGE, C. (1948): Acclimatization in the Andes. The Johns Hopkins Press, Baltimore, Md.: 2-3.

PIIPER, J., CERRETELLI, P., CUTTICA, F. and MANGILI, F. (1966): Energy metabolism and circulation in dogs exercising in hypoxia. J. appl. Physiol., 21, 4: 1143-1149.
PUGH, L. G. C. E. (1958): Muscular exercise on Mount Everest. J. Physiol. (Lond.), 141: 233-261.
ROSKAMM, H., LANDRY, F., SAMEK, L., SCHLAGER, M., WEIDEMANN, H. and REINDELL, H. (1969): Effects of standarized ergometric training program at three different altitudes. J. appl. Physiol., 27: 840-848.
VELÁSQUEZ, T. (1966): Acquired acclimatization: to sea level. In: On Life at High Altitudes. Pan Amer. Hlth Org., sci. Publ. No. 140 (Wash.), 58-63.
VAN LIERE, E. J. and STICKNEY, J. C. (1963): Hypoxia. Univ. of Chicago Press, Chicago, Ill.: 381.

Chapter 3, Section 3: Physiological Effects of Altitude in Wild Mammals (P. R. MORRISON)

ALTMAN, P. L. and DITTMER, D. S. (1961): Blood and other body fluids, Fed. Amer. Soc. exp. Biol. (Wash.): 191.
ASTE-SALAZAR, H. and MORRISON, P. R. (1962): Paper electrophoresis and alkali denaturation of hemoglobins from some Peruvian rodents. Comp. Biochem. Physiol., 6: 257-267.
BULLARD, R. W. and KOLLIAS, J. (1966): Functional characteristics of two high-altitude mammals. Fed. Proc., 25: 1288-1292.
BULLARD, R. W. (1972): Vertebrates at altitudes. In: Physiological Adaptations: Desert and Mountain. M. K. Yousef, S. M. Horvath and R. M. Bullard (ed.). Academic Press, New York: 209-225.
COE, M. J. (1967): The Ecology of the Alpine Zone of Mount Kenya. Monogr. Biol., XVIII. W. Junk, The Hague.
DILL, D. B. (1938): Life, Heat and Altitude. Harvard Univ. Press, Cambridge.
HALL, F. G., DILL, D. B. and BARRON, E. S. G. (1936): Comparative physiology in high altitudes. J. cell. comp. Physiol., 8: 301-313.
HALL, F. G. (1966): Minimal utilizable oxygen and the oxygen dissociation curve of blood of rodents. J. appl. Physiol., 21: 375-378.
HART, J. S. (1957): Climatic and temperature induced changes in the energetics of homeotherms. Rev. Canad. Biol., 16: 133-174.
HART, J. S. (1971): Rodents. In: Comparative Physiology of Thermoregulation. 2. G. C. Whittow (ed.). Academic Press, New York.
HOCK, R. J. (1964): Physiological responses of deer mice to various native altitudes. In: Physiological Effects of High Altitude. W. H. Weihe (ed.). Pergamon Press, London: 59-72.
KALABUKOV, N. (1937): Some physiological adaptations of the mountain and plain forms of the wood mouse (*Apodemus sylvaticus*) and of other species of mouse-like rodents. J. Anim. Ecol., 6: 254-274.
LENFANT, C., TORRANCE, J., ENGLISH, E., FINCH, C. A., REYNAFARJE, C., RAMOS, J. and FAURA, J. (1968): Effect of altitude on oxygen binding by hemoglobin and on organic phosphate levels. J. clin. Invest., 47: 2652-2656.
LENFANT, CLAUDE (1973): High Altitude Adaptation in Mammals. Amer. Zool., 13: 447-456.
MONASTIRSKY, O. A. and POLOVINKINA, P. A. (1970): In: Proc. Symp. Adaptation of Animals and Man to Environmental Extremes. X. High altitude and hypoxia, Acad. Sci. USSR, Novosibirsk: 81-84.
MORRISON, P. R. and ELSNER, R. (1962): Influence of altitude on heart and breathing rates on some Peruvian rodents. J. appl. Physiol., 17: 467-470.
MORRISON, P. R., KERST, K. and ROSENMANN, M. (1963): Hematocrit and hemoglobin levels in some Chilean rodents from high and low altitude. Int. J. Biometeor., 7: 45-50.
MORRISON, P. R., KERST, K., REYNAFARJE, C. and RAMOS, J. (1963): Hematocrit and hemoglobin levels in some Peruvian rodents from high and low altitude. Int. J. Biometeor., 7: 51-58.

MORRISON, P. R. (1964): Wild animals at high altitudes. Symp. zool. Soc. (Lond.), 13: 49-55.
MORRISON, P. R. (1968): Responses of Mountaineers to multiple stressors. Arch. environm. Hlth, 17: 589-602.
PEARSON, O. P. (1951): Mammals in the highlands of Southern Peru. Bull. Mus. comp. Zool. (Harv.), 106: 117-174.
RAND, P. W., NORTON, J. M., BARKER, N. D., LOVELL, M. D. and AUSTIN, W. J. (1973): Responses to graded hypoxia at high and low 2,3-diphosphoglycerate concentrations. J. appl. Physiol., 34: 827-832.
REYNAFARJE, B. and MORRISON, P. R. (1962): Myoglobin levels in some tissues from wild Peruvian rodents native to high altitude. J. biol. Chem., 237: 2861-2864.
ROBERTS, J. C., HOCK, R. J. and SMITH, R. E. (1966): Seasonal metabolic responses of deer mice (*Peromyscus*) to temperature and altitude. Fed. Proc., 25: 1275-1285.
ROSENMANN, M. and MORRISON, P. R. (1974a): Physiological responses to hypoxia in the tundra vole. Amer. J. Physiol., in press.
ROSENMANN, M. and MORRISON, P. R. (1974b): Maximum oxygen consymption and heat loss facilitation in small homeotherms by He-O_2. Amer. J. Physiol., in press.
ROSENMANN, M. and MORRISON, P. R. (1974c): Metabolic response of highland and lowland rodents to simulated high altitude and cold. Comp. Biochem. Physiol., submitted.
SCHMIDT-NIELSEN, K. and LARIMER, J. L. (1958): Oxygen dissociation curves of mammalian blood in relation to size. Amer. J. Physiol., 195: 424-428.
SEGREM, N. P. and HART, J. S. (1967a): Oxygen supply and performance in *Peromyscus*: metabolic and circulatory responses to exercise. Canad. J. Physiol. Pharmacol., 45: 531-542.
SEGREM, N. P. and HART, J. S. (1967b): Oxygen supply and performance in *Peromyscus*: comparison of exercise with cold exposure. Canad. J. Physiol. Pharmacol., 45: 543-549.
SWAN, L. W. (1961): The ecology of the high Himalayas. Sci. Amer., 205, 4: 68-78.
TIMIRAS, P. S., KRUM, A. A. and PACE, N. (1957): Body and organ weights of rats during acclimatization to an altitude of 12,470 feet. Amer. J. Physiol., 191: 598-604.
WOLLASTON, A. F. R. (1922): Natural history. In: Mount Everest, The Reconnaissance. Howard-Bury et al. (eds). W. T. Longmans Green & Co., New York: 281-303, 344-350.

CHAPTER 4

EFFECTS OF NOISE STRESS ON LABORATORY ANIMALS

Chapter 4, Section 1: Endocrine and Neuro-Histochemical Aspects of Noise Stress in Laboratory Animals (A. ANTHONY)

ANTHONY, A and ACKERMAN, E. (1955): Effect of noise on eosinophil levels of audiogenic-seizure-susceptible and seizure-resistant mice. J. acoust. Soc. Amer., 27: 1150-1153.

ANTHONY, A. and ACKERMAN, E. (1955): Effect of noise on the blood eosinophil levels and adrenals of mice. J. acoust. Soc. Amer., 27: 1144-1149.

ANTHONY, A. and ACKERMAN. E. (1957): Biological Effects of Noise in Vertebrate Animals. Wright Air Developm. Center, Techn. Rep. 57-647, Wright Patterson Air Force Base, Dayton Ohio, 118 pp.

ANTHONY, A. and ACKERMAN, E. (1959): Stress Effects of Noise in Vertebrate Animals. Wright Air Developm. Center, Techn. Rep. 58-622, Wright Patterson Air Force Base, Dayton, Ohio, 49 pp.

ANTHONY, A. and ACKERMAN, E. (1959): Noise stress in laboratory rodents. I. Behavioral and endocrine responses of mice, rats and guinea pigs. J. acoust. Soc. Amer., 31: 1430-1437.

ANTHONY, A. (1973): Azure B-RNA changes in the adrenal and cerebral cortex of rats exposed to intense noise. Fed. Proc., 32: 2093-2097.

ANTICAGLIA, J. R. and COHEN, A. (1970): Extra auditory effects of noise as a health hazard. Amer. industr. Hyg. Ass. Quart., 31: 277-281.

BIRÓ, J., SZOKOLAI, V. and KOVICH, A. G. B. (1959): Some effects of sound stimuli on the pituitary-adrenocortical system. Acta Endocr., 31: 542-552.

BUTTE, J. C., KAKIHANA, R. and NOBLE, E. P. (1972): Rat and mouse brain corticosterone. Endocrinology, 90: 1091-1099.

CHAI, C. K. and DICKIE, M. M. (1966): Endocrine variations. In: Biology of the Laboratory Mouse. E. L. Green (ed.). McGraw Hill, New York: 387.

CRAWSHAW, J. E. (1973): Cytophotometric Analysis of Azure B-RNA Changes in Auditory Cortical Neurons of Rats Exposed to Intense Noise, Ph.D. Thesis, The Pennsylvania State Univ., Univ. Park, 73 pp.

DE WIED, D. and WEIJNEN, J. A. W. M. (1970): Pituitary, Adrenal and the Brain. Progress in Brain Research, 32, Elsevier Publ. Co., New York 357 pp.

FULLER, J. L. and WIMER, R. E. (1966): Neural, sensory and motor functions. Biology of the Laboratory Mouse. E. L. Green (ed.). McGraw Hill, New York, 609-628.

GANONG, E. F. and MARTINI, L. (1969): Frontiers in Neuroendocrinology. Oxford Univ. Press, New York, 442 pp.

GEBER, W. F., ANDERSON, T. A. and VAN DYNE, G. (1966): Physiological responses of the albino rat to chronic noise stress. Arch. environm. Hlth, 12: 751-754.

HAMBERGER, C. A. and HYDEN, H. (1945): Cytochemical changes in the cochlear ganglion caused by acoustic stimulation and trauma. Acta oto-laryng., 61: 1-89.

HAMMER, G. (1956): A quantitative study of shock wave effects on spiral ganglion cells. Acta oto-laryng., S. 127: 1-64.

HENKIN, R. K. and KNIGGE, M. (1963): Effect of sound on the pituitary adrenal axis. Amer. J. Physiol., 204: 710-714.

JAMES, V. H. T. (1968): Recent Advances in Endocrinology. Little, Brown and Company, Boston, 334 pp.

JENSEN, M. M. and RASMUSSEN, A. F. (1963): Stress and susceptibility to viral infections. I. Response of adrenals, liver, thymus, spleen and peripheral leucocyte counts to sound stress. J. Immunol., 90: 17-20.

JENSEN, M. M. (1969): Changes in leukocyte counts associated with various stressors. J. reticuloendoth. Soc., 6: 457-465.

KRYTER, K. D. (1970): The Effects of Noise on Man. Academic Press, New York, 633 pp.

SACKLER, A. M., WELTMAN, A., BRADSHAW, M. and JUSTSHUK, R. (1959): Endocrine changes due to auditory stress. Acta endocr. (Kbn), 31: 405-418.
SINGH, K. B. and RAO, P. S. (1970): Studies on polycystic ovaries of rats under continuous auditory stress. Amer. J. Obstet. Gynec., 108: 557-564.
SINGH, K. B. (1972): Effect of sound on the female reproductive system. Amer. J. Obstet. Gynec., 112: 981-991.
TALWAR, G. P. (1969): Brain RNA. In: Biological Nasis of Medicine, 5. E. E. Bittar and N. Bittar (eds). Academic Press, New York: 77-109.
TARMAS, J. and GRZESIK, J. (1963): Morphological and histochemical changes in cerebral nerve cells of guinea pigs evoked by loud noise. Arch. Immunol. Ther. exp. 11: 664-670.
VERNIKOS-DANELLIS, J. (1965): The regulation of the synthesis and release of ACTH Vitam. and Horm. Res. 23: 97-152.
WELCH, B. L. and WELCH, A. S. (1970): Physiological Effects of Noise. Plenum Press, New York, 365 pp.
WOODBURY, D. M. and VERNADAKIS, A. (1967): Influence of hormones on brain activity. In: Neuroendocrinology, 2. L. Martini and W. F. Ganong (eds). Academic Press, New York: 335-375.
YATES, F. E., RUSSELL, S. M. and MARAN, J. W. (1971): Brain adenohypophysial communication in mammals. Ann. Rev. Physiol., 33: 393-444.
ZONDEK, B. and TAMARI, I. (1964): Effect of audiogenic stimulation of genital function and reproduction. Acta endocr., (Kbh.), Suppl. 90: 227-234.

CHAPTER 5

EFFECTS OF ELECTRIC, MAGNETIC AND ELECTRO-MAGNETIC FIELDS AND IONIZATION ON ANIMALS

Chapter 5, Section 1: Biological Effects of Ionization of the Air on Animals (A. P. KRUEGER)

D'ABILARD, T. F. (1752): Letter to Acad. Sci. (3).
ANDERSEN, I. (1969): The effect of ionized air on ciliary activity in trachea. Biometeorology, 4.II. S. W. Tromp. and W. H. Weihe (eds). Proc. 5th int. biometeor. Congr. Swets & Zeitlinger, Amsterdam,: 135.
BERTHOLON, L'ABBÉ P. (1780): De l'Électricité du Corps Humain dans l'État de Santé et de Maladie. Bernuset, Lyon.
CZERMAK, P. (1901): Über Elektrizitätszerstreuung bei Föhn. Mitt. wien. Akad. Wiss., 27.
DANON, A. and SULMAN, F. G. (1969): Ionizing effect of winds of ill repute on serotonin metabolism. Biometeorology, 4, II. J. W. Tromp and W. H. Weihe (eds). Proc. 5th int. biometeor. Congr. Swets & Zeitlinger, Amsterdam: 135-136.
ELSTER, J. and GEITEL, H. (1899): Über die Existenz elektrischer Ionen in der Atmosphere. Terrestr. Mag., 4.
FRANKLIN, B. (1752): Phil. Trans., 47: 289.
GARATTINI, S. and VALZELLI, L. (1965): Serotonin. Elsevier Publ. Co., Amsterdam: 181-182.
GILBERT, G. O. (1971): Effect of negative air ions upon emotionality and brain serotonin levels in isolated rats. Thesis, Pacific Lutheran Univ.
GUALTIEROTTI, R. (1965): The influence of ionization on endocrine glands. In: Aero-ionotherapy. R. Gualtierotti, I. H. Kornblueh and C. Sirtori (eds). Carlo Erba Foundation, Milan, Italy: 56-60.
GUILLERM, R., BADRE, R. and HEE, J. (1966): Effects of light atmospheric ions on the ciliary activity of sheep and rabbit tracheal mucosa in vitro. C. R. Acad. Sci. (Paris), 262: 669-W71.
KRUEGER, A. P. and SMITH, R. F. (1960a): The biological mechanisms of air ion action. I. 5-hydroxytryptamine as the endogenous mediator of positive air ion effects on the mammalian trachea. J. gen. Physiol., 43: 533-540.
KRUEGER, A. P. and SMITH, R. F. (1960b): The biological effects of air ion action. II. Negative air ion effects on the concentration and metabolisms of 5-hydroxytryptamine in the mammalian respiratory tract. J. gen. Physiol., 44: 269-276.
KRUEGER, A. P., KOTAKA, S. and ANDRIESE, P. C. (1966): Studies on the biological effects of gaseous ions. A review. Spec. Monogr. Ser., 1, Biometeor. Res. Centre, Leiden.
KRUEGER, A. P. and LEVINE, H. B. (1967): The effect of unipolar positively ionized air on the course of Coccidioidomycosis in mice. Int. J. Biometeor., 11: 279-288.
KRUEGER, A. P., ANDRIESE, P. C. and KOTAKA, S. (1968): Small air ions: Their effect on blood levels of serotonin in terms of modern physical theory. Int. J. Biometeor., 12: 225-239.
KRUEGER, A. P. and KOTAKA, S. (1969): The effect of air ions on the brain level of serotonin in mice. Int. J. of Biometeor., 13: 61-73.
KRUEGER, A. P., KOTAKA, S., REED, E. J. and TURNER, S. (1970): The effect of air ions on bacterial and viral pneumonia in mice. Int. J. Biometeor., 14: 247-260.
KRUEGER, A. P., KOTAKA, S. and REED, E. J. (1971): The course of experimental influenza in mice maintained in high concentrations of small negative air ions. Int. J. Biometeor., 15, 1: 5-10.
KRUEGER, A. P. and REED, E. J. (1972): Effect of the air ion environment on influenza in the mouse. Int. J. Biometeor., 16, 3: 209-232.
KRUEGER, A. P., REED, E. J., DAY, M. B. and LAZAR, K. A. (1974): Further observations on the effect of air ions on influenza in the mouse. Int. J. Biometeor., 18, 1: 46-56.

KONG, Y. M., LEVINE, H. B., MADIN, S. H. and SMITH, C. E. (1964): Fungal multiplication and histopathological changes in vaccinated mice infected with Coccidioides immites. J. Immunol., 92: 779-790.

LEMMONIER, L. G. (1752): Observations sur l'électricité de l'air. Mem. Acad. Sci., 2, 233.

OLIVEREAU, J. (1971): Incidences psychophysiologiques de l'ionisation atmosphérique. Dissertation, Univ. Paris: 1-187.

PALTI, V., DENOUR, E. and ABRAMOV. A. (1966): The effect of atmospheric ions on the respiratory system of infants. Pediatrics, 38: 404-411.

ROBINSON, N. and DIRNFELD, F. S. (1963): Ionization state of the atmosphere as a function of the meteorological elements and of various sources of ions. Int. J. Biometeor., 6: 101-110.

SULMAN, F. G. (1971): Meteorologische Frontverschiebung und Wetterfühligkeit. Ärztliche Praxis XXIII, 17, 998-999.

WILSON, C. T. R. (1901): On the ionisation of atmospheric air. Proc. roy. Soc. A., 68: 151-161.

WINSOR, T. and BECKETT, J. C. (1958): Biological effects of ionized air in man. Amer. J. phys. Med., 37: 83-89.

Chapter 5, Section 2a: Biological Effects of Magnetic Fields on Neural Behaviour (F. RUSSO and W. E. CALDWELL)

BARNWELL, F. H. and BROWN, F. A., Jr. (1964): Responses of planarians and snails. In: Biological Effects of Magnetic Fields. M. F. Barnothy (ed.). Plenum Books, New York: 263-278.

BECKER, R. O. (1962): Some observations indicating the possibility of longitudinal charge – carrier flow in the peripheral nerves. In: Biological Prototypes and Synthetic Systems. F. Bernard and M. Kare (eds). Plenum Press. New York: 36.

BEISCHER, D. F. (1965): Biomagnetics. Ann. N.Y. Acad. Sci., 134: 454-458.

BROWN, F. A., Jr. (1971): Some orientational influences of nonvisual terrestrial electromagnetic fields. Ann. N.Y. Acad. Sci., 188: 224-241.

CAHILL, L. J., Jr. (1965): The magnetosphere. Sci. Amer., 212: 58-68.

CALDWELL, W. E. and RUSSO, F. (1968): An exploratory study of the effects of an a.c. magnetic field upon the behavior of the Italian honeybee (*Apis mellifica*). J. genet. Psychol., 113: 233-252.

CASSIAMO, O. and AUDISIO, B. (1966): Some human autonomic responses to electromagnetic field activity. Minerva anest., 32: 261-264 (In Italian).

COHEN, DAVID (1967): Magnetic fields around the torso: Production by electrical activity of the human heart. Science, 156: 652-654.

COHEN, DAVID (1968): Magneto encephalography: Evidence of magnetic fields produced by alpha-rhythm currents. Science, 161: 784-786.

COHEN, DAVID and McCAUGHAN, D. V. (1972): Magnetocardiograms and their variation over the chest in normal subjects. Amer. J. Cardiol., 20: 678-685.

FRIEDMAN, HOWARD, BECKER, R. O. and BACHMAN, C. H. (1967): Effect of magnetic fields on reaction time performance. Nature (Lond.), 213, 5079: 949-950.

FRISCH, KARL VON (1955): The Dancing Bees. Harcourt, Brace Co., New York, 183 pp.

GALAMBOS, ROBERT (1961): A glia-neural theory of brain function. Proc. nat. Acad. Sci. (Wash.), 47: 129-136.

GANONG, W. F. (1969): Review of Medical Physiology. 3rd Ed. Lange med. Publ. Los Altos, Cal., 628 pp.

GENGERELLI, J. A., HOLTER, N. J. and GLASSCOCK, W. R. (1961): Magnetic fields accompanying transmission of nerve impulses in the frog's sciatic. J. Psychol., 52: 317-326.

GENGERELLI, J. A., HOLTER, N. J. and GLASSCOCK, W. R. (1964): Further observations on magnetic fields accompanying nerve transmission and tetanus. J. Psychol., 57: 201-212.

GOTTLIEB, N. D. and CALDWELL, W. E. (1967): Magnetic field effects on the compass mechanism and activity level of the snail (*Helisoma duryi endiscus*). J. genet. Psychol., 111: 85-102.

HESS, ARTHUR (1958): The fine structure of nerve cells and fibers, neuroglia, and sheaths of the ganglion chain in the cockroach (*Periplaneta Americana*). J. biophys. biochem. Cytol., 4: 731-742.
HYDÉN, H. and PIGON, A. (1960): A cytophysiological study of the functional relationship between oligodendroglial cells and nerve cells of Dieters' nucleus. J. Neurochem., 6: 57-72.
HYDÉN, H. and McEWEN, B. (1966): A glial protein specific for the nervous system. Proc. nat. Acad. Sci (Wash.), 55: 354-358.
KEETON, W. T. (1971): Magnets interfere with pigeons homing. Proc. nat. Acad. Sci. (Wash.), 68 (1); 102-106.
KHOLODOV, Y. A. (1964): Effects on the central nervous system. In: Biological Effects of Magnetic Fields. M. Barnothy (ed.). Plenum Press, New York: 196-200.
KÖHLER, WOLFGANG (1940): Dynamics in Psychology. Liveright Publ. Co., New York.
LINDAUER, M. and MARTIN, H. (1968): The earth's magnetic field affects the orientation of honeybees in the gravity field. Z. vergl. Physiol., 60: 219-243. (In German).
McCLEAVE, J. D., ROMMEL, S. A., Jr., and CATHCART, C. L. (1971): Weak electric and magnetic fields in fish orientation. Ann. N.Y. Acad. Sci., 188: 270-282.
PICTON, H. D. (1966): Some responses of Drosophilia to weak magnetic and electrostatic fields. Nature (Cond.), 211, 5046: 303-304.
RATNER, S. C. (1973): A structure responsive to magnetic fields in molluscs: Its occurrences and functions. Proc. 13th int. ethol. Conf. Washington, D.C. (Addendum).
RUSSO, FRANK and CALDWELL, W. E. (1971): Biomagnetic phenomena: Some implications for the behavioral and neurophysiological sciences. Genet. Psychol. Monogr., 84: 177-243.
SCHARDT, A. W. and OPP, A. G. (1967): Particles and fields. In: Space Science 1966. NASA pub. SP-155. U.S. Gvt Print. Off. (Wash.): 205-287.
SEIPEL, J. H. and MORROW, R. D. (1960): The magnetic fields accompanying neuronal activity: A new method for the study of the nervous system. J. Wash. Acad. Sci., 50: 1-4.
SOUTHERN, W. E. (1971): Gull orientation by magnetic cues: A hypothesis revisited. Ann. N.Y. Acad. Sci., 188: 295-310.
STUTZ, A. M. (1971): Effects of weak magnetic fields on gerbil spontaneous activity. Ann. N.Y. Acad. Sci., 188: 312-323.
SVAETICHIN, G., LAUFER, M., MITARAI, G., FATEHCHAND, R., VALLECALLE, E., and VILLEGAS, J. (1961): Gilal control of neuronal networks and receptors. In: Neurophysiologie und Psychophysik des visuellen Systems. R. Jung and H. Kornhuber (eds). Springer-Verlag, Berlin, Germany: pp. 445-456.
TASAKI, I. and CHANG, J. J. (1958): Electric response of glia cells in cat brain. Science, 128: 1209-1210.
U.S. DEPARTMENT OF THE ARMY (1959): Basic Theory and Application of Transistors. TM 11-690. Gvt Print. Off. (Wash.).
VANDER, A. J., SHERMAN, J. H. and LUCIANO, D. S. (1970): Human Physiology: The Mechanisms of Body Function. McGraw-Hill Book Co., New York/London, 610 pp.
WALKER, F. D. and HILD, W. J. (1969): Neuroglia electrically coupled to neurons. Science, 165: 602-603.
YAKOVLEVA, M. I. and MEDVEDEVA, M. V. (1972): Conditioned control of cardiac activity and respiration and morphological changes in the brain of pigeons under the action of a constant magnetic field. Zh. Vyssh. nerv. Deyat Pavlova, 22: 288-293 (In Russian).

Chapter 5, Section 2b: Biological Effects of Magnetic Fields on Animal Behaviour (M. A. PERSINGER)

BECKER, G. (1966): On the orientation of diptera according to the geomagnetic field. 3rd int. biomagn. Symp., Univ. Ill., Chicago: 9-11.
BECKER, R. O. (1972): Electromagnetic Forces and Life Processes. Techn. Review, 75, 32-38.
BRANDSCH, R. and JITARIU, P. (1970): The effect of a pulsating electromagnetic field on the development and the 1st divisions of the embryos of Tubifex tubifex. Rev. Roum. biol. Ser. Zool. 15, 6: 431-436.

BROWN, F. A., Jr. (1966): Organismic responsiveness to very weak magnetic fields. 3rd int. biomagn. Symp., Univ. of Ill., Chicago: 6-8.
DE LORGE, J. (1972): Operant behavior of rhesus monkeys in the presence of extremely low frequency-low intensity magnetic and electric fields: Experiment 1. Naval Aerosp. med. Res. Lab., NAMRL-1155, Pensacola.
FRIEDMAN, H. and CAREY, R. J. (1969): The effects of magnetic fields upon rabbit brains. Physiol. and Behav., 4: 539-541.
GRISSETT, J. D. and DELORGE, J. (1971): Central-nervous-system effects as measured by reaction time in squirrel monkeys exposed for short periods to extremely low-frequency magnetic fields. Naval Aerosp. med. Res. Lab., NAMRL-1137, Pensacola.
LAZAR, M. and BARCA, C. (1969): Influence of the electromagnetic field of small intensity and of the homogeneous field of large intensity on the ions of chicken tissue. An. Stiint. Univ. Al. I. Cuza. Iasi. Sect. II (A) Biol., 15 (1): 1-7.
MARR, M. J., RIVERS, W. K., and BURNS, C. P. (1973): The effect of low energy, extremely low frequency (ELF) electromagnetic radiation on operant behavior in the pigeon and the rat. Off. of Naval Res., Arlington.
MESZAROS, I., KUZMANN, E., PELYHE, I. and PUSZTAI, J. (1971): Central nervous effect of a permanent magnetic field. Acta physiol. Acad. Sci. hung., 39: 221-222.
NEAGA, N. and LAZAR, M. (1972): Histological changes in the thyroids of chicks after treatment with a pulsing magnetic field. Stud. Cercet. Biol. Ser. Zool., 24, 2: 119-124.
OSSENKOPP, K-P. and SHAPIRO, L. J. (1972): Effects of prenatal exposure to a 0.5 Hz low-intensity rotating magnetic field on White Peking ducklings. Amer. Zool., 12, 4: 38.
PERSINGER, M. A., LUDWIG, H. W. and OSSENKOPP, K-P. (1973): Psychophysiological effects of extremely low frequency electromagnetic fields: A review. Percept. and Motor Skills, 36: 1131-1159.
PORUMB, S. (1971): The influence of an electromagnetic field on glucose-6-phosphate dehydrogenase activity in rabbit red blood cells. Stud. Cercet. Biol. Ser. Zool., 23: 239-242.
REILLE, A. (1968): Essai de mise en évidence d'une sensibilité du pigeon au champ magnétique à l'aide d'une conditionnement nociceptif. J. Physiol. (Lond.), 60: 85-92.
TRAVKIN, M. P. (1972): Effect of a weak magnetic field on the bioelectric potentials of Tradescantia. Fiziol. Rast., 19, 2: 448-450.
TROMP, S. W. (1968): Review of the possible physiological causes of dowsing. Int. J. Parapsychol., 10: 363-391.
WILTSCHKO, W. (1968): Über den Einfluss statischer Magnetfelder auf die Zugorientierung der Rotkehlchen (*Erithacus rubecula*). Z. Tierpsychol., 25: 537-559.
YAKOVLEVA, M. I. and MEDVEDEVA, M. V. (1972): Conditioned control of cardiac activity and respiration and morphological changes in the brain of pigeons under the effect of a constant magnetic field. ZH. vyssh. nervn. Deyat. Pavlova, 22, 2: 288-293.

Chapter 5, Section 3: Biological Effects of Various Physical Parameters of the Environment on Animals (R. S. POGRUND)

BECKER, R. O. (1963): Relationship of geomagnetic environment to human biology. N.Y. St. J. Med., 63: 2215-2219.
CORRADO, A. E. and BECKETT, J. C. (1960): Air ion density and the allergic individual. Wash. St. Allergy Soc. Meeting, Richland, Wash., 1959. West. Med., March.
DUMANSKIY, Y. D. (1967): Hygienic evaluation of radio-frequency electromagnetic fields in populated areas. Vestn. Akad. med. Nauk, 8: 65-71.
FREY, A. H. and GRANDA, R. E. (1961): Human reactions to air ions. In: Proc. Int. Conf. Ioniz. Air, Philadelphia, Pa. Gen. electr. Advanc. electron. Center, Cornell Univ., Ithaca, N.Y.
FREY, A. (1962): Human auditory system response to modulated electromagnetic energy. J. appl. Physiol., 17: 689-692.
FREY A. (1971): Biological function as influenced by low-power modulated RF energy. Inst. electr. electron. Eng. Trans. Microwave Theory and Techn., M.T.T.-19, 2: 153-164.
HENDERSON, J. J., BENSON, F. B. and CALDWELL, D. E. (1973): Indoor-outdoor air pollution relationships. Environm. Protect. Ag. Off. Res. Developm., Research Triangle Park, N.C., Publ. AP-112b.

INGALLS, C. E. (1967): Sensation of hearing in electromagnetic fields. N.Y. St. J. Med., 67: 2992-2997.
KATO, K. (1964): Ions in air. Clean Air (Tokyo), 2: 48-53.
KORNBLUEH, I. H., SWOPE S. T. and DAVIS, F. K. (1971): Natural ions levels in enclosed spaces. In: Lacustrine Climatology. R. Gualtierotti and U. Solimene (eds). Proc. int. Congr., Como Italy. Publ., Camera di Commercio Industria, Artigianato e Agricoltura di Como.
KRUEGER, A. P., SMITH, R. F., HILDEBRAND, G. V. and MEYERS, C. E. (1959): Further studies of gaseous ion action on trachea. Proc. Soc. exp. Biol. (N.Y.), 102: 355-357.
KRUEGER, A. P., ANDRIESE, P. C. and KOTAKA, S. (1968): Small air ions: Their effects on blood levels of serotonin in terms of modern physical theory. Int. J. Biometeor., 12: 225-239.
MCDONALD, R. D., BACHMAN, C. H. and LORENZ, P. J. (1967): Some psychomotor and physiological tests on humans exposed to air ions, Aerospace Med., 145: 148.
NAZZARO, J. R., JACKSON, D. E. and PERKINS, L. E. (1967): Effects of ionized air on stress behavior. Med. Res. Eng., 6: 25-28.
OLIVERAU, J. M. (1973): Psychophysiological effects of ionization of the atmosphere. Medicina Termale e Climatol., 17: 10-12.
POGRUND, R. S. (1969): Biologic synergisms in rats produced by carbon monoxide and positive ions. Int. J. Biometeor., 13: 123-133.
POGRUND, R. S. (1970): Why rats choose booze. Calif. Hlth, 28: 6-7, 10.
POGRUND, R. S. and THRASHER, J. (1970): Air Pollutants as Carcinogenic and Mutagenic Agents. Project Cleaun Air. Univ. of Calif. Res. Rep., 2 (S-3): 1-17.
SOMMER, H. C. and VON GIERKE, H. E. (1964): Hearing sensations in electric fields. Aerospace Med., Sept., 834-839.
SWENSON, Jr. G. W. and COCHRAN, W. W. (1973): Radio noise from towns: measured from an airplane. Science, 181: 543-544.
VYALOV, A. M. (1967): Magnetic fields as a factor in the industrial environment. Vestn. Akad. med. Nauk, 8: 72-79.
WEIHAUPT, J. G. (1964): Geophysical biology. Bioscience, 14: 18-24.

CHAPTER 6

EFFECTS OF GRAVITATIONAL FIELDS

Chapter 6: Effects of Altered Gravitational Fields on Laboratory Animals (J. OYAMA

ABAKUMOVA, O. Yu., KUTSENKO, N. G., YERTANOV, I. D., SEMINA, Ye. M., and LERMAN, M. I. (1973): Influence of accelerations on activity of the protein-synthesizing system and RNA synthesis in the liver of rats. Space Biol. Med. (USSR), 7: 65-74.
ADEY, W. R. (1964): Effects of gravity on the functions of the central nervous system. (1964) Life Sci., 2: 267-286.
ADEY, W. R. and HAHN, P. M. (1971): Introduction-Biosatellite III results. Aerospace Med., 42: 273-280.
AMTMANN, E. and OYAMA, J. (1973): Changes in functional construction of bone in rats under conditions of simulated increased gravity. Z. Anat. Entwickl.-Gesch., 139: 307-318.
ANDRIANOVA, L. A. (1968): The effect of transverse accelerations on the neurosecretory function of the nuclei of the frontal hypothalamus. Space Biol. Med. (USSR), 2: 38-43.
BARER, A. S. (1958): The effect of centripetal acceleration on the content of acetylcholine, adrenaline, adrenalin-like substances, potassium and sodium in the blood of animals. Bull. exp. Biol. Med., 46: 56-59.
BARR, P.-O., BJURSTEDT, H., and COLERIDGE, J. C. G. (1959): Blood gas changes in the anesthetized dog during prolonged exposure to positive radial acceleration. Acta physiol. scand., 47: 16-27.
BAYBARA, V. S. (1971): Morphological changes in bronchial vessels associated with experimental hypodynamia and hypokinesia. Arkh. Anat. Gistol. Embriol., 11: 95-99.
BELAY, V. Ye., VASIL'YEV, P. V. and GLOD, G. D. (1970): Effects of some drugs on animal tolerance to extreme stress. Space Biol. Med. (USSR), 4: 110-113.
BENGELE, H. H. (1969): Water intake and urine output of rats during chronic centrifugation. Amer. J. Physiol., 216: 659-665.
BESCH, E. L., BURTON, R. R., and SMITH, A. H. (1972): Organ and body mass changes in restrained and fasted domestic fowl. Proc. Soc. exp. Biol. (N.Y.), 141: 456-459.
BIRD, J. W. C., WUNDER, C. C., SANDLER, N., and DODGE, C. H. (1963): Analysis of muscular development of mice at high gravity. Amer. J. Physiol., 204: 523-526.
BRICKER, L. A., JOHNSON, W. A., DAVIES, C. R., and DOTTORE, R. A. (1967): Effect of -7G_x acceleration on renal excretion of solutes in rabbits. Aerospace Med., 38: 35-38.
BRITTON, S. W., COREY, E. L., and STEWART, G. A. (1946): Effects of high acceleratory forces and their alleviation. Amer. J. Physiol., 146: 33-51.
BRITTON, S. W., PERTZOFF, V. A., FRENCH, C. R., and KLINE, R. F. (1947): Circulatory and cerebral changes and protective aids during exposure to acceleratory forces. Amer. J. Physiol., 150: 7-26.
BURTON, R. R., BESCH, E. L., SLUKA, S. J., and SMITH, A. H. (1967): Differential effects of chronic acceleration on skeletal muscles. J. appl. Physiol., 23: 80-84.
BURTON, R. R. and SMITH, A. H. (1969): Hematologic findings associated with chronic centrifugation. Space Life Sci., 1: 501-513.
BURTON, R. R. and SMITH, A. H. (1972): Stress and adaptation responses to repeated acute acceleration. Amer. J. Physiol., 222: 1505-1510.
BURTON, R. R. (1973): Positive ($+G_z$) acceleration tolerances of the miniature swine: application as a human analog. Aerospace Med., 44: 294-298.
DALIGCON, B. C. and OYAMA, J., (1970): In vitro stimulation of glucose uptake and utilization by diaphragm of rats exposed to chronic centrifugation. Physiologist, 13: 174.
DODGE, C. H. and WUNDER, C. C. (1963): Growth of juvenile red-eared turtles as influenced by gravitational field intensity. Nature (Lond.), 197: 922-923.

EVANS, J. W., SMITH, A. H. and BODA, J. M. (1969): Fat metabolism and chronic acceleration. Amer. J. Physiol., 216: 1468-1471.
EVANS, J. W. and BODA, J. M. (1970): Glucose metabolism and chronic centrifugation. Amer. J. Physiol., 219: 893-896.
FELLER, D. D., NEVILLE, E. D., OYAMA, J. and AVERKIN, E. G. (1965): Chemical and metabolic changes of hepatic lipids from rats exposed to chronic radial acceleration. Proc. Soc. exp. Biol. (N.Y.), 119: 522-525.
FELLER, D. D. and NEVILLE, E. D. (1966): Blood glucose and corticosterone changes accompanying altered lipid metabolism induced by exposure to acceleration stress. Proc. Soc. exp. Biol. (N.Y.), 121: 223-227.
FELLER, D. D., TALARICO, K. S. and NEVILLE, E. D. (1967): Liver regeneration in rats exposed to simulated changes in gravity. Proc. Soc. exp. Biol. (N.Y.), 124: 1079-1083.
FELLER, D. D., NEVILLE, E. D. and TALARICO, K. S. (1968): Adipose tissue in altered lipid metabolism of rats exposed to centrifugation stress. Amer. J. Physiol., 214: 1434-1437.
FLUCKIGER, E. VON (1956): Der Sauerstoffverbrauch der Ratte bei vermindertem Sauerstoffpartialdruck. Helv. physiol. pharmacol. Acta, 14: 369-381.
FOSSE, G. (1971): The radiodensity of skeletal parts in animals growing and living in a constant artificially increased gravitational field. Growth, 35: 33-53.
GELL, C. F. and CRANMORE, D. (1956): Dislocation of organs and tissues of rats exposed to acceleration stress. J. Aviat. Med., 27: 497-504.
GELL, C. F. (1971): Descriptive catalog of aerospace medical biodynamic facilities in the United States and Canada. Advisory Group Aerospace Res. and Developm. (AGARD), Publ. 150: 57-100.
GUALTIEROTTI, T., BRACCHI, F., and ROCCA, E. (1972): Orbiting Frog Otolith Experiment (OFO-A). Final Report. Piccin Medical Books, Milan, Italy.
GYURDZHIAN, A. A. (1966): The effect of prolonged acceleration on gas exchange and resistance of rats to hypoxia. Abst. 17th astronaut. Congr., Madrid, NASA TT F-10, 406.
HARRISON, M. H. (1970): Effect of prolonged centrifugation on tissue carbohydrate in fasted rats. J. appl. Physiol., 28: 771-774.
HAYMAKER, W. (1974): The effects of particle radiation on pocket mice aboard Apollo 17, in preparation.*
HENRY, J. P., BALLINGER, E. R., MAHER, P. J., and SIMONS, D. G. (1952): Animal studies of the subgravic state during rocket flight. J. Aviat. Med., 23: 421-432.
HOWARD, P. (1965): The physiology of positive acceleration. In: A Textbook of Aviation Physiology. J. A. Gillies (ed.), Pergamon Press, Oxford; 551.
IGARASHI, M. and NAGABA, M. (1967): Vestibular end organ damage in squirrel monkeys after exposure to intensive linear acceleration. 3rd Symp. Role of the Vestibular Organs in Space Exploration. Pensacola, Florida. NASA SP-152: 63-81.
JANKOVICH, J. P. (1971): Structural development of bone in the rat under earth gravity, simulated weightlessness, hypergravity and mechanical vibration. NASA Contractor Rep., NASA CR-1823; 141.
KEIL, L. C. (1969): Changes in growth and body composition of mice exposed to chronic centrifugation. Growth, 33: 83-88.
KEIL, L. C. (1973): The effect of chronic exposure to hypergravity on bone development and calcium metabolism in the rat. Doctoral thesis, Univ. Calif., Davis.
KUPRIYANOV, V. V. and PETRUKHIN, V. G. (1971): The effect of G forces on the microcirculatory system. Arkh. Anat. Gistol. Embriol., 11: 17-30.
LAWTON, R. W., GREENE, L. C., KYDD, G. H., PETERSON, L. H., and CROSBIE, R. J. (1958); Arterial blood pressure responses to G forces in the monkey. J. Aviat. Med., 29: 97-105.
LEON, H. A. and CHACKERIAN, M. J. (1968): Nutritional and hormonal factors concerned with the regulation of liver protein synthesis during acute centrifugation stress. Endocrinology, 82: 429-435.
MAGARIA, R., GUALTIEROTTI, T., and SPINELLI, D. (1958): Protection against acceleration forces in animals by immersion in water. J. Aviat. Med., 29: 433-437.
MATSYNIN, V. V. (1964): Body temperature and heat generation in white rats under the effect of radial acceleration. Physiol. J. Acad. Sci. Uk. SSR, Kiev, 10: 663-670.

* See page 291.

McCALLY, M. (Ed.) (1968): Hypodynamics and Hypogravics, The Physiology of Inactivity and Weightlessness. Academic Press, New York, pp. 306.
MEEHAN, J. P. (1971): Biosatellite 3: A physiological interpretation. Life Sci. Space Res., 9: 83-98.
MEYERS, J. W. and GODLY, J. A. (1967): Cardiovascular and renal function during total water immersion of dogs. J. appl. Physiol., 22: 573-579.
NEFEDOV, Y. G., GUROVSKIY, N. N., YEGOROV, A. D., YEGOROV, B. B., KISELEV, A. A., NOKOLAYEV, S. O., POLYAKOVA, A. P., and SVISTUNOV, I. B. (1968): Quantitative evaluation of the cardiovascular system in animals during long-term orbital flight. Space Biol. Med. (USSR), 2: 17-27.
NEVILLE, E. D. and FELLER, D. D. (1969): In vivo utilization of acetate-1-C^{14} in centrifuged rats. Aerospace Med., 40: 164-167.
OYAMA, J. and PLATT, W. T. (1964): Carbohydrate metabolism of mice exposed to simulated changes in gravity. Amer. J. Physiol., 207: 411-414.
OYAMA, J. and PLATT, W. T. (1964): Effects of deceleration of rats exposed to prolonged centrifugation. Nature (Lond.), 203: 766-767.
OYAMA, J. and PLATT, W. T. (1965): Effect of prolonged centrifugation on growth and organ development of rats. Amer. J. Physiol., 209: 611-615.
OYAMA, J. and PLATT, W. T. (1965): Metabolic alterations in rats exposed to acute acceleration stress. Endocrinology, 76: 203-209.
OYAMA, J. and PLATT, W. T. (1967): Reproduction and growth of mice and rats under conditions of simulated increased gravity. Amer. J. Physiol., 212: 164-166.
OYAMA, J. and DALIGCON, B. C. (1967): Liver glycogenesis, glycogen synthetase and adrenal responsiveness of rats exposed to acceleration stress. Endocrinology, 80: 707-713.
OYAMA, J. and ZEITMAN, B. (1967): Tissue composition of rats exposed to chronic centrifugation. Amer. J. Physiol., 213: 1305-1310.
OYAMA, J., PLATT, W. T. and HOLLAND, V. B. (1971): Deep-body temperature changes in rats exposed to chronic centrifugation. Amer. J. Physiol., 221: 1271-1277.
OYAMA, J. (1973): Effect of artificial gravity on thermoregulation, respiratory metabolism and intermediary metabolism of animals. NASA Tech. Memo. X-69354, Wallops Station, Wallops Island, Virginia: 27-51.
OYAMA, J. and CHAN, L. (1973): Oxygen consumption and carbon dioxide production in rats during acute centrifugation stress and after adaptation to chronic centrifugation. Fed. Proc., 32: 392 (Abstract).
OYAMA, J. (1974): Drug alteration of brain biogenic amines and effect of acceleration stress-induced hypothermia. Fed. Proc., in press.
PANKOVA, A. S. (1972): Morphological changes in the juxtaglomerular apparatus in the kidneys of rats during multihour exposure to accelerations in different directions. Space Biol. Med. (USSR), 6: 26-30.
PESMAN, G. J. (1971): Acceleration terminology, table of comparative equivalents. In: Principles of Biodynamics Applicable to Manned Aerospace Flight, Prolonged Linear and Radial Acceleration. Techn. Editing and Reprod. Ltd., London; 3.
PITTS, G. C., BULL, L. S., and OYAMA, J. (1972): Effect of chronic centrifugation on body composition in the rat. Amer. J. Physiol., 223: 1044-1048.
PODYMOV, V. K. (1970): Dynamics of morphological changes in the supraoptic nucleus of the hypothalamus during prolonged transverse acceleration. Space Biol. Med. (USSR), 4: 23-29.
POLIS, B. D. (1962): Increase in acceleration tolerance of the rat by 2-dimethylaminoethyl-p-chlorophenoxyacetate (Lucidril). Aerospace Med., 33: 930-934.
POPOVIC, V. (1973): Methods for study of cardiovascular adaptation of small laboratory animals during exposure to altered gravity. Wallops Station, Wallops, Island, Virginia, NASA techn. Mem. X-69354: 158-175.
PREOBRAZHENSKAYA, I. N. (1970): The effect of back-chest G forces on metencephalic and mesencephalic blood vessels. Arkh. Anat. Gistol. Embriol., 11: 60-64.
REEVES, E. (1961): The effect of ageing on the G-tolerance of rats. U.S. Naval Air Developm. Center, Johnsville, Pennsylvania, NADC-MA-6116.
REEVES, E. (1963): The effect of sex on the G tolerance of rats. U.S. Naval Air Developm. Center, NADC-MA-6313.

SANDLER, H. (1966): Angiocardiographic and hemodynamic study of transverse (G_x) acceleration. Aerospace Med., 37: 901-910.

SAITO, I., FUJIHARA, H., and IWANE, M. (1965): The effects of g-force on body temperature. Aerospace Med. Psychol., 2: 78-83.

SAVIN, B. M. (1966): Comparative analysis of the effect of gravitational stress and hypoxia on oxygen tension in brain tissue. Pat. Fiziol. Obsh. Patolog., 61 : 19-23, NASA TT F-10, 288.

SAVINA, Ye. A. and PODYMOV, V. K. (1972): Morphology of the adrenal cortex after exposure to transverse accelerations for up to twenty-four hours. Space Biol. Med., (USSR), 6: 19-26.

SCANO, A. (1971): Biodynamics facilities in NATO European countries. Advisory Group Aerospace Res. and Developm. Publ. 150: 101-124.

SIMMONDS, R. C., BOYD, J. F., and GARRIOTT, O. K. (1974): Skylab inflight fish orientation demonstration, in preparation.

SIPPLE, W. C. and POLIS, B. D. (1959): A physiologic end point for the study of the tolerance of small mammals to high acceleration stress. U.S. Naval Air Developm. Center, Johnsville, Pennsylvania, NADC-MA-5906.

SKINNER, D. B. and CAMP, T. F. (1969): Effects of increased gravity on gastroesophageal reflux in monkeys. Aerospace Med., 40: 14-17.

SMIRNOV, K. V. and SOKOLOVA, T. A. (1973): Effect of accelerations on the active transport of glucose in the small intestines of rats. Space Biol. Med. (USSR), 7: 38-43.

SMITH, A. H., WINGET, C. M., and KELLY, C. F. (1959): Growth and survival of birds under chronic acceleration. Growth, 23: 97-108.

SMITH, A. H., and KELLY, C. F. (1963): Influence of chronic acceleration upon growth and body composition. Ann. N.Y. Acad. Sci., 110: 410-424.

SMITH, A. H. and BURTON, R. R. (1967): Influence of the ambient accelerative force upon mature body size. Growth, 31: 317-329.

SMITH, A. H. and BURTON, R. R. (1971): Chronic acceleration of animals. In: Gravity and the Organism. S. A. Gordon and M. J. Cohen (eds), Univ. Chicago Press, Chicago; 371.

SMITH, A. H., RAHLMAN, D. F., KODAMA, A. M., and PACE, N. (1974): Metabolic responses of monkeys to increased gravitational fields. COSPAR Life Sci. Space Res., Akademie-Verlag, Berlin, XII, in press.

SMITH, A. H. (1974): Principles of gravitational biology. In: The Foundations of Space Biology and Medicine. 2. USA-NASA/USSR-Acad. Sci. Ecological and Physiological Basis of Space Biology and Medicine.

STIEHM, E. R. (1961): Acceleration protection by means of stimulation of the recticuloendothelial system. U.S. Naval Air Developm. Center, Johnsville, Pennsylvania, NADC-MA-6129.

SUTULOV, L. S. et al. (1971): Post-flight histological analysis of turtles aboard Zond 7. Life Sci. Space Res. (USSR) 9: 123-128.

SUVOROV, P. M. (1958): The effect of radial accelerations on salivary and gastric gland secretion and on the periodic contractions of the stomach. Bull. exp. Biol. Med. (USSR), 46: 28-34.

THACH, S. J., Jr. and GRAYBIEL, A. (1968): Behavioral responses of unrestrained normal and labyrinthectomized squirrel monkeys to repeated zero gravity parabolic flights. Aerospace Med., 39: 734-738.

VINOGRADOV, V. M. and PASTUSHENKOV, L. V. (1969): Protective effect of antihypoxic drugs during gravitational accelerations. Space Biol. Med. (USSR), 3: 16-21.

VON BAUMGARTEN, R. J., BALDRIGHI, G. and SHILLINGER, G. L., Jr. (1972): Vestibular behavior of fish during diminished G-force and weightlessness. Aerospace Med. 43: 626-632.*

VON BECKH, H. J. A. (1954): Experiments with animals and human subjects under sub- and zero-gravity conditions during dive and parabolic flight. J. Aviat. Med., 25: 235-241.

VOSE, G. P. and SMITH, A. W. (1968): Effect of water displacement system for gravity counteractionupon skeletal status in the dog. Aerospace Med., 39: 266-270.

WUNDER, C. C., BRINEY, S. R., KRAL, M., and SKAUGSTAD, C. (1960): Growth of mouse femurs during continual centrifugation. Nature (Lond.), 188: 151-152.

WUNDER, C. C. (1962): Survival of mice during chronic centrifugation. Aerospace Med., 33: 866-870.

WUNDER, C. C. (1965): Care and growth of animals during chronic centrifugation. In: Methods of Animal Experimentation, 2. W. I. Gay (ed.), Academic Prcss, New York: 371-448.
WUNDER, C. C., MILOJEVIC, B., and EBERLY, L. (1966): Growth and food consumption of labyrinthectomized hamsters during chronic centrifugation at 5 g and 6 g. Nature (Lond.), 210: 177-179.
WUNDER, C. C., CLARK, M. E., and BENGELE, H. H. (1971): Water intake and urine output during chronic centrifugation. Amer. J. Physiol., 221: 559-563.
WUNDER, C. C. (1971): Chronic acceleration of animals. In: Gravity and the Organism, S. A. Gordon and M. J. Cohen (eds), Univ. of Chicago Press, Chicago; 389-411.

* References received during printing of Progress Series.
HAYMAKER, W., LOOK, B. C., WINTER, D. L., BENTON, E. V. and CRUTY, M. R. (1975): Project BIOCORE M212), a biological cosmic ray experiment: Procedures, summary, and conclusions. Aviat. Space environm. Med., 46:467-481.
ILYIN, E. A., SEROVA, L. V., PORTUGALOV, V. V., TIGRANYAN, R. A., SAVINA, E. A., GAYEVSKAYA, M. S., KONDRATYEV, YU. I., NOSKIN, A. D., MILYAVSKY, V. I. and YUROV, B. N. (1975): Preliminary results of examinations of rats after a 22-day flight aboard the Cosmos-605 biosatellite. Aviat. Space environm. Med., 46: 319-321.
VON BAUMGARTEN, R. J., SIMMONDS, R. C., BOYD, J. F. and GARRIOTT, O. K. (1975): Effects of prolonged weightlessness on the swimming pattern of fish aboard Skylab 3. Aviat. Space environm. Med., 46: 902-906.

CHAPTER 7

EFFETCS OF CHEMICAL AIR POLLUTANTS ON ANIMALS

Chapter 7, Section 1: Effects of Different Air Pollutants on Domestic Animals (F. L. HAYS)

AMERICAN CONFERENCE OF GOVERNMENTAL. INDUSTRIAL HYGIENISTS (1964): Threshold Limit Values Sor. 1964. Arch. Environm. Hlth, 9: 545-554.

AMERICAN CONFERENCE OF GOVERNMENTAL. INDUSTZIAL HYGIENISTS (1966): Documentation of Thresh Limct Values. Rev. Ed. Cinninnati; Ohio.

AMERICAN CONFERENCE OF GOVERNMENTAL. INDUSTRIAL HYGIENISTS (1971): Documentation of the Threshold Limit Values for Substances in Workroom Air. 3rd Ed., Cincinnati, Ohio.

AVIADO, D. M. and SALEM, H. (1968): Acute effects of air pollutants on the lungs. Arch. environm. Hlth, 16: 903-907.

BURRELL, G. A. (1914): The use of mice and birds for detecting carbon monoxide after fires and explosions. U.S. Bur. Mines techn. Paper, 11: 1-23.

CHISOLM, J. J. Sr. (1971): Lead poisoning. Sci. Amer., 224, 2: 15-23.

COTES, J. E. (1968): Lung function: assessment and application in medicine. 2nd Ed. F. A. Davis Co., Philadelphia.

EVANS, L. L. (1967): The toxicity of hydrogen sulphide and other sulphides. Quart. J. exp. Physiol., 52: 231-248.

FRANK, N. R., YODAR, R. E., BRAIN, J. D. and YOKOYAMA, E. (1969): SO^2 (^{35}S labeled) absorption by the nose and mouth under conditions of varying concentration and flow. Arch. environm. Hlth, 18: 315-322.

GOLDWATER, L. S. (1971): Mercury in the environment. Sci. Amer., 224, 5: 15-21.

GOODMAN, L. S. and GILMAN, A. (1970): The pharmacological basis of therapeutics. 4th Ed. The MacMillan Company, New York.

HAYS, F. L. (1972): Studies of the effects of hydrogen sulfide in animals. Ph. D. Thesis. Univ. Missouri, Columbia, Missouri.

HUDSON, J. R. (1954): Mortality and morbidity during the London fog of December, 1952. Publ. Hlth and med. Rep., H.M.S.O. (London) Subj. 95: 45-46. (Appendix A).

LILLIE, R. J. (1970): Air pollutants affecting the performance of domestic animals: A Literature Review. Agric. Res. Ser., Handbook No. 380, U.S. Dept. Agric. (Wash.)

McCABE, L. C. and CLAYTON, G. D. (1952): Air pollution by hydrogen sulfide in Poza Rica, Mexico. Arch. industr. Hyg., 6: 199-213.

MORROW, P. E. (1964): Animals in toxic environments: mammals in polluted air. In: Handbook of Physiology. Sect. 4: Adaptation to the Environment, D. B. Dill, E. F. Adolph and C. G. Wilber (eds). Amer. physiol. Soc. (Wash.): 795-808.

SIEGMUND, O. H. (Ed.) (1967): The Merck Veterinary Manual. Merck and Co., Inc., Rahway, N.Y.

STOKINGER, H. E. (1965): Pollutant gases. In: Handboek of Physiology. Sect. 7: Respiration. W. O. Jenn and H. Rahn, (eds). Amer. physiol. Soc. (Wash.): 1067-1086.

TENNY, S. M. (1970): Respiration in mammals. In: Dukes' Physiology of Domestic Animals, 8th Ed. M. J. Swenson (ed.). Cornell Univ. Press, Ithaca: 293.

WEEDON, F. R., HARTZELL, A. and SETTERSTROM, C. (1940): Toxicity of ammonia, chlorine, hydrogen cyanite, hydrogen sulphide and sulphur dioxide gases. V. Animals. Boyce Thompson Inst. Contrib., II: 365-385.

Chapter 7, Section 2: Effects of Lead on the Growth of the Laboratory Rat (M. A. LESSLER and G. L. WRIGHT)

BAZELL, R. L. (1971): Lead poisoning: Zoo animals may be the first victims. Science, 173: 130-131.

CARDONA, E., LESSLER, M. A. and BRIERLEY, G. P. (1971): Mitochondrial Oxidative Phosphorylation: Interaction of Lead and Inorganic Phosphate. Proc. Soc. exp. Biol. (N.Y.), 136: 300-304.
CARPENTER, S. J., FERM, V. H. and GALE, F. F. (1973): Permeability of the golden hamster placenta to inorganic lead. Experientia (Basel), 29: 311-313.
CHISOLM, H. J. Jr. (1971): Lead poisoning. Sci. Amer., 224: 15-21.
CHOW, T. J. and EARL, J. L. (1970): Lead aerosols in the atmosphere: Increase-concentrations. Science, 169: 577-580.
DELVES, H. T. (1970): Micro sampling method for the rapid determination of lead in blood by atomic absorption spectrometry. Analyst, 95: 431-438.
EMMERSON, B. T. (1970): Atmospheric contamination with lead. Arch. intern. Med., 68: 488-490.
GILFILLAN, S. C. (1965): Lead poisoning and the fall of Rome. J. occup. Med., 7: 53-60.
GOLDSMITH, J. R. and HEXTER, A. C. (1967): Respiratory exposure to lead: epidemiological and experimental dose-response relationships. Science, 158: 132-134.
GORDON, C. C. (1968): East Helena Reports for Director of Air Pollution Abatement. Dec., 1968.
HABIBI, K. (1970): Characterization of particulate lead in vehicle exhaust. Environm. Sci. Techn., 4: 239-248.
HAMMOND, P. B. and ARONSON, H. L. (1964): Lead poisoning in cattle and horses in the vicinity of a smelter. Ann. N.Y. Acad. Sci., 111: 595-611.
KEHOE, R. A. (1964): Normal metabolism of lead. Arch. environm. Hlth, 8: 232-235.
KOSTIAL, K., SIMONOVIC L. and PISONIC, M. (1971): Reduction of lead absorption from the intestine of new born rats. Environm. Res., 4: 360-363.
LAGERWERFF, J. V. and SPECHT, A. W. (1970): Contamination of roadside soil and vegetation with cadmium, nickel, lead and zinc. Environm. sci. Techn., 4: 583-586.
LAZRUS, A. L., LORANGE, E. and LODGE, J. P. (1970): Lead and other metal ions in U.S. precipitation. Environm. sci. Techn., 4: 55-58.
LESSLER, M. A. and CARDONA, E. (1970): Time course of hematologic changes in chronic lead poisoning. Physiologist, 13: 248.
McCLAIN, R. M. and BECKER, N. A. (1970): Placental transport and teratogenicity of lead in rats. Fed. Proc., 29: 347.
MUROZUMI, M., CHOW, T. J. and PATTERSON, C. (1969): Chemical concentrations of pollutant lead aerosols, terrestrial dusts, and sea salts in Greenland and antartic snow strata. Geochim. Cosmochim. Acta, 33: 1247-1294.
PROC. NAT. ACAD. SCIENCES (Wash.) (1970): Lead: Airborne Lead in Perspective.
SIX, K. M. and GOYER, R. A. (1970): Experimental enhancement of lead toxicity by low dietary calcium. J. Lab. clin. Med., 76: 933-942.

Chapter 7, Section 3: Effects of Ozone, Carbon Monoxide, Cadmium and Nitrogen Dioxide on the Cardiovascular System of Animals (Alice Y.S. P'AN)

ASTRUP, P., KJELDSEN, K. and WANSTRUP, J. (1967): Enhancing influence of carbon monoxide on the development of atheromatosis in cholesterol-fed rabbits. J. Atheroscler. Res., 7: 343-354.
AXELSSON, B., and PISCATOR, M. (1966): Renal damage after prolonged exposure to cadmium. Arch. environm. Hlth, 12: 360-372.
AXELSSON, B., DAHLGREN, S. E. and PISCATOR, M. (1968): Renal lesions in the rabbit after long-term exposure to cadmium. Arch. environm. Hlth, 17: 24-28.
BILS, R. F. (1970): Ultrastructural alterations of alveolar tissue of mice: III. Ozone. Arch. environm. Hlth, 20: 468-480.
BRINKMAN, R. and LAMBERTS, H. B. (1958): Ozone as a possible radiomimetic gas. Nature (Lond.), 181: 504.
BRINKMAN, R., LAMBERTS, H. B. and VENINGA, T. S. (1964): Radiomimetic toxicity of ozonised air. Lancet, 1: 133-136.
BUCKLEY, R. D. and LOOSLI, C. G. (1969): Effects of nitrogen dioxide inhalation on germfree mouse. Arch. environm. Hlth, 18: 558-595.

CHOW, C. K. and TAPPEL, A. L. (1973): Activities of pentose shunt and glycolytic enzymes in lungs of ozone exposed rats. Arch. environm. Hlth, 26: 205-208.
DOWELL, A. R., LOHRBAUER, L. A., HURST, D. and LEE, S. D. (1970): Rabbit alveolar macrophage damage caused by in vivo inhalation. Arch. environm. Hlth, 21: 121-127.
FIRKET, J. (1931): The cause of the symptoms found in the Meuse Valley during the fog of December 1930. Bull. roy. Acad. Med., 11: 683-731.
FRIBERG, L. (1959): Chronic cadmium poisoning. Arch. industr. Hlth, 20: 401-407.
GOLDSTEIN, E., EAGLE, M. C. and HOEPRICH, P. D. (1972): Influence of ozone on pulmonary defense mechanisms of silicotic mice. Arch. environm. Hlth, 24: 444-448.
GREENBURG, L., JACOBS, M. B., DROLETTE, B. M. et al. (1962): Report of an air pollution incident in New York, November 1953. Publ. Hlth Rep., 77: 7-16.
KAMATSU, F. (1955): Shinshu myocardiosis. Digest Sci. Lab., 10: 315-318.
LAZZARINI-ROBERTSON, A. (1968): Oxygen requirements of the human arterial intima in atherogenesis. Prog. Biochem. Pharmacol., 4: 305-316.
LOGAN, W. P. D. (1953): Mortality in the London fog incident, 1952. Lancet, 1: 336-338.
LOWRY, T. and SCHUMAN, L. M. (1956): Silo-filler's disease: a syndrome caused by nitrogen dioxide. J. Amer. med. Ass., 162: 153-160.
LUCIS, O. J. and LUCIS, R. (1969): Distribution of cadmium 109 and zinc 65 in mice of inbred strains. Arch. environm. Hlth, 19: 334-336.
MORLOCK, C. G. (1939): Arterioles of the pancreas, liver, gastro-intestinal tract and spleen in hypertension. Arch. intern. Med., 63: 100-118.
MOTLEY, H. L., SMART, R. H. and LEFTWICH, C. I. (1959): Effect of polluted Los Angeles air (smog) on lung volume measurements. J. Amer. med. Ass., 171: 1469-1477.
NIEMI, M. (1968): Cadmium-induced testicular necrosis. Work environm. Hlth, 5: 54-60.
P'AN, A. Y. S. and JEGIER, Z. (1972): Ozone-induced arterial lesions. Arch. environm. Hlth, 24: 229-232.
REMMERS, J. E. and BALCHUM, O. J. (1965): Effects of Los Angeles urban pollution upon respiratory function of emphysematous patients. 58th Ann. Meeting Air Pollution Control Association, Toronto, June 22, 1965.
SCHEEL, L. D., DOBROGORSKI, O. J., MOUNTAIN, J. T. et al. (1959): Physiologic, biochemical immunologic and pathologic changes following ozone exposure. J. appl. Physiol. 14: 67-80.
SCHRENK, H. H., HEIMANN, H., CLAYTON, G. B. et al. (1949): Pollution in Donora, Pa: epidemiology of the unusual smog episods of October 1948. Preliminary Report. Bull. 306, Publ. Hlth Serv.
SCHROEDER, H. A. and VINTON, W. H. Jr (1962): Hypertension induced in rats by small doses of cadmium. Amer. J. Physiol., 202: 515-518.
SCHROEDER, H. A. (1965): Cadmium as a factor in hypertension. J. chronic. Dis., 18: 647-656.
SCHROEDER, H. A., BAKER, J. T., HANSEN, M. N. M. et al. (1970): Vascular reactivity of rats altered by cadmium and zinc chelate. Arch. environm. Hlth, 21: 609-614.
SHAIKH, Z. A. and LUCIS, O. J. (1971): Isolation of cadmium binding proteins. Experientia (Basel), 27: 1024-1025.
SIGGAARD-ANDERSEN, J., BONDE-PETERSEN, F., HANSEN, T. et al. (1968): Plasma volume and vascular permeability during hypoxia and carbon monoxide exposure. Scand. J. clin. Invest., 22: Suppl. 103, 39-48.
WANSTRUP, J., KJELDSEN, K. and ASTRUP, P. (1969): Acceleration of spontaneous intimal-subintimal changes in rabbit aorta by a prolonged moderate carbon monoxide exposure. Acta. path. microbiol. scand., 75: 353-362.

Chapter 7, Section 4: Effects of Sulphur Dioxide and Related Compounds on Monkeys (Y. ALARIE)

ALARIE, Y., ULRICH, C. E., HADDOCK, R. H., JENNINGS, H. J. and DAVID, E. F. (1970): Respiratory system flow resistance with digital computer techniques: Measured in cynomolgus monkeys and guinea pigs. Arch. environm. Hlth, 21: 483-491.

ALARIE, Y., BUSEY, W. M., MacFARLAND, H. N., SWANN, H. E. Jr. and KRUMM, A. A. (1971a): The effects of the chronic inhalation exposure of primates to nitrogen dioxide. Toxicol. appl. Pharmacol., 19: 416-417.

ALARIE, Y., KRUMM, A. A., JENNINGS, H. J. and HADDOCK, R. A. (1971b): Distribution of ventilation in cynomolgus monkeys: Measurement with real-time digital computerization. Arch. environm. Hlth, 22: 633-642.

ALARIE, Y., ULRICH, C. E., KRUMM, A. A., HADDOCK, R. H. and JENNINGS, H. J. (1971c): Mechanical properties of the lung in cynomolgus monkeys: Measurement with real-time digital computerization. Arch. environm. Hlth, 22: 643-654.

ALARIE, Y., ULRICH, C. E., BUSEY, W. M., KRUMM, A. A. and MacFARLAND, H. N. (1972): Long-term continuous exposure to sulfur dioxide in cynomolgus monkeys. Arch. environm. Hlth, 24 :155-128.

ALARIE, Y. (1973a): Sensory irritation by airborne chemicals. Crit. Rev. Toxicol., in press.

ALARIE, Y., BUSEY, W. M., KRUMM, A. A. and ULRICH, C. E. (1973b): Long-term continuous exposure to sulfuric acid mist in cynomolgus monkeys and guinea pigs. Arch. environm. Hlth, 27: 16-24.

ALARIE, Y., KANTZ, R. J., ULRICH, C. E., KRUMM, A. A. and BUSEY, W. M. (1973c): Long-term continuous exposure to sulfur dioxide and fly ash mixtures in cynomolgus monkeys and guinea pigs. Arch. environm. Hlth, in press.

BANERJEE, C. M., ALARIE, Y. and WOOLARD, M. (1968): Gas tensions in conscious monkeys. Proc. Soc. exp. Biol. (N.Y.), 128: 1183-1185.

BINNS, R. and CLARK, G. C. (1973): An experimental model for the assessment of the effects of cigarette smoke inhalation on pulmonary physiology. Ann. occup. Hyg., 15: 237-247.

DeBIAS, D. A., BANERJEE, C. M., BIRKHEAD, N. C., HARRER, W. V. and KAZAL, L. A. (1973): Carbon Monoxide inhalation effects following myocardial infarction in monkeys. Arch. environm. Hlth, 27: 161-167.

ECKARD, R. E., MacFARLAND, H. N., ALARIE, Y. and BUSEY, W. M. (1972): The biologic effect from long-term exposure of primates to carbon monoxide. Arch. environm. Hlth, 25: 381-388.

MacFARLAND, H. N., ULRICH, C. E., MARTIN, A., KRUMM, A. A., BUSEY, W. M. and ALARIE, Y. (1971): Chronic exposure of cynomolgus monkeys to fly ash. In: Inhaled Particles III, 1. W. H. Walton (ed.). Unwin Brothers, Surrey, England: 313-326.

MacKLEM, P. T. (1971): Airway obstruction and collateral ventilation. Physiol. Rev., 51: 368-436.

PALM, P. E., McNERNEY, J. M. and HATCH, T. (1956): Respiratory dust retention in small animals. Arch. industr. Hlth, 13: 355-365.

Chapter 7, Section 5: Effects of Pesticides on Animals (S. C. BHATIA)

AKERA, T., BRODY, T. M., and LEELING, N. (1971): Insecticide inhibition of Na-K-ATPase activity. Biochem. Pharmacol., 20: 471-473.

ARCOS, J. C., CONNEY, A. H., and BUU-HOI, N. P. (1961): Induction of microsomal enzyme synthesis by polycyclic aromatic hydrocarbons of different molecular sizes. J. biol. Chem., 236: 1291-1296.

ASKARI, E. M. and GRABLIKS, J. (1973): DDT and immunological responses. II Altered histamine levels and anaphylactic shock in guinea pigs. Arch. environm. Hlth, 26: 309-312.

BHATIA, S. C., SHARMA, S. C., DAMODARAN, V. N., and VENKITASUBRAMANIAN, T. A. (1971a): Changes in rat ascorbic acid status and adrenocortical activity during acute dieldrin toxicity. Indian J. Biochem. Biophys., 8: 57-58.

BHATIA, S. C., SHARMA, S. C., and VENKITASUBRAMANIAN, T. A. (1971b): Acute dieldrin toxicity: Heptic glycogenic and gluconeogenic response. In: Proc. joint meeting of Biochemical Society of Biological chemists. Bangalore (India). 77-78.

BHATIA, S. C. and VENKITASUBRAMANIAN, T. A. (1972): Machanism of dieldrin-induced fat accumulation in rat liver. J. Agric. Food. Chem., 20: 993-996.

BHATIA, S. C., SHARMA, S. C., and VENKITASUBRAMANIAN, T. A. (1972a): Effect of dieldrin on certain enzyme systems of rat liver. Brit. J. exp. Path., 53: 419-426.

BHATIA, S. C. (1972): Metabolic Studies on the effect of dieldrin. Ph. D. Thesis, Univ. Delhi, India.

BHATIA, S. C., SHARMA, S. C., and VENKITASUBRAMANIAN, T. A. (1972b): Acute dieldrin toxicity: Biochemical Changes in the blood. Arch. environm. Hlth., 24: 369-372.

BHATIA, S. C., SHARMA, S. C., and VENKITASUBRAMANIAN, T. A. (1973): Effect of dieldrin on hepatic Carbohydrate metabolism and protein biosynthesis in vivo. Toxicol. Appl. Pharmacol., 24: 216-229.

BITMAN, J., CECIL, H. C., HARRIS, S. J., and FRIES, G. F. (1968): Esterogenic activity of O,P′-DDT in the mammalian uterus and avian oviduct. Science, 162: 371-372.

BURLINGTON, H., and LINDMAN, V. F. (1950): Effect of DDT on testes and secondary sex characters of white leghorn cockerels. Proc. Soc. exp. Biol. (N.Y.), 74: 48-51.

CUTKOMP, L. K., YAP, H. H., VEA, E. V., and KOCH, R. B. (1971): Inhibition of oligomycin-sensitive (mitochondrial) $Mg^{2\pm}$ Atpase by DDT and selected analogs in fish and insect tissue. Life Sci., 10: 1201-1209.

DEICHMANN, W. B., KEPLINGER, M., DRESSLER, I. and SALA, F. (1969): Retention of dieldrin and DDT in tissues of dogs fed aldrin and DDT individually and as a mixture. Toxicol. appl. Pharmacol., 14: 205-213.

DUBISKI, S. and FRADETTE, K. (1966): The feed back mechanism in immunoglobulin synthesis. Proc. Soc. exp. Biol. (N.Y.), 122: 126-130.

FITZHUGH, O. G., NELSON, A. A. and QUIFE, M. L. (1964): Chronic oral toxicity of aldrin and dieldrin in rats and dogs. Food cosmet. Toxicol., 2: 551-552.

GABLIKS, J., ASKARI, E. M., YOLEN, N. (1973): DDT and Immunological responses. I Serum antibodies and anaphylactic shock in guinea pigs. Arch. environm. Hlth, 26: 305-308.

GHAZAL, A. (1963): The action of insecticidal chlorinated hydrocarbons on the central nervous system of rats. Arch. expt. Pathol. Pharmacol., 244: 504-509.

GILLETT, J. W., CHAN, T. M. and TERRIERE, L. C. (1966): Interactions between DDT analogs and microsomal epoxidase systems. Agric. Food Chem., 14: 540-545.

HATHWAY, D. E., MALLINSON, A. and AKINTONWA, D. A. A. (1965): Effects of dieldrin picrotoxin and telodrin on the metabolism of ammonia in brain. Biochem. J., 94: 676-686.

HODGE, H. C., BOYCE, A. M., DEICHMANN, W. B. and KRABILL, H. F. (1967): Toxicology and no effect levels of aldrin and dieldrin. Toxicol. appl. Pharmacol., 10: 613-615.

HOSEIN, E. A., and PROULX, P. (1960): Chemical and biochemical analysis of brain tissue preparations during the epileptiform like activity of dieldrin and other cerebral convulsants. J. Agric. Food Chem., 8: 428-431.

HUTTERER, F., SCHAFFNER, F., KLION, F. M., and POPPER, H. (1968): Hypertrophic, hypoactive smooth endoplasmic reticulum: A sensitive indicater of hepatotoxicity exemplified by dieldrin. Science, 161: 1017-1020.

JACKSON, D. A., GARDNER, D. R. (1973): The effect of some organochlorine pesticide analogs on Salmonid Brain ATPases. Pestic. Biochem. Physiol., 2: 377-382.

JEFFERIES, D. J. and FRENCH, M. C. (1969): Avian thyroid: Effect of p,p′-DDT on size and activity. Science, 166: 1278-1280.

KACEW, S. and SINGHAL, R. L. (1973a) Adaptive response of hepatic Carbohydrate metabolism to oral administration of p,p′-1,1,1-Trichloro-2,2-Bis (p-chlorophenyl) ethane in rats. Biochem. Pharmacol., 22: 41-57.

KACEW, S. and SINGHAL, R. L. (1973b): Metabolic alterations after chronic exposure to α-chlordane. Toxicol. appl. Pharmacol., 24: 539-544.

KIMBROUGH, R. D., GAINES, T. B. and LINDER, R. E. (1971): The ultrastructure of livers of rats fed DDT and dieldrin. Arch. environm. Hlth, 22: 460-467.

KITABCHI, A. E. (1967): Ascorbic acid in steroidogenesis. Nature (Lond.), 215: 1385-1386.

KOCH, R. B., CUTKOMP, L. K. and DO, F. M. (1969): Chlorinated hydrocarbon insecticide inhibition of cockroach and honeybee ATPases. Life Sci., 8: 289-297.

KOCH, R. B., CUTKOMP, L. K. and YAP, H. H. (1971): Inhibition of oligomycin sensitive fish adenosine triphosphatase activity by chlorinated hydrocarbon insecticides. Biochem. Pharmacol., 20: 3243-3245.

LUCIER, G. W. and McDANIEL, O. S. (1972): Alterations in rat liver microsomal and

lysosomal beta glucoronisidase by compounds which induce hepatic drug-metabolizing enzymes. Biochim. Biophys. Acta 261: 168-176.
McFARLAND, L. Z. and LACY, P. B. (1969): Physiologic and endocrinologic effects of the insecticide kepone in the Japanese Quail. Toxicol. appl. Pharmacol., 15: 441-450.
ONIKIENKO, F. A. (1963): Enzymic changes from early stages of intoxication with small doses of chloroorganic insecticides. Gig. Fiziol. Tr. Proizv. Toksikol. (Kiev. Gos Izd. Med. Lit. Ukr. SSR): 77-80 (In Russian).
PETERS, D. A. V., HRDINA, P. D., SINGHAL, R. L. and LING, G. M. (1972): Role of brain serotonin in DDT-induced hyperpyrexia. J. Neurochem., 19: 1131-1136.
RATCLIFFE, D. A. (1967): Decrease in eggshell weight in certain birds of prey. Nature (Lond.), 215: 208-210.
SAYERS, G. (1950): The adrenal cortex and homeostasis. Physiol. Rev., 30: 241-320.
SCHWABE, U. (1964): The effect of DDT on the metabolism of unesterified fatty acids. Arch. exp. Pathol. Pharmacol., 248: 195-205.
SINGHAL, R. L., VALADARES, J. R. E. and SCHWARK, W. S. (1970): Metabolic control mechanisms in mammalian systems-IX Estrogen-like stimulation by o.p′-1,1,1-trichloro-2,2-bis (p-chlorophenyl) ethane. Biochem. Pharmacol., 19: 2145-2155.
SOLARINO, G. and TRIPODO, C. (1954): Plasmocitisi spermentale da DDT e reazioni immunobiologiche: Nota 2. Compartments del quadro proteico eletsoforetico. Boll. Ist Sieroter Milan, 33: 482-488 (In Italian).
WASSERMANN, M., WASSERMANN, D., GERSHON, Z. and ZELLERMAYER, L. (1969): Effects of organochlorine insecticides on body defense systems. Ann. N.Y. Acad. Sci., 160: 393-401.
WASSERMANN, M., WASSERMANN, D., KEDAR, E. and DJAVAHERIAN, M. (1971): Bull. environm. Cont. Toxicol., 6: 426-435.
WASSERMANN, M., WASSERMANN, D., KEDAR, E. and CUCOS, S. (1972): Effects of dieldrin and gamma BHC on serum proteins and PB1. Bull. environm. Cont. Toxicol., 8: 177-185.
WELCH, R. M., LEVIN, W., and CONNEY, A. H. (1969): Estrogenic action of DDT and its analogs. Toxicol. appl. Pharmacol., 14: 358-367.
WHATLEY, S. and GLICK, B. (1966): The effect of o,p-DDD in the chicken. Experientia (Basel), 22: 179-180.
WITTER, R. F. and FARRIOR, W. L., Jr. (1964): Effects of dieldrin or DDT in vivo on alpha-alanine, gamma-amino-butyric acid, glutamine and glutamate in rat brain. Proc. Soc. exp. Biol. (N.Y.), 115: 487-490.
WOODWELL, G. M. (1967): Toxic substances and ecological cycles. Sci. Amer., 216: 24-31.
WURSTER, C. F., Jr. and WINGATE, D. B. (1968): DDT residues and declining reproduction in the Bermuda petrel. Science, 159: 979-981.

Author Index

Subject Index